# The Theory of Categorial Conversion
## Rational Foundations of Nkrumaism in Socio-natural Systemicity and Complexity

Adonis & Abbey Publishers Ltd
St James House
13 Kensington Square,
London, W8 5HD
United Kingdom

Website: http://www.adonis-abbey.com
E-mail Address: editor@adonis-abbey.com

Nigeria:
Suites C4 & C5 J-Plus Plaza
Asokoro, Abuja, Nigeria
Tel: +234 (0) 7058078841/08052035034

British Library Cataloguing-in-Publication Data
A catalogue record for this book is available from the British Library

ISBN: 978-1-909112-67-4(Paperback)
        978-1-909112-72-8(Hard back)

# The Theory of Categorial Conversion
## Rational Foundations of Nkrumaism in Socio-natural Systemicity and Complexity

Kofi Kissi Dompere

ADONIS & ABBEY
PUBLISHERS LTD

# Table of Contents

## CHAPTER SIX
### The Theory of Categorial Conversion: The Analytical Building Blocks in Socio-Natural Systemicity

# DEDICATION

To all theorists who seek methodological paths for explaining the general mechanism of complex socio-natural transformations where the old ones are destroyed and new ones are created through the internal dynamics of actual-potential polarities in systemicity and complexity in relational continuity and unity.

To all members in thinking societies with cognitive independence
In thought, reasoning and practice of ideas with the understanding of the power of principle of doubt, interdependence of ideas, its practices and history;

Where new ideas are locked in audacity, curiosity and creativity;
With a full recognition of the nature of knowledge-production system,
Where most ideas need continual corrections towards perfections as the Information-knowledge system expands with increasing complexity of relational continuum and unity toward greater and greater understanding of the relational structure of matter, energy and information with the hope that such understanding will enhance the view that knowledge is self-exciting, self-correcting and decision-choice determined with a continual exposure of human ignorance.

In this respect, the monograph is a dedication to all multidisciplinary theorists at the frontier of sciences of systemicity, complexity and their interactions.

# ACKNOWLEDGEMENTS

I wish to express my thanks to all my friends, critics and admirers who have given me encouragements and emotional support. I am grateful to the staff of the Legon Hall at the University of Ghana for accommodations, when I was editing and refining my views. This monograph has benefited from the contributions of the references on the African conceptual system of the principle of opposites as well as those multidisciplinary references that are in the monograph. I express my thanks to my admirers and critics whose positive and negative reflections on my works have enhanced my theoretical and philosophical convictions and made this work enjoyable to the end. My thanks also go to all my potential admirers and critics, and those whose will take time to research on static and dynamic states of qualitative dispositions in system's complexity where a complex form of qualitative mathematics is required for analytical works on defining and analyzing qualitative motions. This monograph has also benefited from my seminar works on the current methodological frontiers of the fuzzy paradigm and principle of opposites with some students in the Computer Science Department and African Studies Department at Howard University. I thank Professor Kwabena Osafo-Gyimah for his continual encouragement, inspiration and motivation. Special thanks go to Ms. Jasmine Blackman for her proofreading suggestions on the first draft. Controversial ideas and terminologies are intentional, and intentionally directed to restructure the paradigm of thinking to account for new explanations in the theory of socio-natural transformations. I accept all responsibilities for errors that may logically arise.

# PREAMBLE

Insofar as Egypt is the distant mother of Western culture and sciences, as it will emerge from the reading of this book, most of the ideas that we call foreign are oftentimes nothing but mixed up, reversed, modified, elaborated images of creations of our African ancestors, such as Judaism, Christianity, Islam, dialectics, the theory of being, exact sciences, arithmetic, geometry, mechanical engineering, astronomy, medicine, literature (novel, poetry, drama), architecture, the arts, etc.

One can see then how fundamentally improper is the notion, so often repeated, of the importation of foreign ideologies in Africa. It stems from a perfect ignorance of the African past. Just as modern technologies and sciences come from Europe, so did, in antiquity, universal knowledge stream from the Nile Valley to the rest of the world, particularly to Greece, which will serve as a link [R1.86, p. 3]

The outcome of his [R1.243] study of Pharaonic mathematics confirms and surpasses what we already know through the work of his predecessors, and the spontaneous collaboration of his stepdaughter, Lucie Lamy, enabled him to present this mathematical thought in all its details. This achievement is all the more astonishing in the light of claims that philosophy and science as we know them were invented by the Greeks.

Indeed, it is easy to forget that Moses and Pythagoras, among others, received their entire culture from the Egyptian temple, but much more difficult, it seem to me, is categorically to deny this fact.[ R1.243] p. 4 ].

Foundations: The lesson of immediate interest to us in the general study of the Pharaonic mind can be epitomized as follows:

1. Faith in an origin that cannot be situated in time and space. This is reality absolute, not to be grasped by our intelligence. This cannot be regarded as a mystery: it is the eternal Present Moment, indivisible Unity.

2. Through an internal act, the irrational source undergoes a polarization that manifests itself in spiritual substance. This substance appears as the energy of which the universe is constituted. Such is the mystery of the division into two, which, with the irrational origin, comprises the mystic ternary.

3. The phenomenon of universe in all its aspects is made up of this energy substance to the various degrees of its positive (north) polarity going toward its negative (south) polarity.

This becoming is accomplished by alteration, a positive, negative and negative-positive oscillation. Here the point of equilibrium must be the return to the nonpolarized source, the present Moment, which cannot be situated.

4. Therefore the universe is but a struggling search for the predominance of each polarity, one provoking the other, but with the negative unable to predominate, that is unable to give rise to phenomena except by reactively becoming positive in negative: active inertia, annihilation (death) is thus surpassed by the new polarization, new (second) mystery| that of reversal. This double play continues until all negative residue is reabsorbed into the Present moment.

5  The cause, the first mystery (called the primary category in this monograph), can have but one aim: the second mystery, and then the final equilibrium (called categorial equilibrium in this monograph) in the Present Moment. This can be attained through integral activation of the negative, whatever the means: natural, unconscious, or artificial, or conscious in the sense of psychological consciousness.

6  Everything in the universe maintains continuity (reproduces itself) through a polarization, in the image of the mystery of original polarization; and the alternation of polarities makes for existence (apparent life) growth, maturity, and aging.

7. Proportionality (called relative characteristic set in this monograph) produces form or variety (called categories in this monograph), which animates and names (specifies) the being (call categorial identity in this monograph). There are twelve essential forms, of which five are double and two single.

8. Becoming, up to the mystery of reversal (when negative inertia in turn divides so as to be positive reactivity), constitutes genesis. It is sole-singular, and in its part in similar to the whole.

The terms "positive" and "negative" are general notions that take different names according to the category to which they are applied [R1.243, p.11-12].

In order to interpret a script, the meaning of the characters must be known. The West, lead astray by Greek thinking, which is concerned with appearances only, must once again learn the meaning of a "natural symbol" which is never deceptive [R1.243 p. 11]

After the indigenous Africans of Egypt and other Nile Valley High-culture developed the fundamental "PRINCIPLES OF THE LAW OF OPPOSITES" as "THE UNDERLYING FACTOR OF LIFE IN THE UNIVERSE and applied it to natural phenomena Plato, another so-called "GREEK PHILOSOPHER," came forth with his alleged "THEORY OF IDEAS," which he culminated with "THE REPUBLIC: THE IDEAL STATE." However, this work, allegedly written by Plato himself, is still being disputed as to its authorship, just as it was with Plato's contemporaries when he first published it [R1.37, p. 319].

Massey pioneered the effort to connect Old Kamite thought to its origins in Africa's hoary antiquity. He showed that it was a mode of consciousness exhibiting multiple levels of meaning and connecting in unexpected ways to system s of thought widely separated in time and space. This paradigm of thought was the direct progenitor to the philosophy, metaphysics, religion and science that eventually shaped Western civilization. As death approached in 1907, Massey had his epitaph inscribed as follows: "Born 1828, Re-born 1907," an implicit recognition of the unbreakable links connecting life, death and rebirth. It is the same with the incalculable wisdom of Old Kam: what was lost is now being found; what was dead is now being re-born, and Massey has had a hand in this process of recovery, reconstruction and rebirth (Charles S. Finch III, in[R1.179 Vol. I, p. ix]).

This (European) concept of separate being, of substance (to use the scholastic term again which find themselves side by side, entirely independent one of another, is foreign to Bantu thought. Bantu hold that created beings preserve a bound one with another, an intimate ontological relationship, comparable with the causal tie which binds creature and Creator... Just as Bantu ontology is opposed to the European concept of individuated things, existing in themselves, isolated from others, so Bantu psychology cannot conceive of man as an individual, as a force existing by itself and apart from its ontological relationship with other living things and from its connection with animals, or intimate forces around it (Placid Temples, Bantu Philosophy quoted in [R1.8, p10-11]).

Certainly scholars have since provided dictionaries and grammars of the Ancient Egyptian language, but that is not enough to grasp the Egyptian way of thinking, so utterly different from our Western logical mind. There is first of all what has been called the multiplicity of approaches: statements (and answers to problems) which to us would seem absolutely contradictory appear side by side and did not in the least disturb the Egyptian, on the contrary, our own logical abstractions would probably have appeared to the Egyptian mind as an impoverishment and falsification of the fullness of significant truth … All this would lead—as a modern scholar has put it—to the impossibility of translating Egyptian thoughts into modern language, for the distinctions we cannot avoid making did not exist for the Egyptians (A.A. Barb, The Legacy of Egypt, quoted in [R1.8 p.10-12]).

# PREFACE

## The Theory of Categorial Conversion: Rational Foundations of Nkrumaism

And if the changes on the earth's surface during the last few millions of years appear to our present ethical notions to be in the nature of a progress that gives no ground for believing that progress is a general law of the universe. Except under the influence of desire, no one would admit for a moment so crude a generalisation from such a tiny selection of facts. What does result, not specially from biology, but from all the sciences which deal with what exists, is that we cannot understand the world unless we can understand change and continuity [R12.24b, p. 111]

To conceive the universe as essentially progressive or essentially deteriorating, for example, is to give to our hopes and fears a cosmic importance which may, of course, be justified, but which we have as yet no reason to suppose justified. Until we have learnt to think of it in ethically neutral terms, we have not arrive at a scientific attitude in philosophy; and until we have arrive at such an attitude, it is hardly to be hoped that philosophy will achieve any solid results [R12.24b, p.86]

## I. THE MONOGRAPH

There is always a motivation for the development of a theory as either a new one on the same phenomenon or on a new phenomenon. The motivation may also be involved in extension or refinement of existing theory of a known phenomenon. The motivation may also be driven by the practice of epistemic methodology of doubt where one cultivates cognitive creativity, logical imagination and intellectual curiosity of importance in the development of either explanatory or prescriptive structures in specific theory. These require the development of strong will to evade the victimization by the general slavery of cognitive familiarity where one stays in an epistemic box of thought. The motivation coming out of the practice of the principle of doubt is an epistemic recognition that knowledge production system is self-exiting and self-correcting system which depends on doubt, curiosity and perseverance within the truth-falsity duality of learning and knowing. It is the principle of doubt, cognitive curiosity, perseverance and the fight against being a victim of slave walls of familiar that are the motivation for the development of this monograph that deals with a search of general theory of transformation. The development of any theory of ontological and epistemological transformations require special care since it must deal with simultaneous time-processes of qualitative and quantitative dispositions of elements. It is the existence of the simultaneity of behaviors of qualitative and quantitative dispositions that renders the current theories of economic development and social transformation inadequate and substantially wanting. Social transformations and economic development are not time-space phenomena. They are quality-quantity-time phenomena where quantity-quality duality exist in a relational continuum and unity such that every qualitative disposition has a corresponding quantitative disposition with a relational structure between quantitative and qualitative equations of motion that determine the states of categorial systems.

To understand social self-transformation in economics, politics and law and their interrelated effects, one must understand natural transformations and how the concepts of quantity and quality are analytically combined to structure the human understanding of the general mechanism of socio-natural transformations. The understanding of self-transformations in nature will be an important cognitive instrument to construct either an explanatory theory or prescriptive theory of social transformation whose foundation is in crisis. Any attempt to find a foundation of the theory of social transformation from society itself seems to produce epistemic frustration and cognitive confusion. The foundation of any scientific theory of explanation or prescription of social transformation within either explanatory science or prescriptive science must be based on combined interactions of matter, energy and information that will allow various combinations of quantity, quality and time categorial elements. Here, one is confronted with the problems and difficulties within subjective-objective duality with relational continuum and unity, where the subjective-objective duality is mapped into quality-quantity duality. In other words, the subjective-objective duality is viewed as a function defined over the quality-quantity duality in the epistemological space for understanding by cognitive agents.

By the relational continuum and unity as characteristics of the fundamental principle of opposites, duality and polarity manifest themselves in all areas of the universe, in all ontological elements as well as all levels of organizing power and force of transformation and creation

xvii

of new forms. In order to understand transformations as it relates to the universal elements, our world, and in fact human life, it is necessary to identify and understand opposites and in reference to polarity, duality, negative and positive duals and how they are expressed in ontological elements under principles of relational continuum and unity that are the characteristics of the universal order. Individually, polarity and duality are internally indivisible, where the poles of the polarity, and the duals of the duality exist in relational continuum and unity. The polarities and dualities are also indivisible from each other, where every pole of a polarity has a residing duality. In this respect, subjective-objective duality is indivisible from quality-quantity duality, where there are relational structures, experiential information structure, dualistic preferences, matter and energy. Here, one cannot understand social self-transformation in a relational continuum and unity if one does not understand ontological self-transformation as a process defined in a space of relational continuum and unity that involve the dynamics of qualitative and quantitative dispositions. The self-transformation is governed by interplays of conflicts and contradictions as produced by the opposites. In respect of this process, one of the greatest contributions of science would be a development of provisions of either an explanatory structure or a prescriptive structure of how contradictions and conflicts are internally produced by matter and turned into energy, and then to power for qualitative instability and transformations of categorial identities of ontological elements.

The Africentric conceptions of the principles of opposites that are captured by concepts of polarity, duality, relational continuum, integrated unity and continual creation as epistemic foundations of Nkrumah's Consciencism are not understood and related to the Africentric philosophical unity on states, processes, category, change and transformations in the quality-quantity-time space. This lack of critical understanding has created an important cognitive gap in appreciating Nkrumah's policy choices and the applications required to guide actions. Furthermore, this lack of understanding has created a situation where people emphasize the ideological component and completely overlook the scientific character of Nkrumah's contribution to the understanding of social transformations and economic development in all planes including freedom and justice. The cognitive gap must be closed by epistemic actions that will connect ontology to epistemology as well as to the relational unity of socio-natural transformations. The closing of this cognitive gap includes reconstruction of the relational structure between the concepts of qualitative and quantitative dispositions, and how the concept of qualitative disposition is related to category formation and language. Here, a new epistemic look will be required to enhance the structure as part of the structure of the development of this monograph. Nkrumah's scientific explanation and policy prescriptive actions of social change may be abstracted from all his writings and speeches. The central and the basic philosophical-scientific pieces of the theory of social transformation, however, are contained in his book Consciencism [R1.203].

In developing the scientific theory of social transformation in the Consciencism, and faced by the African experiential information structure that contains African traditions, colonialism, slavery, European oppression, exploitation and complete political disenfranchisement, Nkrumah's central question may be stated as how can Africa transform herself from within to restate her traditional greatness and advance new social frontiers. The search for answers to this social question lead Nkrumah to ask another questions as to how do ontological elements transform themselves from within by destroying what they are (the actual) to construct what they are not (the potential) through destruction-construction process from within without external impression. There are two questions to answer. They are social transformation question and natural transformation question which may be aggregated into socio-natural transformation questions. Certain analytical difficulties must be overcome by the development of answers to the socio-natural transformation questions. To overcome these analytical difficulties, two important and interrelated epistemic sub-systemic blocks in a sequential form are developed. The first systemic building block is *categorial conversion* followed by a second building block of *philosophical Consciencism*. The categorial conversion is an intellectually honest and serious knowledge development to first establish a *general mechanism of internal qualitative and quantitative self-motions* of socio-natural processes within categories to bring about intra-categorial and inter-categorial transformations and changes. The intra-categorial conversion is evolutionary in the sense of producing gradual qualitative changes which induce increasing internal conflicts to create internal crisis, while the inter-categorial conversion is revolutionary in the sense that it resolve the crisis and bring about a new order. The universality of this categorial-conversion approach to quality-quantity dynamics is yet to be understood, appreciated and made applicable across scientific spectrum in theory and applications. The concept of categorial conversion and the logic of its use allow the establishment of the necessary internal conditions of convertibility of categories by discussing the *qualitative laws of motion* which are implicit in Consciencism. These laws of motion must be developed. This development is taken up in this monograph. The problem of socio-natural transformations are conceived in pyramidal relations of qualitative disposition, quantitative disposition and time as they relate to matter, energy and information, where at the level of knowing, the subjective and objective dispositions must be integrated with qualitative and quantitative dispositions as it will be developed in this monograph to establish an epistemic unity required for the understanding of socio-natural transformations. The foundations of this categorial-conversion approach are presented in [R1.92][R3.7][R3.13]

The establishment and the understanding of the mechanism of transformations and changes in Nkrumah's conceptual system require the development of the theory of categorial conversion to examine conditions of qualitative and quantitative states for change and convertibility of categories. The theory of categorial conversion is not explicitly developed by Nkrumah. He, however, provides reasonable conceptual and analytical materials for one to develop the needed theory for the establishment of the internal mechanism for qualitative transformation and change if one has good knowledge of category theory, theory of formal and informal languages and foundations of fuzzy paradigm with its logic and mathematics. The good knowledge must be related to the principle of opposites and the basic Africentric conceptual system [R1.8] [R1.35] [R1.37] [R1.86] [R1.92] [R1.243]. This monograph is about the development of the required theory which constitutes a point of entry into Nkrumah's conceptual system. The analytical foundations and directions of application of Nkrumah's conceptual system are provided in his book Consciencism [R1.203].

## II The Objective of the Monograph, its Entry Points and Points of Departure

The objective of this monograph is the construction of the needed organic theory that provides the mechanism for general transformations along the lines of Nkrumah's conceptual system. The development of the needed organic theory requires first to establish the logic and conditions of *categorial formation*. This is the *theory of category formation* where such a theory establishes the identities of socio-natural elements and then places them in their respective categories to meet the conditions of *nominalism* in the epistemological space and in relations to the ontological space. This is a first entry point into the structure of the organic theory in the process of developing a general mechanism of all transformations. One cannot speak of categorial conversion when there are no categories. The second requirement of the organic theory is to establish internal conditions for convertibility of categories by specifying qualitative laws of motion that must be related to quantitative and qualitative dispositions of categorial elements given time. This is the second entry point into the construction of the organic theory in understanding categorial conversions of socio-natural categorial elements.

The theory for the establishment of mechanism of categorial formation must be developed from characteristic sets as qualitative properties that tend to establish nominalism which is linked with vocabulary and the development of languages. To be epistemologically meaningful, analytically useful, communicationally understandable, and logically constructible and computable, the theory of categorial formation must establish *identities* of socio-natural elements and induce partitions on actual and potential spaces. The elemental and categorial identities must have the capacity to be related to formal and informal languages [R3.13]. Generally, therefore, the theory of categorial formation must establish conditions of socio-natural identities for epistemic identification and the development of experiential information structure on the basis of human limitations in the space of know-hows within the epistemological space for cognitive actions. The results of category formation are analytically connected to a general time set and a set of processes where every entity at any time point is seen as belonging to a category in such a way that time-dated elements with given quantity-time dispositions have their own identities in a fuzzily defined organic category which is an enveloping category. It is useful to familiarize oneself, at this point, with the development of mathematical theory of categories and philosophical reflection on the concepts of category [R1.92] [R3.7] [R3.10] [R2] [R2.18 [R2.22]. The familiarity of the theory of time set and its relation to processes will be useful. It will become clear that internal qualitative changes are defined over *continuous time set*, and associated with *continuous processes* and intra-categorial conversions. The establishments of categorial changes are defined over *discrete time set* as categorial crises are resolved in a particular time point, and associated with *discrete processes* and inter-categorial conversions. These processes will be combined in the development of the theory of categorial conversion that we seek. The categorial-conversion approach is another point of departure as well as a point of entry into the development of a framework for the study socio-natural development on the basis of qualitative mathematics and reasoning.

The concepts of identity, time, characteristic sets, negative and positive characteristic sub-sets and category are the main analytical blocks from which the theory of category formation is developed by methodological constructionism and reductionism. The mutual interconnections of these analytical blocks allow distinctions to be placed on the socio-natural elements, for examination of entities individually or collectively to be undertaken, for comparison and cross-sectional and time-series studies of contents of history of elemental existence of general forms of space-time and non-space-time phenomena to be undertaken. By the properties of these concepts, every stage, for any given element, is a category defined by its own characteristic set which is defined to exist as relative combinations of negative and positive characteristics sub-sets. In other words, given an element with a defined qualitative disposition, the relative quantitative combinations of the negative and positive characteristic sub-sets provides a categorial *relative characteristic set* that establishes the elemental identity over infinite time domain. A categorial element in two time positions is not necessary the same; it is said to be the same if its categorial relative characteristic set is the same under the same logical construction.

The introduction of characteristic set and its division into negative and positive characteristics subsets as defining every element places every element as duality in in time and as polarity over time under the principle of opposites in a relational continuum, unity, internal tension and struggle for the production of plenum of forces for the acquisition of internal power and dominance by either the negative or positive characteristic sub-set (dual). In this respect, the theory of category formation extends the concept of duality into polarity and relates to both duality and polarity, the contents of actual-potential dispositions of elements in the ontological space to establish forces of qualitative transformation on the relative behavior of the positive and negative characteristic sub-sets of the elements. The introduction of the negative and positive characteristic sub-sets, and the manner through which they work to relate their behaviors to the dualities and polarities under the Africentric principle of opposites is another important point of departure from other searches for a unified theory of mechanism for transformations and changes. It is also a point of entry into the construction of the theory of transformation.

The theory of categorial conversion takes as its analytical development the existence of ontological categories in time and over time and are transcribed onto epistemological categories by cognitive mapping, and then made intelligible in the epistemological space through the principle of nominalism. The existence of ontological and epistemological categories is provided by the theory of categorial formation. The use of these categories in the theoretical construct is as an important point of departure and entry into the search for a general mechanism of change. The concept of category is a foundational pillar in the transformational analytics of intra-state and inter-state behaviors of elements. In order to show intra-categorial and inter-categorial conversions, Nkrumah acknowledges the existence of transformations from within elements that bring about internal self-changes of qualitative properties of elements. Such internal qualitative changes are referred to in this monograph as intra-categorial conversions involving qualitative equations of motion, transient process and transversality conditions. The qualitative changes are induced by internal qualitative self-motions which proceed to induce inter-categorial conversions as explained in the monograph.

Qualitative changes take place by agencies of action which Nkrumah associated with internal forces. By claiming that matter is a plenum of forces which are in antithesis to one another, Nkrumah evokes the Africentric principles of opposites [R1.8] [R1.37] [R1.48]R1.56] [R1.86][R1.92][R1.243][R1.249]. This claim extends to all socio-natural categories with primary and derived categories, where the derived category emerges from the primary category by a process. The introduction of primary category and derived category is important departure as well as an entry point of transformation analytics on the basis of the principle of opposites. The socio-natural categories, whether primary or derived, are asserted to be internally active and hence endowed with powers of self-motion. The principle of the powers of self-motion is a derivative of the principle of internal tension on the basis of contradictions and tension as seen from the fundamental principle of internal opposites within elements and between categories. The fundamental principle of opposites constitutes the primary logical category from which all other logical categories are derived under categorial-conversion analytics. The unified theory of categorial conversion, as providing explanation to the mechanism of all transformations, must show how forces are created from within. Here, it is useful to keep in mind that motion requires force and force generation requires energy, where energy and matter are intersupportive. It is here that the power of the principle of opposite provides us with logical possibility to develop the needed theory. The manner in which the construct of the theory combines categories and the principle of opposites is another point of departure for the development of the general theory of mechanism of socio-natural transformations. Here, as an example, one may conceive the universe as one element composed of negative and positive characteristic sets under the principle of opposites.

The principle of opposites under a relational continuum and unity constitutes the epistemic conditions to develop the theory of categorial conversion by the use of methodological constructionism and reductionism to understand qualitative and quantitative motions of categories under the principle of relational interconnectedness. Here, the analytical argument of Nkrumah in placing a distinction between quality and quantity and as well as their relational unity provide a powerful analytical process to abstract the necessary conditions of intra-categorial and inter-categorial convertibility of elements. The principle of opposites must be understood from the Africentric conceptual system in terms of duality that exists in relation continuum and unity but not in excluded middle without relational connectivity [R1.74] [R1.86]. For universal transformations, the principle of opposites must not be defined in a multiplicity of forces in a multiplicity of directional changes. It must be defined as a collection of dualities where every element exists as a duality that preserves and presents its identity. The nature of this duality must be specified in relation to qualitative and quantitative dispositions of elements in order to activate the internal contradictions and tensions which will produce the required internal energy and force for self-motion. The nature of the opposites, the structure of the antithesis, and the methods of the production of mutual negation are not provided by Nkrumah.

At the level of general abstraction, the agencies of decision-choice actions to produce the sufficient conditions to support the necessary conditions for categorial transformations are not provided. These agencies of decision-choice actions are discussed when Nkrumah applies categorial conversion to transform Africa's socio-political reality which was his main concern given Africa's experiential information structure.

The abstract analytical work is left to be developed to solve the problem of general mechanism of socio-natural dynamics. The general mechanism for socio-natural transformations and how they are related to the Africentric conceptual system that is logically connected to never-ending evolution-creational phenomenon is extensively discussed in [R1.92]. Here, there is an emphasis on actual-potential polarity as the universal polarity. The analytical structure begins with *nothingness-somethingness* polarity as the primary category of polarity from which all other polarities are derived. Nothingness is the actual pole with its residing duality and somthingness is the potential pole with its residing duality of the actual-potential polarity. The methodological analytics has been named as *polyrhythmics* and the general conceptual framework as *polyrhythmicity*. Every transformation dynamic must specify the nature of the specific polarity by identifying the actual pole and the residing duality as well as identifying potential pole and the residing duality. The elements of the system of opposites exist in relational continuum and unity. The use of this methodological approach allows the examination of the Nkrumah's conceptual system to be expanded and developed into the required theory of categorial conversion and show its applicable areas of thought and establish the foundation for a solution to the transformation problem in socio-natural categories. It is here that the African conceptual system takes precedence in providing a powerful understanding of negation of negation of transformational enveloping of  categorial elements and evolution of never-ending forms.

At the level of general theoretical abstraction, Nkrumah presents the problems of categorial conversion and the search for conditions of convertibility of socio-natural categories. The problem of categorial conversion is general across all scientific boundaries and socio-natural elements. The existence of the logic of convertibility is general but the set of conditions of convertibility are area and object specific with common logical links for epistemic construct. Toward the construction of the general theory relevant to the solution to the problem of categorial conversion Nkrumah states:

> Today, however, philosophy has little need to resort to crypticism. Speaking, in general terms, I may say that philosophy has fashioned two branches of study which enable it to solve the problem of categorial conversion in a satisfying way. These tools are logic and science, both

of which owe their origin and early development to the demands of philosophy. The conceptual tools which philosophy has fashioned in Logic, and by means of which it can cope with the formal problems of categorial conversion, are contained in nominalism, constructionism and reductionism. For philosophy's model of categorial conversion, it turns to science [R1.203NK, p. 21].

The manner in which nominalism, constructionism and reductionism operate in science, logic, knowledge-construction and the supporting mathematical system is discussed in [R3.7] [R3.10] [R3.13]. Taking into account the advice of Nkrumah, the current monograph is developed as a unified theory of dynamics of qualitative and quantitative dispositions of socio-natural elements from within the elements and categories, on the basis of methodological nominalism, constructionism and reductionism. The methodological nominalism allows one to establish categories, assign names and distinguish between primary and derived categories. The methodological constructionism is a *forward logical process* that allows one to show the mechanism through which the derived categories emerge from the primary category. The methodological reductionism is *backward logical process* that allows one to reveal the primary category as the source of the derived category. In other words methodological reductionism is a test of validity of methodological constructionism. The relational unity of this pyramidal logic of knowing, combined with the principle of opposite, provides the necessary instruments for the development of the theory of categorial conversion as a unified conceptual system. The unified theory that is sought is to create a general theory for the understanding of the mechanism of transformation-creational phenomenon of socio-natural elements. The dynamics of the qualitative dispositions relate to transformations of the nature of the internal characteristics of elements through their internal qualitative self-motions that bring about intra-categorial conversions, where crises are continually manufacture at increasing degrees of intensities to bring in categorial moments. These degrees of intensities, the *categorial moments* reach breaking points (tipping points), where inter-categorial conversions take over to manufacture *categorial transfer functions* through their amplifications of the dynamics of quantitative dispositions to satisfy the sets of *categorial transversality conditions*. The development of this unified theory is not only the task of this monograph, but a support and amplification of Nkrumah's conceptual system as a conceptual map to integrate thought and action. Let us keep in mind that transformation is process in destruction-construction duality and polarity with relational continuum and unity. Transformation is impossible without destruction and construction (creation) is impossible without transformation. The destruction-construction duality is an alternative logical way of viewing the cost-benefit duality that implies the transformation-substitution phenomena under the principle of opposites with relational continuum and unity in the universal system. The universe is simply a family of infinite sets of transformation-substitution processes operating on matter-energy-information configurations with decision-choice actions under a system of cost-benefit rationalities. In all processes of categorial conversion, the actual is the cost and the potential is the benefit since the actual is destroyed and the potential is created. The potential is the substitution for the actual as such the potential is the cost and the actual is the benefit forgone. There is inter-substitution process of costs and benefits in destruction-construction process that the theory of categorial conversion reveals.

To Nkrumah, the logical foundation of the theory must acknowledge the principle that every element is composed of plenum of forces which are opposed to one another to generate energy for internal self-motion. Here, Nkrumah makes an appeal to the Africentric concept of opposites where duality with relational continuum is a basic property of elements with internal dynamics of mutual negations as seen in the *Dogonic system* of thought, *adinkralogy* or adinkramatics of the Akan conceptual system, *Pharaohnic-Nubian-Ethiopian conceptual* system and African traditional thought system in general. To activate the core of the principle of duality, Nkrumah introduces its application through the dualistic concepts of negative and positive actions in its application under the development of Philosophical Consciencism as applied to social transformation. The concepts of negative and positive actions will be shown to be foundational and universal to dynamics of qualitative and quantitative dispositions of all socio-natural elements and categories. The objective of Nkrumah in his analysis of general philosophy and the discussion of categorial conversion is to provide a reasonable logical justification in support of internal self-motion and change to transform the African societies from within by decolonization and complete emancipation through socio-economic development. The objective is not to develop a unified theory of categorial conversion. This task of development of the unified theory is undertaken in this monograph, where the theory of categorial conversion is placed on general philosophical and scientific levels.

Socio-natural transformations are categorial conversions which are internally induced in all socio-natural elements. To bring about any categorial conversion requires the understanding of convertibility conditions of socio-natural categories. This understanding requires specifying from within each element and category the attributes that constitute the duality and tend to generate negative and positive actions. The attributes are seen as negative characteristic sub-set that constitutes the negative dual, and the positive characteristic sub-set that constitutes the positive dual to combine to form the duality of any element. Another way of understanding the duality of any element is in the cost-benefit plane where every element is simultaneously cost and benefit to constitute cost-benefit duality in a relational continuum and unity under the asntrofi-anoma decision-choice rationality. The principle of mutual negation of never-ending process of qualitative transformation is seen through the principle of relational continuum and unity where every negative characteristic sub-set has an inseparable supporting positive characteristic sub-set and vice versa. The same principle reveals the principle of creative destruction in relational cost-benefit duality. It is this relational continuum and unity that provides the foundational claim that in *great panorama of creation, nobody ever lived saw its beginning, and nobody will ever live to see its end* which is given the symbolic representation of the *African wisdom knot* that expresses the non-distinguishability of nothingness and somethingness or beginning and end [R1.74][R1.92] [R1.234]. The principle of continuum and unity is fully expressed by Nkrumah in the appendix to his book Consciencism [R1.203].

The universal duality qualitatively appears in the categorial conversion as a negative duality with the power of the negative dual dominating that of the positive dual, and positive duality with the power of the positive dual dominating that of the negative dual These negative and positive dualities must be related to the universal polarity where the negative pole is defined by the negative duality and the positive pole is defined by the positive duality. All the conceptual elements, under the principle of opposites, where polarities and dualities are central, reveal themselves in relational continua and unity to define categorial existence and transformations. The elements of the negative characteristic sub-sets generate internal *negative actions* that form the aggregate negative action which is transmitted through the negative duality to the negative pole of the polarity for internal polar action. Similarly, the elements of the positive characteristic sub-set generate internal *positive actions* that form the aggregate positive action which is transmitted through the positive duality to the positive pole of the polarity for an internal polar action. The theory of categorial conversion acknowledges the existence of the space of actual and the space of potential without which transformations are not possible. The space of actual contains actual elements with defined identities whether cognitively known or not. The space of potential contains potential elements with defined identities whether cognitively conceived or not. To develop the theory of categorial conversion process, the polarity is associated with the actual and potential elements and named the *actual-potential polarity*. All these taken together constitute the Africentric concepts of the principle of opposite from which tension, energy, force and power of qualitative internal self-motion of elements and categories may be abstracted from matter. The philosophical basis of this claim may also be found in [R1.92].

## III Reflections on Theories of Transformations

The current theories of transformation at the levels of nature and society are viewed in terms of epistemic separation without logical connectivity and unity. The fact is that both nature and society exist in the general ontological space, where their behaviors are governed by the same universal laws of states, changes and processes. When one holds the conceptual system of either explanation or prescription of behaviors of nature and society then one must view the universe as an awesome factory for continual destruction and creation where old forms are destroyed and new forms are created in a never-ending organic process under transformation-substitution principle. In this organic universal factory of continual destruction-construction process, nothing is quantitatively lost or gained within its internal self-transformation dynamics except continual transformation of qualitative characteristics of elements and categories as well as categorial distribution of elements for a continual emergence of *differentiation-unification dualities* with the destruction of one polarity and emergence of another polarity leading to new *distribution of polarities*. *Differentiation* means destruction of existing forms; and *unification* means the creation of new forms in the universal system while the distribution means allocation and reallocation of matter, energy and information while transformation means the combination of all of them in the family of sets of categories.

As cognitively viewed and logically conceived, the current theories of socio-natural transformation dynamics in quality-quantity plane, including the whole corpus of economic development theory are plague with epistemic problems, misleading notions and weak theoretical structures which create wanting elements of scientific character. Here, the lack of explicit and implicit mechanisms of decision-choice processes and the value system that motivates action and inaction prevent important critical and insightful analyses as well as the understanding of socio-natural transformations as an ontological elements are placed in the quality-quantity space with neutrality of time. The neutrality of time means that the passage of time alone cannot bring transformation in the quality-quantity space. In order for transformation to take place the duals must organize, mobilize, develop control instruments that must be managed to change terms of relation between the negative and positive duals and dualities. The concept of socio-natural elements as a unit of theoretical construct in the theory of categorial conversion brings us to an important question of whether explanatory and prescriptive theories of transformations of social elements are the same for natural elements in their dynamic settings. An additional question involves around how to abstract the similarities and differences of these theories in exact and inexact sciences.

In all areas of transformation of ontological elements, the theory of categorial conversion is composed of two sub-theories of categorial formation and categorial dynamics. The objective of the theory of category of category formation is to establish meaningful categories and create a family of sets of categories on the basis of their internal characteristics. The objective of the theory of categorial dynamics is to derive the necessary conditions of convertibility of elements among categories where relational structures are established among quality, quantity, subjectivity, objectivity and their relational connections to exactness and inexactness in all areas of socio-natural decision-choice activities. To understand these relational structures and connections in theoretical constructs of categorial conversions, *quality and quantity* are viewed as *ontological phenomena. Objectivity and subjectivity* are viewed as *epistemological phenomena. Exactness and inexactness* are viewed as *information phenomena.* The epistemic challenge of the construction of any theory of categorial conversion is the development of self-contained logical system that integrates these essential attributes of ontology, epistemology and information linkages which will bring into focus the qualitative and quantitative general laws of categorial conversion where time is taken as neutral in the matter-energy-information relational interplays.

In the search for a unifying epistemic structure which will connect the epistemological space to the ontological space, the central concepts of critical interest are quality, quantity, category, primary category, and derived category, laws of motion and, transmission mechanism. These must be supported by a paradigm composed of its logic and mathematics for the construction of the needed unified epistemic structure. These central concepts when operated on with the selected paradigm must be related to thought and find usefulness in explanation, prescription and prediction in socio-natural transformations. The success of the unified epistemic construct lies in the meaningful relational structures among language, information and paradigm of thought over the epistemological space. Here, language and paradigm occupy a special position in the knowing process and the connection to the theory of categorial conversion by their impacts on what can be abstracted from the information over the epistemological space relative to the ontological elements.

## IV The Motivational Objective of the Monograph

The central motivational objective of the monograph is the development of the theory of categorial conversion from its analytical foundations of philosophy and mathematics. The theory must be a unified one that provides an understanding of general mechanism of quality-quality dynamics that is equally applicable to social and natural systems. The unified theory must identify categories and show how the categories are related to nominalism. It must also show the relationship between actual and potential categorial elements as well as the mechanism where the actual element fades into the potential and potential elements are transformed into the actual through qualitative motions to alter the relative negative-positive characteristics of categorial elements. To claim a categorial conversion of a categorial element, the motivation is to construct a theory that must indicate the general conditions of the *points of transition* to either actual or potential. It must also indicate the categorial transversality conditions. The conditions of points of categorial transition are designated as *Nkrumah Delta* which may also be called *categorial-conversion delta*. The categorial conversion (Nkrumah) Delta is classified as *negative categorial-conversion Delta* or *positive categorial-conversion Delta* in the transient points. The *negative categorial-conversion delta* corresponds to negative categorial conversion where the negative dual and duality have the power of dominance over the positive dual and duality in the new actual. The *positive categorial-conversion delta* corresponds to the positive categorial conversion where the positive dual and duality have the power of dominance over the negative dual and duality in the new actual. The added motivation of the monograph of is to show the conceptual generality of the categorial-conversion Delta where Nkrumah Delta varies from category to category depending on the transversality conditions. This Delta is the demarcation force required for the effectiveness of the transfer function to move an old actual into potential and a potential element to a new actual in the categorial-conversion process.

The set of conditions that ensures the conversion is the set of convertibility conditions which also varies from category to category. These convertibility conditions must be created from within categories if sustainability of categorial conversion is to be maintained. In natural categories, these conditions are the creation of internal natural decision-choice dynamics of conflicts of duals and dualities in nature. In social categories, these convertibility conditions are the complicated works of the members of the society through human information-decision-choice processes of balancing the individual and collective preferences and selections in the individual-collective duality under a relational continuum and unity. The development of the unified theory requires the abstraction from multiple areas of knowledge such as category theory, the theory of formal languages, mathematics, philosophy and many others. The task of the development of unified theory of categorial conversion as an analytical foundation of Nkrumaism begins with a preamble and followed by a prologue where the works of critics and admirers of Nkrumah are examined in a philosophical-mathematical space. The motivation of the examination is to define the conceptual problems and analytical pillars of Nkrumaism and how these are related to policy options that may be required of decision-choice actions. The introduction of the notions of characteristics, identity, opposites, duality, polarity, continuum, tension, energy and information are motivated in relation to the roles they play in the categorial conversion which is then related to socio-natural development and growth. The general nature of the theory is indicated and the areas of application in sciences are suggested. The basic motivation, here, is to present a foundation and analytical morphology of a general theory of socio-natural transformations as well as present the conditions that may be extended to general theory of engineering sciences. Another important motivation in the construct of theory of categorial conversion within the constructioninism-reductioninism duality is to show how the epistemic construct draws its weapons and analytical tools of reasoning from the African roots and African conceptual system under the principles of opposites composed of polarity, duality, negative-positive characteristic subsets and actual-potential relational structural distribution over categories established by primary category and derived categories[R1.83b] [R1.84] [R1.92] [R1.203] [R1.218a,b,c][R1.243]. In this framework, the members of African intelligentsia and educated class are charged with the greatest responsibility to move the African conceptual system forward. In this endeavor, no traditional stone from the ancient to the present and connected to the future must remain unturned where a continual epistemic success can only come if all the members of African intelligentsia and educated class shoulder this intellectual responsibility through relational unity and continuum of traditional efforts to decolonize the African mind. The effort to decolonize the African mind is another important motivation for the development of the monograph. It is interesting to note that this monograph is based on the complete traces of the African creativity in thinking as a thought system without significant disruption of African epistemic origins from the ancient to now. This reveals the African epistemic unity in creative thinking. The African thoughts that are attributed to Greeks must be challenged as to their historic validity as has been pointed out by a number of writers.

## V Organization of the Monograph

The main body of the monograph is organized into eight chapters. These chapters are introduced with a preamble and prologue. The preamble provides some reflections on paradigm, laws of thought and African traditions as they relate to its paradigm and the structure of the universe. The prologue presents some key concepts within the African traditions, and the general and specific epistemic structure of the monograph, providing exist and entry points into the general logical framework in the constructionism-reductionism methodological duality with relational continuum and unity. The contents of the preamble and prologue introduce the entry points of category formation and the principle of opposites that proves us both an exist and entry points to the development of polarity and duality as logical instruments to deal with the dynamics of qualitative and quantitative dispositions as well as categorial dynamics within the African conceptual system. The objective, here, is to connect the theoretical developments of both categorial conversion and Philosophical Consciencism to the fundamentals of African cognitive unity as it has been presented in [R1.92] in support of the African cultural unity as it has been presented in [R1.84] as well as African socio-political unity as presented in [R1.91] [R1.202]. The driving force of this methodological approach is simply the notion that the an integrated African conceptual system, its paradigm of thought, logical, philosophical and mathematical contributions will remain in epistemic obscurity and cannot be developed until the members of African intelligentsia become awakened to connect it with the conceptual foundation of the principle of opposites that forms the core of African thought system as it was in Kamet (Ancient Egypt and Nile Valley Civilization). In this monograph of the theory of categorial conversion and the follow up monograph on the theory of Philosophical Consciencism, the interest is not only on Nkrumah, but on the on the development of an African-centered foundation of his thinking and reconstruction of African social formation from within Africa.

Chapter one deals with the problem of abstract ideas and the practice of ideas in social settings. It brings into focus the relational structure of philosophy, ideology, freedom, liberation and human action as derived from within on the basis of decision-choice activities to resolve conflicts of preferences in the  individual-collective duality. Chapter Two deals with the problem of mental structure of the colonized and oppressed people in general, and with special reference to Africa as it relates to mental colonization. Contending ideologies and opposing philosophies are discussed where the task of Nkrumah on the road to Africa's decolonization, independence and emancipation is analytically presented in relation to the development of a theory that justify internal self-motion. Chapter Three deals with the intellectual tasks in creating thought to guide the Africa's decolonization and emancipation from within under the principle of self-motion that relates to experiential information structure. Reflections that are made by some Pan-African personalities on the intellectual tasks for Africa are discussed in relation to ontological-epistemological duality and polarity. These reflections are generalized across boundaries of thinking and knowing.

Chapters Four to Eight are used to develop the general unified theory of categorial conversion. Philosophical foundations and extensions of

Nkrumah's conceptual system are discussed in Chapter Four. Here, the theory of category formation is presented to initialize the conditions and the logic of qualitative disposition in relation to quantitative disposition. From the theory of categorial formation, the general conditions that define elemental identities, and establish categories under the principle of relational continuum and unity are derived. The categories are overlapping categories that are distinguished in the epistemological space by the application of fuzzy logic and the methods of fuzzy decision. The methods of fuzzy decision are related to the principle of opposites composed of actual-potential polarity and supported by a system of dualities.

Chapter Five is used to present the axiomatic foundations of the theory of categorial conversion. The chapter is used to discuss the concepts of *time trinity* of the past-present-future phenomenon represented in *adinkralogy* as the *sankofa-anoma*, the axiomatic relations among the actual, reality and potential in the categorial-conversion process. The discussion of the system of language, vocabulary and grammar of the theory of the categorial conversion is analytically presented to the reader. The system is linked to information-knowledge structure and further linked to the Africentric conceptual system with special reference to wisdom and intelligence in the controllability of the system of categorial conversions. A set of Africentric postulates, axioms and principles are presented to show their relational foundation to the development of the theory of categorial conversion.

Chapter Six is used to present the conditions and relevance of the analytical building blocks of the theory, and how the analytical building blocks are defined in the ontological and epistemological spaces. The development of this chapter provides a framework to present a relational continuum and unity of matter, energy, information and self-motion in relation to the principle of opposites. The mathematical foundation and symbolic representation of the categorial-conversion process is presented and discussed in Chapter Seven. The categorial-conversion problem is mathematically stated with a discussion on its structure and how it relates to fuzzy mathematics and fuzzy laws of thought. Chapter Eight concludes the monograph by relating the mathematical problem of the categorial conversion to the problem of general transformations. A solution to the problem is abstracted. Additionally, the method of *categorial-conversion analytics* and conflicts are relationally presented and linked to the theory of games under the conditions of polarity, duality, relational continuum and unity where the games are zero-sum, two-entity games. The zero-sum-two-entity game allows the formulation of the problem of the conflicts between the negative dual and duality on one hand  and the positive dual and duality on the other hand, and then examine the strategies and  tactics with their effects on the poles of any actual-potential polarity.

# PROLOGUE

## I. Relections on Critics and Admirers of Nkrumah

There are many people that have laid claims to be Nkrumaist and subscribe to some ideas of Nkrumah. These people may be separated into different levels of intellectual groups with differential understanding and orientation to the ideas of Nkrumah. Some are drawn to Nkrumah's political position in general and specific. Others are influenced by Nkrumah's anti-colonialist and anti-imperialist and anti-neocolonialist stands. There are others who are attracted by Nkrumah's position on African Unity and its relation to non-imperialist and anti-neocolonialist world. Others are attracted to Nkrumah's orientation and program for complete political and economic emancipation on the basis of self-reliance, communalism and social welfare improvement, which he termed scientific socialism from the basis of African traditions in thought, organization and culture brought to modernity with improvements in science of organization and production. Besides his admirers, there are many who criticize Nkrumah's organic views from one position or more without a full or reasonable understanding their relationships and their intellectual unity. The admirers and the critics fail to fully grasp the organic system of Nkrumaism and how this system is rooted in the African intellectual traditions. The Nkrumaist positions may be grouped under politics, law and economics with one unified laws of thought philosophical positions and science of social organization. These lines of intellectual divisions affect the preferences of the supporters, admirers and the critics.

The appreciation of the respective positions requires that we understand an important guiding principle of the lines of reasoning which forms the foundations of Nkrumah's philosophical position in general. Here, Nkrumah understood that philosophy, like science. is about something, and this something is the relationship between thought and human experience and how this experience translates into social progress and understanding of forces of nature. From this conceptual position he states:

> Philosophy, in understanding human society, calls for an analysis of facts and events, and an attempt to see how they fit into human life, and so how they make up human experience. In this way, philosophy, like history, can come to enrich, indeed to define, the experience of man. [R1.203, p.2]

It is through this understanding, that philosophy is related to society. This philosophical understanding must be related to experiential information structure, conditions of existence and the social environment which helps to define identities, national interest and social vision.

In this way, Nkrumah came to understand philosophy as an interpretation of experiential information and explanation of the interpreted results where the experiential information is rooted in the conditions of particular social set-ups. The philosophical interpretations must be coordinated with the experiential information and then related to social circumstances and social paradigms of thought. The social paradigms of thought are culturally specific, in some sense, but general to human experience through the time trinity of past, present and future as conceptually embedded in the *philosophical sankofa-anoma* [R1.92]. Different experiential information structures give rise to different paths of evolving cultures where each culture is spanned by information on institutions of economics, politics and law broadly defined to understand social problems as they relate to the natural order of things. The conceptual system of Nkrumah and its rational foundations, as being advanced, deal with a complex system of socio-natural organisms in their totality of existence. At the level of social system, they involves more than politics, economics and law as the triangular pillars of societies. They include the analysis of culture, ideology, moral conduct, social conflicts, war, peace, social expression and the methodology of knowledge production in time and over time. The knowledge production includes the system of information collection, processing and interpretation under a particular paradigm of thought, logic and philosophical systems and scientific know-hows. The Nkrumah's analysis and synthesis of the whole of the social structure and its evolutionary dynamics are guided by a particular philosophical foundation that provides a scientific *map of thinking*. This map of thinking is drawing from the understanding of dynamics of natural qualitative disposition and projected onto the social space. The map of thinking and its application to social science and natural sciences are neglected by his admirers and critics. The neglect has produced important distortions in the interpretations of the Nkrumah's conceptual system and how it is intimately related to scientific theories of social and natural motions as seen in the dynamics of both qualitative and quantitative dispositions, and how the qualitative and quantitative dispositions provide continuum and unity of relational structures to matter, energy and information with neutrality of time. The critical objective of the scientific development of this map of thinking in Nkrumaism is to establish the principle that lasting and sustainable change is internal to the object of change. The theory of categorial conversion that is presented here is to provide the rational foundations for the principle of transformation from within categorial elements and hence give explanation to the internal dynamics of qualitative and quantitative dispositions. In this respect a scientific foundation is established for explaining social transformations as well as to engineer a change.

The exposition of scientifically relational structures of quantity, quality and time as applied to socio-natural transformations are completely neglected by critics, thus destroying the application strength of the Nkrumah's conceptual system to their understanding of socioeconomic and politico-legal dynamics of social formation and Nkrumah's policy behavior in African decision-choice space. In this way, both the admirers and critics thus overlook a great philosophical and scientific contribution by Nkrumah to knowledge. The failure, particularly on the side of the African leadership, of the understanding of the character of the Nkrumah's logical departure from the Eurocentric system of thinking has destroyed the development of the needed transformative character required of the African leadership. The link of the Nkrumah's conceptual structure to African-centered system of thinking on the basis of the principles of the opposites with relational continuum and unity abstracted from African experiential information structure is yet to be understood by both the critics, admirers and African leadership who are living in the space of neocolonial slave-ships with pure love for neocolonialists whose ideology has forced the African leaderships at all important social fronts to live in the comfort zone of familiarity of neocolonial mind set thus destroying the essential development of their useful transformative character. The destruction of the development of the transformative character of the African leadership has lead to the creation of unthinking society that fails to understand the meaning of independence and its relationship to Africa social vision, national interest, the relevant social goal-objective configuration and nation-building. The important interactions of transformations and qualitative motions at the levels of nature and society are not understood by both the casual and non-casual reader of the Nkrumah's works which are extensive and intensive. The interactions of science, technology, engineering, mathematics and philosophy are neglected in Nkrumah's conceptual system especially in *Consciencism* [R1.203]. The meaning of category in relation to transformations is misunderstood and the analytical requirements of category formation in mathematics, philosophy, engineering and science are not appreciated and connected to the general system of knowledge production in understanding qualitative and quantitative changes in nature and society. The importance of Nkrumah's review of the essential elements of philosophy with the problem of mind-body duality with relational continuum and unity is completely misplaced in terms of how it relates to categories and transformative migration of elements from one category to the other. The foundational roots of the African traditional thinking, philosophy of life and social organization in the Nkrumah's conceptual system are not understood and appreciated. Perhaps, both the admirers and critics fail to understand the analytical foundations of the African thought system and its logic of reasoning on the basis of which Nkrumah advanced his conceptual system. It was the recognition of this deficiency on the part of the critics and admirers that the work on *polyrhythmicity* was produced [R1.92]. The goal in this monograph is not directed to a fight with the Nkrumah's critics and admires. It is being presented to develop the understanding of what is simply a great initial contribution of Nkrumah's conceptual system to philosophy, mathematics, science, engineering and social theory in dynamic space involving the interactions of quantitative and qualitative dispositions. In the development of this monograph, the driving force of the Nkrumah's seminar book will be made explicit and related to the theory of categorial conversion that is being presented here. This monograph is devoted the development of the theory of categorial conversion as rational foundations of Nkrumaism. This rational foundation is part of general theory of transformation. A minimal expositional extension of the relationship of the theory categorial conversion to the understanding of a possible construct of unified theory of engineering sciences and in relation to the justification of the theory of knowing will be sketched as the one found in [R3.10] [R3.13].

## II The Conceptual Problems and Analytical Pillars of Nkrumaism

The central analytically explicit pillars in the Nkrumah's conceptual system are category, nominalism, constructionism, reductionism and categorial conversion on the basis of self-motion. Nominalism relates to category formation and the principles of epistemic identity. The Implicit pillars are the principle of opposites, polarity, duality defined by negative and positive characteristic sets with relational continuum under unity, conflicts and categorial resolution. The structure of nominalism, as it relates to shades of meaning and how the shades of meaning relates to analog and continuous process are not understood in terms of vocabulary, grammar, language and cognition. There is an implicit or explicit failure in understanding the relational structure between vocabulary and categories which constitutes the essential basis of any language whether formal or ordinary. The universe is viewed in a never ending continuous transformation under the principle of continual creation and recreation of the universal order where the human understanding of this continuous transformation through the methods of digital and discrete processes as approximations of point-to-point qualitative changes is not understood in terms of Nkrumah's analytical tool of categorial conversion on the basis of internal self-motion since explicit connections were not made. Within the language of thought in Nkrumah's conceptual system, the analytical and conceptual roles of categorial conversion and their importance to scientific and philosophical reasoning to planning and engineering are not appreciated. The methodological constructionism allows the derivations of propositions from the primary conditions while the methodological reductionism allows one to terrace the derived propositions to the primary conditions for verification. This is an important methodological foundation in knowledge acquisition and verification under both limited and vague information structure that lead to expressions of probability and possibility in thought.

At the level of history of thought, Nkrumah's analytical perspectives on Western philosophical development, its strengths and weaknesses are neglected and not related to Nkrumah's search for an alternative developmental path from within the African experiential information and Africentric conceptual foundation to the creation of a revolutionary philosophy in support of a revolutionary ideology for decolonization and complete emancipation of Africa and African people. In other words, the basic reasoning and objective on the account of questions and answers of the histographical structure of Western philosophy and methods of reasoning are brushed aside by both critics and admires who merely concentrate on the familiar grounds of conceptual simplicity without critical thinking and reflection on the problems which Nkrumah defined and search for analytical solutions and practices consistent with his vision that practice must be guided by thought and thought must lead to practice. Nkrumah's central problem was about qualitative transformations of the African society on the principles of Africa historic existence where African has African root and bear true African fruits but not Eurocentric roots that give rise to strange fruits. This problem is of two types. Type I involves the qualitative transformation of sovereignty institutions from external control to domestic control on the basis of domestic internal self-motion. Type II involves qualitative and quantitative transformations of socioeconomic institutions and activities within the domestic sovereignty through domestic internal self-motion. Both of these types of the problem are viewed within the African thought system of with conceptual foundations of opposites, polarities, dualities, relational continuum and unity with conflicts involving actual-potential processes. The genius of Nkrumah may be seen in the manner in which he combines the African principle of opposites, with category and self-motion through categorial conversion to craft a philosophy that draws it weapons from the African traditions in thought where every situation or every category is defined to exist as an actual-potential polarity.

The African roots of Nkrumah's revolutionary philosophy and the corresponding revolutionary ideology must reveal their African conceptual foundations. For example, the existence of the universe while partially and experientially obtained does not allow itself to experimental confirmation of any claims of its creation. The African conceptual system affirms the primacy of the universe by the postulate of *no-beginning and no-end* in the sense that the beginning is in the end and the end is in the beginning. In other word, it affirms the ontology as the identity in the knowing process over the epistemological space. For example, in the Akan philosophical system of thought, the postulate of permanency of the universe is affirmed by the statement that: *in the great panorama of creation, there was no living-body that saw its beginning and no living-body will see its end.* Its corresponding statement in duality with a continuum and unity is simply *the end is the beginning and the beginning is the end.* Nkrumah was aware of these postulates and the thought system in which they are embedded which are extremely articulated and discussed by Danquah [R1.73] [R1.74] [R1.75], Diop [R1.83b] [R1.85][R1.86] and Ben-Jochannan [R1.35] [R1.37]. The postulate of the permanency of the universe in Africentric thought system is connected to the principles of opposites containing polarity, duality with relational continuum, unity and conflicts. In this framework, the universal creation and its permanency are defined under the principles of nothingness-somethingness polarity and duality with continua in actual-potential space where analytically actual resides in the potential and the potential resides in the actual. Alternatively stated, every actual has a supporting potential and every potential has a supporting actual in the continual internal transformation of the universal elements. Each polarity has a residing duality while each duality is composed of positive and negative characteristics sets under active tensions that relate to internal transformations [R1.92] [R3.10] [R3.13].

The postulate of the permanency of the universe, combined with the principles of nothingness-somthingness polarity and duality with continua in the actual-potential space requires explanations of the contents of the universe and continual creation in terms of qualitative transformations. The contents of the universe are simply *what there is* and must be linked to its knowability under some logic of knowing. The process of identifying *what there is* leads to *category formation* where the collection of these categories of the universe is an infinite set in its human conception. The process of explaining *what there is* leads to the understanding of continual categorial conversion without end. The continual categorial conversion leads to the postulate that the universe is infinitely self-contained and closed under the principle of continual self-transformation without a loss of matter and energy. What are in the universe are forms of matter and energy which are relationally connected by information characteristics. The permanency of the universe is maintained but there is no permanency of the qualitative dispositions of the elements and categories. The only thing that is lost in the universal system of transformations is categorial quality where an element in one category changes its qualitative characteristics and categorically migrates into another category through qualitative self-motion. It is this never-ending process that the concept of infinity finds expression and meaning both ontological and epistemic spaces.

## III The Notions of Characteristics and Identity

The qualitative self-motion of any element is seen in terms of changes in qualitative positions in the quality-quantity space with time acting as neutral. The identity of each element is defined by its qualitative characteristics that place it into a category. Changes in the qualitative position are related to quantitative position which is reflected in changes in qualitative characteristics. Within each qualitative category there are derived quantitative categories that reflect quantitative disposition. Here, the analytical work must first establish the qualitative characteristics of the different elements in order to specify their differences and similarities. These differences and similarities allow the establishment of the conditions of epistemic identities of *what there is* and the category of belonging. The defining conditions of the category of belonging involve the development of the *theory of category formation*. The required theory of category formation was not developed in the Nkrumah's conceptual system. The concept of category as an instrument of reasoning is assumed and used by Nkrumah. The manner in which categories are formed in relation to nature and society is left to the reader in Nkrumah's analytical works. The relationships between category and nominalism and then to vocabulary and language are implicit that must be abstracted from Nkrumah's work in *Consciencism* [R1.203]. These may be seen as analytical weakness, but they are not. A number of examples are given by Nkrumah to establish the existence of categories. Among such examples are matter, mind and spirit. A full theory of category formation is provided in [R1.92] [R3.7] and the connection to vocabulary is discussed in [R3.10]. The relevant conceptual structure for the development of the theory of category formations will be presented here. There are two types of categories that will be of interest in this analytical process. They are categories in the ontological space which by the principle of nominalism will be called *ontological categories*. The elements in each ontological category will be referred to as *ontological elements*. There are categories in the epistemological space which by the principle of nominalism will be called *epistemological categories*. The elements in each epistemological category will be referred to as *epistemological elements*.

The question that arises, then, is whether ontological categories are independent of one another in the sense that they exist in discrete systems, or they exist in a continuum in the sense that one emerges from another or the existence of one is traceable to another by a process where such a process must be defined and analytically shown to be possible within the actual-potential polarity. The category from which an element emerges is specified as the parent category which is called by Nkrumah as the *primary category*. The category into which an element enters by a process is called *derived category* which is a baby from the parent category. The Universe is seen as a partitioned into a collection of primary and derived categories in a relational continuum. The concept of partition of the universe simply means that every element in both the ontological and epistemological spaces belongs to a category without a residual but in an interconnected and relational mode which is specified in terms of relationality. The interconnected and relational mode of categories preserves the universal diversity and unity under the principle of continuum with continual transformation.

## IV The Concept of Categorial Conversion and Qualitative Dynamics in Nkrumaism

From the discussion in the previous section, it became clear that at any moment of time the universe is partitioned into categories of elements. Every category is either a parent (primary category) or a successor (derived category) in the sense that it either emerges from one of the categories in the universal partition or it is the category from which others emerge through an ontological process. The successors of the primary category are called the derived categories which are obtained by internal changes in their characteristics and arrangements. In order for changes in the characteristics of elements to occur, it must be the case that the primary category is endowed from the beginning with a capacity of qualitative self-motion which becomes the property of all successors. For this qualitative internal self-motion to be operative, the primary category must be endowed with a capacity to generate energy as an internal characteristic in order to produce a force.

The process where one category emerges from another category is what Nkrumah calls *categorial conversion*. The understanding of the working mechanism of the categorial conversion requires the development of the theory of categorial conversion. Such a theory is not explicitly provided by Nkrumah. It is important, however, to grasp the basic structure of the logical and analytical mechanism of the categorial conversion in the universal system of categories. The understanding of this mechanism is important to appreciate the work of Nkrumah in his Consciencism as a revolutionary philosophy and ideology for decolonization and development. The notion of special reference to Africa finds expression in its application to the conditions of the African people.

Nkrumah does not provide us with an explicit logical foundation of the possible theory of categorial conversion; neither does he provide us with the conditions of the shift of paradigm from that of the classical paradigm whose laws of thought is Aristotelian in nature. His work in Consciencism, however, provides *necessary conceptual materials* to create such a theory. The current works in [R1.92] [R3.7] [R3.9][R3.10][R3.13] and Fuzzy logic [R3]provide us with non-Aristotelian paradigm that represents Africa's relational thinking to create sufficient conceptual materials for the development of the theory of categorial conversion. The *concept of nominalism* allows the linking of category formation to vocabulary, grammar and language to science and non-science in a continuous process. The concept of *methodological constructionism* in broad general sense, supported by the African thought of relationality allows the relational continuity of categories in qualitative forward progressive changes to show the possible enveloping of the categorial conversion in scientific and-mathematical reasoning. The concept of *methodological reductionism* supported by the principle of the African conceptual relationality provides us with logical instruments to trace each derived category in a qualitative backward motion to its parent and then in a successive way to the primary category for the test of *categorial validity*. In the process, categorial convertibility conditions are logically derived. The convertibility conditions are made up of conditions of the concept of *categorial moment*, the concept of categorial transfer function and the concept of categorial transversality conditions. The categorial moment is derived from the examination of the conditions of intra-categorial conversion by relating it to the energy field, transfer function and *quantum characteristic set*. The categorial transfer function is derived from the examination of the conditions of inter-categorial conversion by relating it to matter-energy field under information structure. The conditions of categorial transversality are derived from the categorial moment and categorial transfer function to define the point categorial entery. The theory of categorial conversion holds the universe to be closed under transformations of categories. The number of categories is under infinite transformation within the closed universe qualitative transformation given the initial nothingness-somethingness polarity, duality, relational unity under continuum in the actual-potential space. The theory of categorial conversion presents a general theory of mechanism of transformation in nature and society.

## V Opposites, Polarity, Duality, Unity, Continuum, Tension, Energy Information, Actual and Potential in Categorial Conversion

In constructing the theory of categorial conversion, every ontological element is seen as composing of matter and energy that are relationally connected by information and this holds for all categories. This principle of the relational structure of ontological unity and continuum is logically extended to the epistemological space. In the ontological space, energy and energy generation are part of the internal characteristics of any elemental matter in the universe. Any ontological element is both matter and energy. To demonstrate the possibility of energy generation for the internal self-motion, it is necessary to appeal to the general Africentic notion of the principles of opposites. Here, the relevant analytical concepts and tools are polarity, duality and relational continuum under the principle of unity [R1.8] [R1.37] [R1.60b] [R1.92] [R1.124] [R1.243][R1.251a]. All these concepts and tools find expressions in the actual-potential space. Each universal element stands as state and process in relational continuum and unity. As a state and a process, it is defined by actual-potential polarity where the state finds an expression in the actual pole and the process finds an expression in the potential pole. Each pole of the polarity is seen as a duality composed of characteristics set which is divided into negative and positive characteristics subsets that generate negative and positive actions within it to create relational tension under the internal information structure. The tension produces energy for conversion by creating the needed conditions of categorial convertibility. By combining the concepts of opposites with continua and unity, a possibility is opened for a theory of categorial conversion to be developed to show the process of how derived categories emerge from the primary category, and how the derived categories are logically reducible to their parents and then to the primary category.

The identity of each actual-potential polarity depends on the relational structure of its residing duality which in turn depends on the corresponding duals and their relational structure and internal organization. The duals of any duality in relational morphology are seen in terms of internal process of negation of the characteristics where the negative characteristics subset with its negative actions seeks to reduce the quantity of the positive characteristics set in order to maintain itself or alter the identity of the duality to become what it is not. The same process is also taking place with the positive characteristics subset. In general, there are the negative duality associated with the negative pole and positive duality associated with the positive pole in the actual-potential polarity. The identification of the actual or potential pole with either negative or positive pole of the polarity will depend is an epistemic decision-choice action that will also depend on circumstances and conditions. This internal process generates conflicts and negative-positive battles for control within the duality which is relationally transmitted to the poles of the actual-potential polarity for qualitative transformation. The battle for control of the identity of the duality generates the needed energy for the qualitative motion to change the internal structure and organizational arrangement of the internal characteristic set that produces the relative negative-positive characteristic set The internal categorial dynamics leads to the production of the needed transfer function that moves an element from one category to another. The categorial transversility conditions will show the changes in outward expression for cognition. This process is made possible under the principle of relational continuum and unity of both the duality and polarity. The unity and continuum in each duality and each polarity are maintained by an information structure that helps to indicate the direction of the relational change and qualitative self-motion and the effects of the negation on internal quantitative disposition. For the theory of categorial conversion to be successful in its endeavor it must relate quality-quantity duality to *categorial difference*. It also must show how a qualitative motion is abstracted from the quantity-quality-time spaces [R3.13].

## VI Quality-Quantity Duality, Categorial Conversion, Development and Growth

Categorial differences of categories are defined in terms of *qualitative dispositions* and related to *quantitative dispositions* and then to vocabulary, grammar, language of knowing and communication in the epistemological space. Within each category, there is quantitative disposition that affect the categorial quality which is basically its surrogate of its internal arrangement of the quantitative disposition. From the viewpoint of the theory of categorial formation, development is seen as categorial conversion passing through qualitative stages where each stage is placed in a category with defined qualitative characteristics; growth is seen in quantitative stages where each growth is placed in a qualitative state. In respect of the theory of category formation, there is a set of *natural categories* and a set of *social categories*. Two categories are seen to constitute a potential-actual polarity in which resides negative-positive duality. The general dynamics is seen in terms of quantitative and qualitative motions or simultaneity of the two motions [R3.10] [R3.13]. The categorial differences are maintained by the respective qualitative dispositions. Given the development stages we must define *categorial transfer functions* that take one element from one category of development (for example less developed) to another (for example less developed). In other words, we must specify conditions of convertibility for categorial transfer with the potential-actual polarity. It is here, that the study and analytical work of how possibility and probability are connected to actual and potential acquire critical importance in epistemic understanding in the knowledge-production systems under various constraints of information structure. It is also here that that the development of possibility theory with possibilistic reasoning and probability theory with probabilistic reasoning acquires analytical significance in the theory of knowing.

By defining the developmental stages as categories and relating them to categorial moments and categorial transfer functions, conditions and relevant grammar are made possible to work out theories of socioeconomic development and politico-legal transformations. It was on this theoretical basis that, Nkrumah subtitled his Consciencism ,as philosophy and ideology for decolonization and development. The decolonization is in reference to socio-political transformation from colonialism to decolonized states for domestic control of sovereignty while the socioeconomic development is in reference to material welfare changes including nation building and production increases within domestic control of African sovereignty. The two combine qualitative and quantitative dispositions of relevant variables in the dynamics of the social setup. The logical structure for understanding sees social transformation as mimicking natural transformations where independence can emerge from colonialism, freedom from imperialist oppression, more development from less development and similar structures by internal processes of categorial conversion. The categories are distinguished by qualitative differences. The movement of one element from one category to another is a categorial transfer which is induced by categorial conversion. The energy for the categorial transfer is produced by the people who constitute the matter while the social energy and social matter are linked in relational continuum and unity by social information and society and nature are relationally connected in continuum and unity also by information. The social information is produced in the social system as defective and deceptive information structures which influence the decision-choice activities and the controllability of the categorial conversion. It is useful, at this point to understand the relational structure between ideology and deceptive information structure and how it influences decision and control; and the relational structure between dogmatism and defective information structure and how it influences decision and control. The relational structure of quality-quantity duality and development-growth duality is such that there can be development without growth and there can be growth without development as seen in discrete points in relational continuum and unity of the categorial-conversion process for any ontological element.

## VII Polarity, Duality, Conflict, Games and Categorial Conversion

The theory of categorial conversion is about transformations due to the internal dynamics of elements under the principles of opposite. The principles of opposites appear as polarity and duality under relational continuum and unity with negative and positive characteristic subsets. The polarity and duality exist in relational continua and unity with conflicts characterized by fuzzily defined opposite characteristic sets of negative and positive sets. The opposites are related to mutual negation that is induced by transformation processes which pass through penumbral regions of tactics and strategies defined in terms of actions and counter actions. The opposites work under strategies and counter strategies for any given information structure that may be asymmetric. The internal conflicts are between the positive and negative characteristic subsets of any given entity in a fuzzily defined category. The energy required to produce qualitative motion and then categorial conversion is generated by the internal contradictions and conflicts from the strategic and tactical actions by the negative and positive characteristic sets for the negation and negation of negation to bring about categorial conversions. The strategic actions come as positive and negative action and realized as games in penumbral regions defined by fuzzy information structure. The relational structure of the behavior of the negative and positive characteristic sets towards mutual negations may be seen as socio-natural games in the actual-potential polarity. The game is between actual and potential poles to create dominance where the strategies are transcribed through the dualities under information structures and the activities of negative and positive characteristic subsets.

## Viii The General Nature of the Theory of Categorial Conversion

In the development of the theory of categorial conversion, it is important to bring into focus the general nature of the theory and its areas of application. This general nature will define the boundaries of theoretical and logical claims of validities of derived propositions and applications that may be required of them. The theory must be applicable to qualitative and quantitative dynamics in nature and society. Here, the central questions must be generally framed in the quality-quantity-time process as it relates to matter, energy and information to produce qualitative and quantitative motions.

Three types of questions tend to arise in the general conceptual system of categorial-conversion dynamics. The first question relates to whether there are general laws for creating elemental and categorial entities in the ontological space and how these entities are transcribed into the epistemological space for knowing. This question relates to the problem of existence of categories and the mechanism for category formation which also relates to the general ontological problem of *what there is*, which is the actual. The problem is simply defined as the *problem of category formation* in the epistemological space. The problem also relates to the problem of nominalism and languages. The second question relates to whether, given categorial identities, there are general laws of transformation of socio-natural elements where categorial identities may be altered. The implication, here, is that there is a potential space with category formation from which new actuals may be manifested and into which old actuals are absolved into the potential space by a process. This potential space relates to possibility space but must be distinguished from the possible-world space [R3.7] [R3.10] [R10] [R10.1] [R1025]. This is the categorial conversion problem. The third question involves the application of the existence of the general laws of mechanism in the theory of categorial conversion to specific situations. In other words, to what extent does the theory provide a general mechanism of transformation and change in quality-quality-time dimensions. The transformation involves the problem of dynamics of qualitative disposition in quality-time space. To be general the theory must deal with the dynamics of both qualitative and quantitative dispositions of all categories. The first and second questions are the problem of existence of categories and general laws of change. The third question is the problem of the optimal application to the understanding of the general qualitative and quantitative dynamics.

## IX Specific Areas of Application in Sciences

The scientific claim of the theory of categorial conversion is its universality and general application to the understanding of mechanism of qualitative-quantitative changes of socio-natural elements in all areas of ontology and its epistemological understanding. Characteristics are general to all socio-natural elements. It is these characteristics and their internal organization that allow category formation and define *categorial identity* of ontological elements. The existence of differences and similarities of socio-natural elements and human cognitive ability to acknowledge differences and similarities of actual and potential socio-natural elements by acquaintances present an important justification to the claim of universality of category formation. In other words, categories of elements constitute the foundation of knowing of elements, states, processes and transformations that they may artificially and naturally encounter. As it has been pointed out, vocabulary, language and derivations of knowledge are impossible if there are no categories that define similarities and distinction among ontological elements. Claims of scientific discoveries are only possible under categories from which primary and derived categories present the direction and evolution of scientific and other knowledge claims which present knowledge evolution of which Max Planck refers to as scientific world pictures [R12.75][R12.76]. The general process is referred to as enveloping of categorial derivatives [R3.10] [R3.13]. Generally, knowledge production resides in space of epistemological actual-potential polarities relative to the space of ontological actual-potential polarities where actual is what is claimed to be known and potential defines cognitive ignorance in the knowledge-ignorance space. In all applications the actual must be identified so also the preferred element of the potential. The existence of information vagueness and limitation may lead to unintended result relative to the preferred.

To claim a universality for the theory of categorial conversion, as it is being done here, the mechanism of change derived from the theory must help to explain socio-natural transformations of elements in all subject areas including the knowledge production itself and the dynamics of qualitative and quantitative dispositions as well as the actual-potential processes. The application of this theory to knowledge production is provided in [R3.7] [R3.10] [R3.13]. The theory of categorial conversion is applicable to the general understanding of the socio-political transformations at the international politico-economic arrangements of domination, colonialism, neocolonialism, imperialism freedom, justice and related phenomena of oppression and liberty. The theory must also explain politico-domestic arrangements of any social setup as well as its developmental process. It must also constitute a general theory of unified engineering sciences including social engineering, and physical engineering such as biological, medical, chemical and other areas of actual and potential engineering where engineering is viewed as destruction-creational process. This line of thought is left to be developed at an appropriate time.

## X The Path of Reasoning in Applying the Theory of Categorial Conversion

The application of the framework of the theory of categorial conversion requires that one starts with the concept and principle of opposites. The primary category of opposites is the category of polarities which initializes the path of reasoning. From the primary category of opposites, all other opposites are derived by a process and reasoning. The method of reasoning allows one to build epistemic bridges between the ontological space and the epistemological space for understanding and knowing. The category of polarities must be constructed by epistemic actions. Within the category of polarities, a primary category is identified. In the theory of categorial conversion, the primary polarity is actual-potential polarity composed of the actual pole and the potential pole. In all areas of application, the actual pole must be specified. This actual pole may have many contending potential poles in non-controlled socio-natural processes. In a controlled socio-natural process, the potential pole must be specified. The identity of each pole of the actual-potential polarity is defined by the residing duality which must be specified. The actual pole is defined by actual negative duality and hence may also be called negative pole. The potential pole is defined by potential positive duality and hence may be called positive pole. The dualities are defined by the negative and positive characteristic subsets whose union defines the identity of the duality. In the negative duality, the negative characteristic subset dominates the positive characteristic subset while in the positive duality the positive characteristic subset dominates the negative characteristic subset. Within each duality, there are conflicts and tension that generate energy for negative action on the part of the negative characteristic subset and positive action on the part of the positive characteristic subset. The relative net effect of negative-positive actions of the negative-positive dualities will indicate the direction of categorial conversion and the degree of efficiency of the conversion relative to the convertibility conditions of categorial moment, categorial transfer function. There are two crossing points of the categorial-conversion process whose conditions are called categorial-transversality conditions. The two are the negative and positive categorial-conversion crossing points with negative and positive categorial-transversality conditions. The analytical path may is shown as an epistemic geometry in Figure X.1.

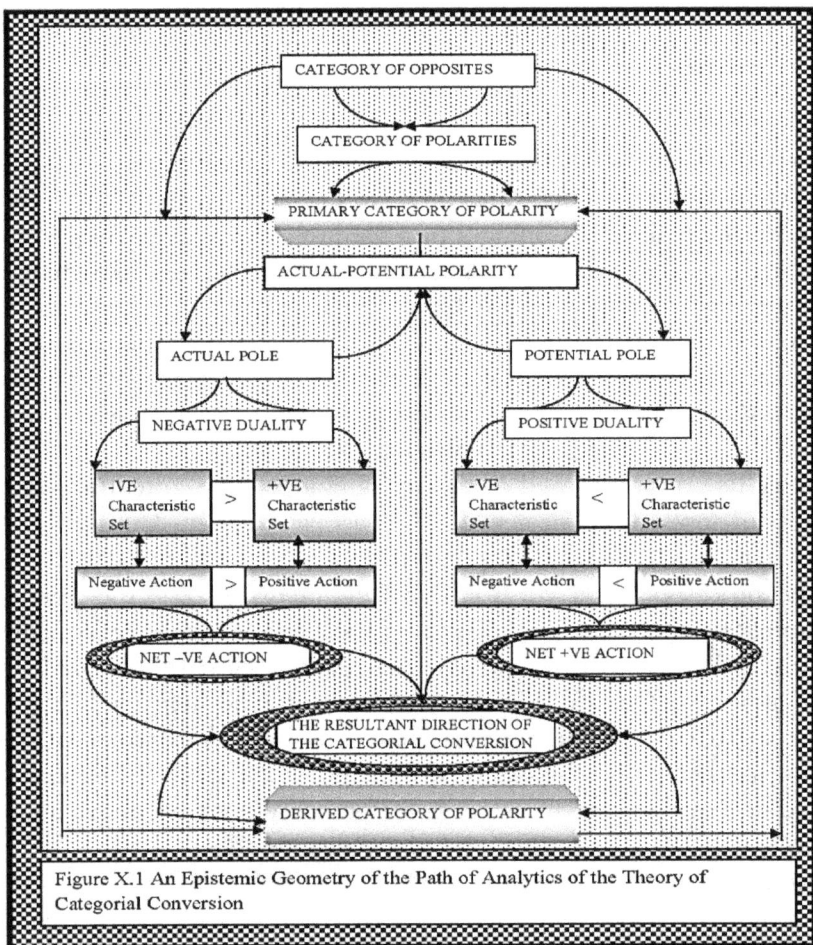

Figure X.1 An Epistemic Geometry of the Path of Analytics of the Theory of Categorial Conversion

The Figure X.1 is neutral to all categorial-conversion analytics across all elements in the ontological space. In specific applications, the nature of the polarity must be specified with well identified dualities. The residing negative and positive dualities must be defined with clearly specified negative and positive characteristic subsets. The nature of the inherent conflicts must be made explicit with clearly specified possible directions of possible conversions that may give rise to a possible derived category of polarities. The natures of the matter, energy and information structure must be indicated in a way that allows the specification of required dynamics of the qualitative-quantitative dispositions and their relational continuum and structural unity. In the analytics of categorial conversion of natural categories, these steps are naturally induced on the basis of information flows and matter-energy relations in the force manufacturing. It is here that the studies of physics, chemistry, biology, energetics, their combinations, synergetic structures, and other forms sciences provide derived useful information structures to assist multiplicity of human actions such as engineering and technological creation of all forms in the interpretive and mimicking modes of natural processes and states. As it has been discussed in [R1.92][R3.13], the theory of categorial conversion explains continual creation and socio-natural transformations where in the natural categorial conversion, the primary category of the actual-potential polarity with the system of supporting dualities is *nothingness-somethingness polarity* where nothingness constitutes the actual pole with a corresponding set of dualities, and somethingness constitutes the potential pole with corresponding set of dualities. Similarly, at the level of knowing, the theory of categorial conversion explains the path of continual reduction of ignorance with transformations of increasing knowing. Here the primary category is *ignorance-knowledge polarity* with a system of dualities where ignorance constitutes the actual pole and knowledge constitutes the potential pole. An increasing reduction of degree of ignorance implies an increasing degree of knowing through information-knowledge processes. The cognitive stages in physics are what Max Planck calls scientific world pictures [Max Planck]. This is generalized under categorial conversion as an epistemic enveloping in construction-reduction processes in knowledge accumulation within the epistemological space [R3.7] [R3.10] [R3.13].

# XI Categorial Conversion and the General and Specific Transformation Problems

The universe is composed of objects and things called *ontological elements*. Each ontological element is defined by a qualitative characteristic set which presents itself as *negative* and *positive characteristic* subsets establishing a duality in relational continuum and unity under the principle of opposites. Together, the negative and positive subsets define a quantitatively relative characteristic set which establishes the *ontological identity* of each element. The ontological identity is independent of the awareness of cognitive agents and other elements in the universal system. The ontological identities of universal elements allow groupings into ontological categories. Each element in the ontological category exists not only as a duality but exists with internal conflict in the duality under the African conceptual principle of opposite [R1.8][R1.92]KKD][R3.13] [R1.35] [R1.R19] [R1.243][R1.247]. The universe is governed by this principle of opposite under relational continuum and unity that maintains continual creation and recreation of never-ending structural changes through qualitative self-transformations of actual-potential polarities in the *identity process* which is simply the *ontological process* which presents a continual transformation of ontologically elemental and categorial identities. At the level of ontology, the ontological process is not seen as a problem by nature. It is simply *what there is*.

The understanding and explanation of the ontological process by cognitive agents become a problem only at the epistemological level which may be called the *transformation problem* where an epistemic search of an epistemic process is geared in creating a derived identity from the information about the ontological identity process. The epistemic process is to construct an epistemological identity process as a solution to the problem of understanding the ontological process. The epistemological transformation problem is to establish a general mechanism of the dynamics of qualitative disposition of categorial element and this qualitative disposition is relationally connected to quantitative disposition in the quality-quantity duality. In establishing the general dynamic mechanism of epistemic process, the epistemological transformation problem is structured into the problem of categorial conversion of actual-potential polarities. The epistemological process as a derived identity process is what is referred to as the theory of categorial conversion. The theory of categorial conversion must be worked with the construction-reduction duality with relational continuum and unity as part of the theory of knowledge. To be general the theory must be able to explain natural qualitative transformation, social transformation and knowledge transformation. The first important theoretical construct is to deal with the epistemological process of ontological identity process of ontological categories and then examine its applications to social and knowledge transformations. The theory of categorial conversion of knowledge transformation has been dealt with in [R3.7] [RR3.10 [R3.13].The general transformation problem is in relation to seeking a unified theory of explanation dynamics of qualitative disposition of all ontological elements. The specific transformation problem applies to human designed systems such as society and engineering of all forms.

Every epistemic development within construction-reduction duality must have an established foundation that constitutes its core.

1. Every element in the universe whether abstract and or concrete is characterized by its qualitative and quantitative characteristic set which is partitioned into negative and positive characteristic subsets whose relative structure is the negative-positive relative characteristic set that defines the identity of the ontological element in terms of its qualitative disposition

2. The simultaneous existence of the negative and positive characteristic subsets defines a duality in relational continuum and unity that have contradiction, tension and uneasy coexistence where both subsets need each other for mutual existence and inhalation through self-destruction. The nature of the duality is defined by the negative-positive relative characteristic set. The duality is said to be negative if the negative characteristic subset dominates the positive characteristic subset and vice versa

3. Every actual existence has a corresponding competing potential existence to create an actual-potential polarity defined in terms of relational structure of negative and positive poles under the principle of opposites. Every pole has a residing duality whose behavior is produced by the conflicting activities between the duals seeking to negate each other in a game-theoretic setting.

4. The conflicting activities of the duals are transmitted by the residing duality to the home pole of the actual-potential polarity seeking to destroy the actual pole and to replace it with a potential through. The conflicts in the dualities produce internal energies that bring about internal qualitative self-motion in terms

of changes of internal organization and arrangement changing the elemental and categorial identity to produce categorial conversion.

5.   Categorial conversion produces epistemic structure of general internal dynamics of qualitative disposition and hence explains internal qualitative development of universal elements and categorial transformations where self-development implies qualitative self-motion.

6.   The processes of constructionism and reductionism  are seen in terms of *fuzzy logic and fuzzy mathematics* where conflicts and contradictions are accepted as existing in the ontological space as reality as well in the epistemological space as thought but not in terms of classical logic and its mathematics, where the system is viewed in terms of the *Africentric principles of opposites* which encompass polarity, duality, and negative and positive characteristic subsets in terms of *qualitative disposition*, with an additional negative-positive relative characteristic set in terms of *quantitative disposition*.

7.   The theory of categorial conversion presents a complete and dynamic epistemic system that provides most general logical framework for understanding and explaining qualitative transformations of nature, society and knowledge production through the internal conflicts and contradictions between the negative and positive characteristic subsets in relational continuum and unity under the principles of opposites. The toolbox of the theory contains, matter, energy, information, qualitative disposition, quantitative disposition, category, category-formation, actual-potential polarity,  duality, negative and positive characteristic subsets, conflicts, contradictions, possibility, probability, fuzzy logic, fuzzy mathematics and fuzzy-stochastic reasoning [R3][R4][R5].

## XII  Logical Challenges of Understanding the Relational Structure of Qualitative Disposition and Quantitative Disposition in Categorial Conversion

The establishment of categories of elements in universal system is essential in the theory of categorial conversion. The theory of categorial conversion initializes its construct on the principles of existence of categories of matter, energy and information. The claim of existence of categories leads to the development of the theory of category formation where there are qualitative categories and quantitative categories. The question that emerges is simply what are the differences and similarities if any. The differences and similarities involve the concepts of quality and quantity and whether these concepts are ontological or epistemological. At the level of ontology, categories are defined and established by their *qualitative disposition* that specifies distinct attributes of elements. The collection of these distinct attributes of elements is called qualitative characteristic sets. The differences of these qualitative characteristic sets place elements in qualitative categories with neutrality of time. Quantitative characteristics can be established in any category of been. In other words, within each qualitative category another category can be defined and established by quantitative disposition that specifies distinct numerical properties of elements in a category. The collection of these distinct numerical properties for any element in a qualitative category is called quantitative characteristic sets. The differences of these quantitative characteristic sets place elements in any qualitative category into quantitative category with neutrality of time. There is a distribution of qualitative categories in the universal system and a distribution of quantitative categories in any qualitative category. The numerical properties may be defined either as fuzzy numbers, linguistic numbers or classical numbers [R3][R3.7].

The concept of neutrality of time means that both quality and quantity can independently change as well as simultaneously change for any phenomenon with forces and energy from within, and that the passage of time alone cannot induce a change. However the speed and size of change can be examine against time where qualitative motion and quantitative motion are defined. In this monograph, qualitative motion is related to transformation of quality and related to categorial conversion that establishes a new category or places an element in a different category in terms of relation-relation alteration. Similarly, quantitative motion is related to change in space in terms of space-time alteration of an existence of objects of the same qualitative characteristics. Complete dynamics requires the combination of qualitative and quantitative motions. In categorial conversions, the method of complete dynamics is required. These statements relate to conceptual similarities and differences of qualitative and quantitative dispositions, in terms of things that unite them and things that set them apart.

Quality and quantity in the theory of categorial conversion is seen as defining duality with relational continuum and unity in transformations where the activities in the duality are related to actual-potential polarity that shows the nature and direction of conversions. Here, a logical problem arises if quality is seen as separate from quantity in terms of general mechanism of elemental movement from one category to the other. Every element in the universal system is simultaneously qualitative and quantitative in existence and transformability. Every qualitative disposition is supported by a quantitative disposition without which their existence is indefinable by any logical means. The qualitative and quantitative dispositions are established by characteristics that exist in negative and positive characteristics that give meaning to identities of elements. In an important logical sequence every element in the universal system is simultaneously negative and positive in existence and transformability that give meaning to self-correction, self-improvement and self-excitement. Here, the negative qualitative (quantitative) disposition is supported by a positive (negative) quantitative disposition without which their existence has no logical mean in application and theory.

In the inner level of things, the ontological categories are formed by their inner characteristic sets in negative and positive proportions that project signals as information to the epistemological space to be used by cognitive agents with a given logical structure to develop epistemological categories at the level linguistic process and cognition. Inside any quality is a quantity just as inside any subjectivity there is an objectivity, both of which exist in relational continuum and unity. The relativity of the negative and positive characteristic sets defines the quantitative disposition in the game of negation of negation under the principle of categorial conversion. Negation either decreases or increases the number of negative or positive qualitative characteristic subsets for the game of dominance toward a *categorial crises zone* where a quantitative dominance occurs changing the structure of the negative-positive relative characteristic subset to a new quantitative relation of the negative and positive qualitative characteristic subsets to bring about a conversion by revealing new qualitative appearance by establishing *categorial difference* and *categorial distance* whose existence must be explained by the theory of categorial conversion. The negation is about both qualitative and quantitative dispositions in two interrelated stages. The first stage is either transforming some negative qualitative characteristics into positive characteristics. This is *intra-categorial conversion* that requires the development of *categorial moment*. The second stage is to change the magnitude of the negative-positive relative qualitative characteristic subset that defines *categorial relative dominance*. This reveals itself as *inter-categorial conversion* through the development of *categorial transfer function* which must satisfy the *categorial transversality conditions*. In other words, the intra-categorial conversion is propelled by a categorial moment to alter qualitative characteristics to create relative categorial dominance of either negative over positive or positive over negative to define the required conditions for elemental self-manufacturing of categorial transfer function to bring about inter-categorial conversion

## XIII The Concepts of Controllability and Convertibilty and their Relational Structure in Categorial Conversion

Categorial-conversion processes in socio-natural elements are defined and maintained by *categorial controllability* and *categorial convertibility*. The analytical framework of the explanation of categorial conversion and its applications in socio-natural processes are built on two interrelated theoretical structures. One theoretical structure relates to convertibility conditions while the other structure relates to controllability conditions. The convertibility conditions are studied by *the theory of categorial conversion* after the establishment of the theory of category formation which provides a logical explanation and justification of the existence of categories that relate to *matter, energy* and *information* under the principles of relational continuum and unity. The controllability conditions are studied by the *theory of Philosophical Consciencism*. The convertibility conditions characterize the internal dynamic conditions of *forces at work* and *production of energy* under a defined information structure that together establish qualitative motion which moves one element from one category to another or transform one category to a different category in the universal socio-natural elements and categories. The controllability conditions characterize the internal control mechanisms of strategies and tactics, and counter strategies and tactics that produce the dynamic conditions of *forces at work* and *production of energy* under a defined information structure under principles of opposites in a game space with negative and positive characteristic sets that together shapes the direction and produce the resultant qualitative motion which then moves one element from one category to another or transform one category to a different category in the universal socio-natural elements and categories according to the control and counter-control decision-choice structure.

The theory of category formation is essential to the development of the theory of categorial conversion which is essential for the development of the theory of Philosophical Consciencism. Without the justified existence of categories one cannot even think of language and categorial transformations. This monograph deals with both the theories of category formation and the theory of categorial conversion. The theory of Philosophical Consciencism is reserved for another monograph. The convertibility and controllability conditions are the foundation for understanding the dynamics of change in natural categories and the applications of which this understanding can be brought bear on society and human artificial creations such as all kinds of engineering, production-consumption systems under the principles of opposites where the focus is always on matter, energy and information structures, all in relational continuum and unity. The principles of opposites in relational continuum and unity on the basis of which self-transformation is explained depends on the existence of *actual-potential polarity* that gives meaning to primary and derived categories where the primary category is the actual and the derived category is an actualized potential in a never-ending process of continual transformation. In this case every actual and every potential are in temporary qualitative equilibria.

# CHAPTER ONE

## Abstract Ideas and Practice of Ideas in Social Settings: Extentions and Reflections on Nkrumah and Africa under Systems Thinking

In a logical frame regarding the theory of knowing, categories of philosophy arise to define a family of philosophical categories that reflects the multiplicity of complex social formations and experiences which also give rise to the culture of defined boundaries of reasoning to establish knowledge representation. By examining the representatives from the philosophical categories, one may learn the basic needs of a particular philosophy and the socio-cultural foundation of the philosophical category and its accepted logic of creation, as well as its utility in the social settings. From the critical examination, a possibility may be opened to discover the inherent contradictions, logical inconsistencies and the difficulties of reconciling the inner structure of the system of ideas with the existing social practice. From this discovery, one may proceed to construct the fundamental objections and design a logical process for avoiding them. The discovery of the inherent contradictions and inconsistencies in the system of ideas are illuminations of the philosophical category and the problem or set of problems dealt with. Such a discovery may point to a direction of a logical construct to avoid them.

The point here is simple, in that philosophical categories provide us with accounts of the world as seen in cultural settings and social experiences. The framework of the cultural settings, and social experiences is consistent with the position of W.E. Abraham in that:

> All events of larger significance take place within the setting of some culture, and indeed derive their significance from the culture in which they find themselves. It could therefore happen, and does indeed happen, that the same event, occurring as it were between the frontiers of two different cultures, should be invested with differing significance, with different capacities for arousing strong reaction, and with different capacities for determining the direction of policies arising therefrom. This immediately raises problems for a number of disciplines including, above all, history and social anthropology. The writing of the history of one culture from the milieu of another culture, which is not – relevant

to the events and situations concerned – isomorphic, raises serious questions of cultural bias and distortion. It does not necessarily offer objectivity, and indeed could not offer it in any sense in which this involved freedom from cultural colour. In terms of objectivity, where it touches evaluation of facts and events, a cultural alien can only offer an alternative set of prejudices [R1.1, p.11].

There are some importantly conceptual key elements in Abraham's statement. These key elements are related to interpretations of what constitutes significant events in experiential information structure and the development of response to the events contained in it under differential cultural setting. They acquire increasing importance in individual and collective decision-choice space regarding national affairs. They will help to understand the argument on different types of Philosophical Consciencism and social transformations as they relate to categorial instrumentations and the management of command and control to bring about socio-political and economic transformations. The complexity of the problems associated with these will be extensively discussed in [R1.90b], where the cultural differences will be used to partition the space of consciousness and conscience into corresponding sub-spaces.

The cultural settings define particular ideologies that tend to influence thought [R3.7, pp. 153-165]. The point, here again, is that the acceptance of a philosophical social universalism neglects the basic fact of philosophical reflections of life in specific experiential settings. This basic fact of philosophical reflections and the impact of cultural milieu on thought were clearly understood by Nkrumah in his attempts to conceptualize the African colonial conditions relative to the African traditions within the Western social set-up, African cultural conditions, and the need to liberate the African masses from the shackles of imperial oppression and the colonial mindset. It was necessary to understand the foundation of social philosophy that gave rise to colonialism, imperial aspirations and the cultural confines that gave birth to it as well as maintains it. To embark on the African liberation, the logic of the Western philosophical system of subjugation must be analyzed to undermine the pillars of its intellectual credibility. The African liberation and a search for freedom must be guided by a philosophical *liberationism*. This philosophical liberationism must be constructed from African social and cultural conditions as Nkrumah observed and as it is examined in [R1.92]. The African philosophical liberationism must be conceptually and analytically freed from the philosophical colonialism, neocolonialism and imperialism who's derived ideological justified slavery and nonhuman essence of the Africans. The African social progress demanded an immediate decolonization of the colonial mindset which had evolved through colonial education and indoctrination to enslave the African mind in order to de-Africanize the African personality.

The understanding of the process through which the African mind had been colonized and enslaved belongs to critical *history*. The liberation of the African mind from the mental colonialism and imperial enslavement belongs to *critical philosophy*. Both the critical history and the critical philosophy must be combined with the support of a critical logic to liberate the African mind from the shackles of epistemic colonialism, subjugation and slavery. These critical works are observed by Nkrumah's statement: *The critical study of the philosophies of the past should lead to the study of modern theories, for these latter, born of the fire of contemporary struggles, are militant and alive* [R1.203 p.5]. In this statement, Nkrumah observed the militancy of philosophical liberationism that is needed for decolonization, emancipation, and development on the mental, spiritual and physical levels of the African people. He also clearly understood the interdependent relationships among philosophy, ideology and social ideas, and the practice that may be required of them. It is within the understanding of the relationships of ideas and practice that lead Nkrumah to state that: *Practice without thought is blind; thought without practice is empty* [R1.203, p. 78]. This is equivalent to the exposition on the concepts and logico-mathematical symbolism. Symbols without underlying concepts leave us no room of knowledge and concepts without symbols are very difficult to process complex ideas, some of which are dependable and some of which are not. This discussion brings us to the problem of formation of abstract ideas, and their relationships to language, logic, mathematics and paradigms of thought [R3.10] [R3.13].

## 1.1 Abstract Ideas, Possibility Space and the Possible-World Space

Every philosophical system is a language which is composed of building blocks of vocabulary, grammar and logical rules of forward and backward inferences. These three constitute the pillars of the formation of abstract ideas that flow from experiential information structure which reflects the material world and inter-human relations as well as relations with the elements in the actual world as they are interpreted. Ideas are formed from: 1) experiential information acquired through human encounters in socio-natural interactions, and 2) cognitive interpretations of the elements in the experiential information structure. In other words, abstract ideas are generated by the use of an experiential information structure as an input into epistemic processing machine composed of logical rules and decision-choice actions. Formed ideas, both abstract and non-abstract, reflect the subjective connections of people with one another in societies and with the external world, which projects the conditions of human existence within the social and natural environments in which people find themselves. The experiential information structure is differently reflected upon by oppressed people relative to the oppressors from the same socio-natural environment. The people under occupation, colonialism and subjugation develop an experiential information structure different from the people that are doing the occupation, colonialism and oppression under imperial aspirations. These differential information structures become inputs for the knowledge-production system that yields different results for the oppressed-cognitive agents and the oppressor-cognitive agents. In fact, social truths about the same phenomena are often revealed as opposites without relational continuum and as if the universe is operating in disunity. These results and truth asymmetries not only generate conflicts in perception but become inputs into the social decision-making system. The experiential information structure has as many information sub-structures as there are social classes. Each information sub-structure spins a knowledge sub-space. The set of the knowledge sub-structures creates a conflict sub-space in the social decision-choice space where freedom and justice become important casuaties. It will become clear that the importance of the conflict space is to generate energy for social self-transformation induced by the social decision-choice system. Every society, like matter, is endowed with the property of qualitative self-motion that induces internal changes.

Given the experiential information structure, various interpretive combinations are constructed to produce relational elements in the knowledge-possibility space and further amplified into elements in the possible-world space as has been discussed in [R3.13]. These relations are not open to immediate observations from acquaintance as a first principle of knowing. Their conceptualizations require a complex system of definitions in terms of other natural and social relations and are supported with created rules of combination. The complex system of social relations and intercourses leads to the development of ideas, to which no directly conceivable object corresponds in order to justify advantages and benefits. An example of this is the complexity of the relationship among colony, slavery, imperialism, occupation, terror, and resource exploitation, racism, rich and poor. The ideas of race inferiority and superiority are derived by abstract generalization with *definite intentions* but not by comparing a number of conceivable objects. They simply flow from a process in which racial characteristics are formed, and then claimed for advantage and exploitation. The claim is the point of intentions to which Amo Afer elegantly discussed in relation to knowledge-construction and learning in the period of enlightenment [R1.11][R1.12].

All abstract ideas and the conceptual system that holds them, are rooted in experiential information structure through acquaintance as registered by the senses [R1.11] [R3.10] [R3.13] [R8.44] [R8.54] [R12.85].The experiential information structure is basically the same as the sense data. The experiential information structure is developed in accordance with human responses to the needs of a complex system of socio-natural intercourses which constitute the primary category of ideas, while the abstract ideas in various hierarchical forms constitute the derived category of ideas. The experiential information is the epistemological information structure which serves as the primary input in idea-formation. The socio-cultural intercourse abstractly separates into social-to-social intercourse and social-to-natural intercourse. The abstract ideas and the derived knowledge system are obtained through an epistemic processing of the experiential information structure. In other words, the experiential information structure is transformed by a process into the various stages of deferential degrees of abstract ideas, where each level of abstract idea constitutes a primary category for the next levels of derived abstract ideas. Each level of abstract idea has a seed of its own destruction as well as a seed of the germination of the next abstract ideas in a *cognitive categorial conversion*, where every abstract idea belongs to a category with a degree of social acceptance. Every abstract idea has a parent and a successor. The parent is the preceding idea. Each parent serves as a primary category while each successor serves as a derived category of abstract ideas with an attached degree of abstraction at that level. The collection of these levels of abstract ideas together shall be referred to as a family of derived categories of abstract ideas relative to the primary category of the experiential information structure. The toolbox for the development of the abstract ideas is obtained from a particular paradigm of thought and the accepted culture of social-knowledge development. In this context, a paradigm is viewed from the logical plane in terms of laws of thought and their defined inferential uses as it has been explained in [R3.7] [R3.8] [R12.85] [R12.88].

At the levels of experiential and abstract ideas, the claims of knowledge items are encoded into words and symbols and disseminated to the relevant general public as inputs into individual and collective decision-choice actions and implementations. These abstract ideas that become inputs into the individual and collective decision-choice systems do not have to reflect any of the elements in the experiential information structure, which is the primary category of cognitive reality called the sense data. In the processes of epistemic categorial transformation of the primary category of ideas into the sequence of derived category of ideas, misinformation and disinformation are intentionally introduced to create ideology and ideological illusions. At the level of ideological illusions, truthful revelations become casualties, and liars are moved to the ascendency in the epistemic categorial conversion process with further derivatives of the initial liars. The derived system of abstract ideas contains elements of ideology of disinformation, misinformation and elements that are reasonably believed to be true, in the sense of being supported by some experiential information elements, and justified by the methodological reductionism. The abstract ideas are generated from the experiential information structure by the use of *methodological constructionism* as a forward logical process for any given paradigm of thought. They become elements of an *inferential information structure*. The beliefs in the epistemic reality of the abstract ideas as derived from the experiential data are verified by the *methodological reductionism* as a backward logical process for the same given paradigm of thought.

Philosophical systems of abstract ideas and the supporting ideologies, whether applied to nature or society, have implied *intensions* since these ideas become inputs into individual and collective decision-choice systems. The intensions are to influence the directions not only of thought, but also the individual and social decision-choice actions. The effects of such a structural intentionality in the knowledge production and learning are also discussed by Amo Afer [R1.11]. The disinformation and misinformation characteristics of abstract ideas in the socio-political information structure constitute what has been referred to as the *deceptive social information structure* in [R13.8] [R13.9]. These abstract ideas that become inputs into the social decision-choice system do not have to have a correspondence with any of the elements in the experiential information structure that forms the primary information of the cognitive reality obtained by some form of acquaintance [R12.84]. The construction of the abstract ideas may lead to the creation of epistemic elements that simply belong to the *possible-world space* where such elements have no reference to the information basis of their creation. In other words, some of the derived abstract ideas that belong to the possible-world space may simply be *phantom epistemic elements*. When an abstract idea is a phantom epistemic element, the problem of its reductionism basically becomes a phantom problem in the knowledge production in the sense that it has no element in the primary category of reality to which it may be reduced. The elements that belong to the possible-world space may be taken to represent truth and reality without their possible *epistemic conditionality* [R3.10]. They may enter as credible inputs into the information-knowledge and decision-choice systems. It is here that the view is always held that what matters is not the reality in the decision-choice process but the perception of reality. It is useful to keep in mind that the derived knowledge structure that contains proven and unproven abstract ideas enters either as accepted knowledge which needs no further transformation by some individuals, or, as primary information that needs an epistemic processing into an idea as an input into the decision-choice system by some individuals. Let us keep in mind that every derived abstract idea in the epistemological space has a parent and a successor where the successor may be taken as the final stage which needs no further processing, and then used as an input into the social decision-choice system. It may also be taken as an intermediate stage that needs further processing and refinement before taken as a knowledge input into the social decision-choice system.

The abstract ideas are generated with intentions where such intentions are indirectly couched as the criteria of the social decision-choice system for an individual or social purpose. The derived epistemic elements that belong to the possible-world space may be taken to represent proven ideas and reality without their possible indicative epistemic conditionality, and used as input into the decision-knowledge system. Examples of such ideas are racial superiority, racial inferiority, characteristics of beauty, the correct path to God and moral codes. The relational structure of ideas and logic may be presented as a pyramidal epistemic system in the epistemological space.

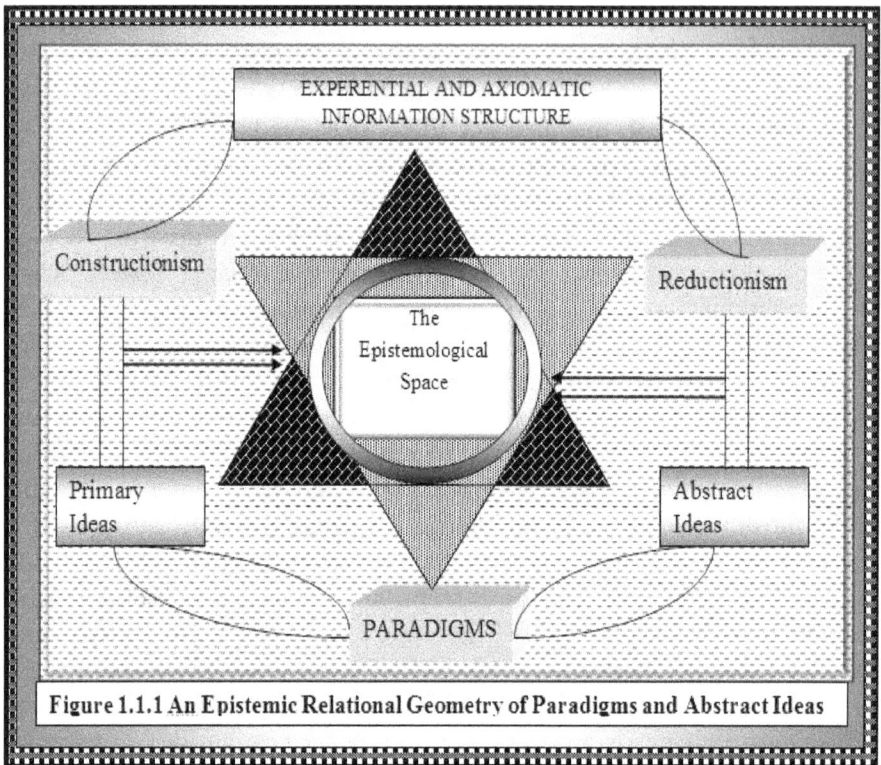

**Figure 1.1.1 An Epistemic Relational Geometry of Paradigms and Abstract Ideas**

The relational structure represents the internal structure of construction-reduction duality with a logical continuum in the knowledge production. The methodological constructionism is used to derive abstract ideas from the experiential information structure. The methodological reductionism is used to reduce the abstract ideas to their basic essentials to test their correspondence with the elements in the primary category. The logical validity of the construction-reduction reality is always in relation to a paradigm and the accepted culture of knowledge production in the society. It is here that a knowledge system may be viewed as a language composed of vocabulary, grammar, definitions and interpretive rules of meaning that may be expanded as human acquaintance space expands.

The knowledge system is embedded in a given culture that gives it legitimacy and meaning. In general, the accepted culture of knowledge production in a society drives the interpretations of the results of the paradigm used for processing the experiential information structure, in order to obtain a structure of knowledge as an input into the social

70

decision-choice system. In social settings, the cultures of knowledge production contain the goals and objectives of the dominant culture which may come to distort the interpretations and then generate interpretive conditions, where *misinformation* and *disinformation structures* are intentionally integrated into the social knowledge structure as natural and scientifically justified. These misinformation and disinformation structures may contain elements from the phantom space and become integral parts of the social ideology [R1.1][R1.203] [R3.7] [R3.13] [R7] [R7.7][R7.11] [R7.12] [R7.21] [R7.24]. The needed actions of the uses of the elements of the knowledge structure drive the *interpretive intentionality* of the results of the information processing. The interpretive intentionality is also driven by the *decision-choice intentionality* both of which acquire their meaning and legitimacy from the dominant culture. Corresponding to each society, in this respect, there will be different ideologies with illusory components. We shall speak of a system of ideologies which corresponds to a system of social formations. The non-illusory components of the social ideology may be viewed as scientific and may correspond to activities to produce further scientific knowledge. It is on this analytical framework that one may logically appreciate Nkrumah's foundational work of Consciencism and its applications to the liberation movement of Africa and Africa's complete emancipation. It is also here, that one may appreciate Nkrumah's contribution to the theory of transformations seen in terms of conflicts between the quantitative and qualitative dispositions that generate energy for the needed categorial conversion moment for motions with the neutrality of time.

## 1.2 Philosophy and Ideology in Social Information and Knowledge

The whole notion and activities of a social decision-choice system is driven by interactive forces of cognitive structure and information structure to produce a knowledge structure that becomes an input into the social decision-choice system. The forces for information transformation to knowledge are the *epistemic energies* that vary from one individual to another individual to produce the space of individual epistemic energy. The aggregation of these individual epistemic energies leads to the collective epistemic energies that transform social information to social knowledge. In the analytical and synthetic framework, we have the individual information-knowledge structure and the collective information-knowledge structure. The individual information-knowledge structure performs two functions. One function is directed to actions in individual decision-choice space where the benefit-cost outcomes are individual specific. The other function is directed to actions in the collective decision-choice space where the benefit-cost outcomes affect the social collectivity and are differentially distributed over the members in the collective. In this framework, there is the individual cognitive capacity and the collective cognitive capacity, which is the aggregation of the individual cognitive capacities towards information processing and decision-choice problem-solving. Given the individual cognitive capacity, these individual and collective epistemic information-knowledge processes leading to the knowledge outcomes that go into the individual and collective decision-choice processes are essential to the integrity of the whole of the social decision-choice activities of the individual and the collective, the colonized and the colonizer, the imperialist and the oppressed, the master and the slaved, and of the structural nature of social dualities. In view of these complexities and this antagonistic system of opposites, social exigencies are crafted with the differential cost-benefit distribution over the individuals depending on the category of the opposite that one belongs to. These social exigencies give rise to the differential experiential social information structure of social classes for processing into the knowledge input of all social decision-choice actions. The experiential information structure of each class is composed of defective and deceptive information substructures that have been explained in the above sections. The deceptive information structure and the socially acceptable paradigm of information processing or knowledge production may be ideologically affected, leading to different knowledge outcomes as inputs into the social decision-choice process. The ruling ideology is dependent on the *dominant philosophy* of the social organism where such a dominant philosophy may be supported by interpretive conditions of the results of science. In some cases, new areas of science are created to provide justified scientific inputs into the maintenance of the dominant philosophy in support of the dominant ideology.

Ideology affects the content of the social information structure through its manipulating impacts on the deceptive information structure. It also affects the paradigm of knowledge production through its impact on the laws of thought for thinking and learning to arrive at conclusions and decision-choice actions in the social set-up. In other words, ideology affects the individual and collective rationalities as have been explained in

[R3.7, pp. 143 – 165]. It may also help to create social stability and general compliance by setting the boundaries of individual and collective mindset. Philosophy and ideology are always in relational continuum and unity over the epistemological space not only in terms of their cognitive effects but also their inter-supportive reinforcements. In this respect, simple working definitions of the concepts of ideology and philosophy will be useful for our discussion.

## Definition 1.2.1: Ideology

Ideology is a system of justified and unjustified abstract and non-abstract ideas, beliefs, and concepts crafted from an accepted dominant *philosophy*, given the individual and collective experiential information structure by the ruling class at any point in time, to reduce social decision-choice complexities in all fields of behaviors of human endeavors to favor the aspirations, social intentionality and social vision of the ruling class as seen in a particular social set-up, where such behaviors are defined in the social decision-choice space for the individual and the collective.

## Definition 1.2.2: Philosophy

Philosophy is a critical investigation into knowledge propositions and conclusions about nature and society, by formulating crude theoretical content of the ontological reality (*what there is*) from an experiential information structure, on the basis of which different branches of knowledge sectors tend to emerge and grow in the cognitive process. When such branches of knowledge sectors emerge, the nature of their contents, claims and appropriate methodological guidance for their development and growth then becomes important objects of philosophical inquiry under the ambit of *philosophical rationality*, which provides a set of critical rules to assist the conduct and methodologies of exact and inexact science. Any knowledge sector, on the other hand, may investigate philosophical claims and critiques in terms of a *knowledge-area rationality* that provides a set of admissible rules in the acceptance of philosophical reflections.

Similarly, the path of philosophical reflections is shaped and being shaped by the progress of science. In fact, one may look at the relationship in terms of science-philosophy duality in a continuum under a tension in cognitive unity, where philosophy tends to negate and challenge science, and science tends to negate and challenge philosophy.

Of course, we can speak of sociology, economics and history of science, but these can be treated under the ambit of philosophy and science broadly defined. The nature of the relationship between science and philosophy in the process of cognition has produced disagreements between scientists on one hand and philosophers on the other, thereby creating intra-group conflicts among scientists and philosophers. Such disagreements, in the unity of cognition, are the energy sources for maintaining a creative tension between science and philosophy, and between science and philosophy on one hand, and the growth of knowledge on the other. All these take place under conditions of rational inquiry in the epistemological space at the guidance of particular social ideology.

Thus, science and philosophy exist in a dynamic unity and relational continuum. The unity and the relational continuum find an expression in the idea that some initial philosophical understanding may be obtained from critical reflections upon the content and nature of science and scientific methodology. In this way, philosophical reflections on scientific methodology, mode of thinking, content of science, and utility of scientific knowledge are essential to the growth of philosophical knowledge and critical inquiry into scientific contents and claims.

From the position of theory construction and unity of knowledge accumulation, the development of philosophy tends to increase the domain of scientific inquiry and cognition, while scientific investigation and its results increase the domain of philosophical inquiry as they provide new fertile grounds of knowledge verification. In general, therefore, philosophical interpretations and critique of contents and results of exact and inexact sciences become fertile grounds for further scientific inquiry, theory and the discovery of that which is humanly new in nature and society. Philosophy draws on achievements of science while science draws on the gains of philosophy. In this way, dynamic and intellectual intercourses and creative conflicts are maintained between philosophy and science as well as between philosophers and scientists in relational continuum and unity. On the road to human cognition and discovery of *what there is* (the actual), conditions of its existence, to change *what there is* and to actualize *what would be* (potential) from *what there is not* (space of potentials), philosophy guides the scientist while on the road to philosophical generalization and interpretation of scientific data, science guides the content of philosophy. This is the trajectory of the growth and development of human mastery of nature and society through the reduction of ignorance. It is this dynamic behavior in science-philosophy duality that places critical demands on: a) scientists to be familiar and possibly acquire the skills of philosophical rationality, b) on philosophers to also be familiar and acquire the skills of scientific rationality, and c) on both scientists and philosophers to search for cognitive rationality in the enterprise of knowledge production which is generated by the decision-choice process. It has been argued in [R3.7] [R3.10] that knowledge production is a decision-choice process as well as self-correcting and self-exiting control system..

The mastery of nature and society is embedded in the dynamics of actual-potential polarity with residing ignorance-knowledge duality on the road to cognition and decision-choice actions. From the view point of general cognition or theory of knowledge, philosophy may be viewed as an organized system of generalized theoretical views of the world, of nature, of society and of people. Philosophy is thus a special type of science. It seeks to confirm a definite orientation of reality. A question may, therefore, be asked as to what are the set of conditions under which philosophy may be regarded as science? The question puts us in the domain of science in the sense that we are forced to look for conditions of existence and the content of philosophy. To look for conditions that define the boundaries of philosophy, an important corresponding question tends to arise. What is science? Any attempt to answer this question places us in a category of philosophy. The answers that may be given to the dual questions will depend on whether one subscribes to idealism or materialism as the primary category of philosophical inquiry, and within them, whether one views the world metaphysically or dialectically. It is also admissible that a mixture of these may be followed to abstract answers to these questions.

In spite of all these demanding questions, the history of the development and growth of knowledge suggests that the intellectual intercourse and creative conflicts between philosophy and science have increased the power of philosophy and science to delve deeply into the structure of natural and social phenomena in the process of answering the question of *what there is* and its knowability on one hand, and *what would be* and its actualization on the other hand. This growth in knowledge has been done through the answers given by various aspects and areas of science to questions of their own specificities and with further reflectively philosophical understanding of various paths of methodology and contents of science. The importance of whether

75

philosophy is science and science is philosophy vanishes, or reduces to minimum, if one views science and philosophy as duality in the categories where categorial transformations occur between philosophy and the specific area of science with dialectical unity in cognition.

Philosophical generalizations and interpretations of scientific methodology and data are fed back creatively into critical science. Such generalizations and interpretations become fertile grounds for further scientific inquiry as well as objects of scientific enterprise. In this way, creative conflict is maintained between science and philosophy and between scientific rationality and philosophical rationality. Science tends to investigate philosophical claims and philosophy reflects on the meta-theoretic ways of science and the enterprise of science. There is, therefore, a living unity between science and philosophy at every moment of the development of cognition and the reasoning path for knowledge accumulation, where philosophy examines the logic of reasoning in terms of induction and deduction or construction and reduction. As such, a mechanical separation of science from philosophy leads to an important denial of any possibly dynamic process of cognition, fruitful theoretical transformations, corrective-feedback processes and effective social realization of knowledge where every claim to truth is temporary and open-ended that may be amplified by ideology.

Every investigator in any specific area of science directly or indirectly works with some degree of *philosophical rationality*. Similarly, every investigator, in any specific area of philosophy directly or indirectly works with some degree of *scientific rationality*. Both investigators in the enterprise of knowledge production are constrained by socially acceptable conditions of *cognitive rationality* in the sense of providing an admissible set of rules of general knowledge production that provides guidance for both philosophical and scientific rationalities. All rationalities are driven by conditions of a decision-choice process which provides discriminatory measures for its usefulness in the knowledge enterprise. The discovery of scientific truth or a knowledge item is always on the path of approximations through uncertainties and degrees of rationality. It will become clear in the later discussions that the degrees of rationality are best classified under conditions of human ignorance characterized by knowledge limitativeness and vagueness, and analyzed by *stochastic* and *fuzzy rationalities,* which tend to define the structure and the boundaries of cognitive rationality. Knowledge limitativeness places us under conditions of *stochastic rationality*. Knowledge vagueness places us under conditions of *fuzzy rationality*. When the two are combined we find ourselves placed under conditions of *fuzzy-stochastic* or *stochastic-fuzzy* rationalities.

The history of human cognition may be seen as an enveloping of the dynamic behavior of philosophy-science duality induced by the collective decision-choice process. In this respect, unless one realizes the power of such an enveloping, relational continuum and living unity of science and philosophy, then on the road to *scientific truth* or *what there is,* one may tend to deal with superficialities of either a conflict or harmony in the cognitive process. Alternatively, if one has a profound intuitive sense of cognitive motion and process unity between the structural developments of categories of science and philosophy, we are inclined to think that, one is likely to get to the roots of reality or *truth* if it is not found. The dynamics of science-philosophy duality may be viewed in terms of Kuhn's *structure of scientific revolution* [R12.54] or Kedrov's *theory of scientific discovery* [R12.46], [R15.23].Here, the creative conflict between philosophy and science is that scientific discoveries outside the prevailing scientific culture are presented as paradigm shifts by Kuhn, while Kedrov presents them as overcoming cognitive barriers which are generated in terms of changes in scientific culture [R12.54] [R15.23]. In terms of creating a conflict between science and philosophy, Nkrumah sees it in terms of social milieu [R1.203]; and Amo sees it in terms of intentionality [R1.11]. It will become clear in these discussions that paradigm and cognitive barriers are generated by a collective search for conformity. They become constraints on scientific rationality. Both of them are elements of institutional ideology of science, where the set of characteristics of institutional ideology constitutes a sub-set of the set of characteristics of general social ideology that guides decisions and intentions. As defined, ideology may be seen as an instrument to abstract social conformity within the domain of individual and collective behavior.

## 1.3 Illusory Ideology and the Possible-World Space

In discussions on rationality, a distinction was made between illusory ideology and scientific ideology and their impacts on the integrity of the knowledge-production system and its effects in the social decision-choice system on the basis of some acceptable rationality and the claims of knowledge [R3.7][R3.10] [R7] [R7.13] [R7.25]]. In reference to ideology and phantom problems in knowledge production, a distinction was also made between the *possibility space* and the *possible-world space* as the

77

knowledge production was put on the analytical focus in terms of the theory of the knowledge square and paradigms of thought. The possible-world space was shown to contain the possibility space as a sub-space. The other subs-pace of the possible-world space shall be referred to as a combination of a *phantom-possible-world sub-space* and *artificial possible-world sub-space*. The possibility space and the possible-world space have been defined and discussed in [R3.10][R3.13]. Similarly, illusory and scientific ideologies have been defined in [R3.7]. It is, therefore, useful to define the phantom possible-world space and artificial possible-world space. The concept of the possible-world space in relation to the possibility space has been defined and discussed in [R3.10].

## Definition 1.3.1: Phantom Possible-World Space

The phantom possible-world space as a sub-set of the possible-world space is a collection of cognitive elements or abstract ideas that have no possibilistic epistemic justification. They are elements that exist as apparitions such as "fauns". They may have a mental construct with a combination of two or more elements from the space of epistemic reality, the possibility space or both. They do not meet the methodological test conditions of the logical transformation in the construction-reduction duality.

## Definition 1.3.2: Artificial Possible-World Space

The artificial possible-world space, as a sub-space of the possible-world space, is a collection of cognitive elements or abstract ideas, the constructs of which have foundations in the possibilistic epistemic justifications. The cognitive elements exist not as apparitions but as actual or possible constructs of mimicries of elements from the space of epistemic reality or possibility space or both. They meet the methodological test conditions of the logical transformation of the construction-reduction duality.

The phantom epistemic elements are abstract ideas which cannot be demonstrated within a paradigm to correspond to an element in the experiential information structure by the methodology of reductionism, even though they might have been created by some form of methodological constructionism. Any given phantom epistemic element is an abstract idea and has a successor and a parent that cannot, by a backward logical process (methodological reductionism), be shown to belong to the experiential information structure. In other words, it is a derived category which has no primary category. To appreciate the analytical process, let us notice that the possible-world space is a derivative from either the possibility space or the space of the epistemic actual or both, which then constitutes its primary category if it meets the test conditions of construction-reduction duality in a relational continuum and unity. The phantom epistemic elements, within the required test conditions, are abstract ideas in an illusory and ideological space. These phantom epistemic elements always have social intentionality in the sense of Amo [R1.11], and arise from social exigencies in the sense of Nkrumah [R1.203], and are developed under cultures in the sense of Abraham [R1.1]. Social intentionality, exigencies and culture contribute to the creation of the phantom epistemic elements in acceptable social knowledge structure. The problems that are defined around these phantom ideas in the illusory-ideological space, as a sub-space of the possible-world space, belong to the *space of phantom problems* which have no solutions in the logical space of their creation. Analytically, such phantom problems are ill-posed and are raised in wrong epistemological sub-spaces with some paradoxical absurdity. One reason is that these phantom ideas and corresponding problems are simply social mirages which are defined in epistemic sub-spaces of unsolvable problems, no matter how hard one tries. The available epistemic toolboxes are outside the limits of the space of the phantom problems.

The point of importance is that the phantom epistemic ideas and problems are actually promoted as real and justified with some crafty social philosophy or religion, where the theological space is connected by a non-logical cognitive mapping to the space of the mirage of ideas and phantom problems. The mirage of the phantom abstract ideas and phantom problems is advanced to create intensions and conditions of rent-resource-seeking environments that make it easy for the oppressors to abstract rent from the oppressed. There are many examples in human societies, such as master-slave relations, imperialist-colonized relations, different forms of race relations, oppressor-oppressed relations and many more, all of which are justified on the basis of phantom abstract ideas. Notice that in all these situations, the relationships exist in dualities with continuum where the existence of the master requires the existence of the slave. This can be seen in terms of negative-positive dualities where the meaning of the negative and the positive depends on the

79

position that one occupies. Similarly, the existence of conditions of master requires the existence of conditions of a slave, where such conditions are justified by a particular ideological system of thought that may contain some phantom abstract ideas.

The understanding of any ideology is also an understanding of a particular philosophy that gives support to the abstract and non-abstract ideas contained in the ideology. It is this philosophy that maintains the conditions of the existence of the negative-positive duality, and defines stabilities in the continuum for any given relational structure. To destroy the existing relational structure, one must first understand the philosophical basis of the ideology and the methods of its development, and then dismantle them in order to do away with the ideological foundations of the conditions that produce the existence of negative-positive dualistic relational structures. It is here that the concept of the battle of ideas acquires a meaning and produces conditions of qualitative equation of motion within dualities in relational continua under the principle of unity. A new and opposing philosophy with a corresponding ideology is then developed to challenge the existing one for a relational restructuring. The new philosophy with the corresponding ideology must create polar neutrality between negative and positive poles in the epistemological space as initialized by the experiential information structure.

This understanding of the battle of ideas and the methodology of their creation in relation to cultural specificity were important driving forces behind Nkrumah's activities in examining the aspects of knowledge production and its impact on society [R1.203]. Every ideology must have its own supporting philosophy [R1.92][R9.10][R9.24][R9.26]. The philosophy constructed by oppressors to create basic and abstract ideas from their experiential information structure  draws its weapons of oppression from its environment, where both the oppressors' experiential information and philosophy are promoted and sold to the oppressed as universal. The conditions of the universality of the abstract ideas, phantom and non-phantom in nature, are promoted to the oppressed to create complicated intellectual dependency, where the oppressed person justifies his or her conditions of oppression with the same philosophical tools of his or her oppressor without critically examining the foundations and historically cultural conditions of the philosophical tools and the underlying abstract ideas. An example of this is the imperialist-colonized duality or independence-neocolonial duality where social stability is maintained in the continuum with imperialist philosophical logical tools and neocolonialist ideological tools which are justified as universal for analyzing human qualitative and quantitative conditions. The essential framework points to Nkrumah's observation that:

> It is my opinion that when we study a philosophy which is not ours, we must see it in the context of the intellectual history to which it belongs, and we must see it in the context of the milieu in which it was born. That way, we can use it in the furtherance of cultural development and in the strengthening of our human society [R1.203, p. 55].

The imperialists and colonialists, to justify their atrocities, violence, torture and terrorism against other people in the process of resource stealing, have developed a social philosophy and a corresponding ideology that supports social oppression, war and "just war", humanitarian killings under various ideological covers such as Christianization, civilization, democracy, development, global peace and many more. The more they commit such atrocities against other humans, the deeper the pool of praises that they swim in, and the more rewards they give themselves. In view of this philosophical brand and the anticipation of resistance, they have created and continue to create instruments of war-making and instruments of mass destruction which are supported by instruments of mass deception under an effective deceptive information structure. To the imperialist and colonialist, their social philosophy must be supported by a deceptively crafted philosophy of war and violence. The imperialist ideology has mutated and continually mutates to justify imperialist aspirations to make allowance for their human and non-human exploitation without regard to the lives of the oppressed. These imperialist and colonialist aspirations are especially particular to the so called *Western nations* that place claims on what they call *Western civilization*. The history of this Western civilization contains the epicenter of terror and violence against humanity. This Western civilization is given a philosophical content, cultural universality and an ideological protective belt. From these philosophical ideas and historical claims, the intellectual lives of the people in the Western nations are transformed into conditions of epistemic slavery, and then encapsulated by the shackles of the imperialist philosophy with its corresponding ideology to accept the practice of injustice and terror against non-members at face value, as these injustices are committed by

81

their leaders. In other words, the Western philosophy with its corresponding ideology found its weapons from the context of the intellectual history and the cultural confines to which it belongs. Justice is seen only in terms of the eyes of benefits for the imperialists and colonialists. The costs for the subjugated people are seen from the contents of their oppressor's philosophies and their ideological covering which they claim to contain the endowment of perfection, absolute truth and universality of ideas. This is consistent with Nkrumah's philosophical retrospect as he tries to show that there is nothing universal about the *Western philosophy*. The Western philosophy grew out from Western social conditions, culture and experiential information structure that provided the environment and the material input for Western philosophical views.

These philosophical views containing phantom ideas are then implanted into the minds of the people under occupation through a well-organized education program which is sometimes supported by force or threat of force. The intention is to fully colonize the collective and individual minds to accept occupation, subjugation and exploitation as the norm. In this respect, colonialism proceeds at the two levels of the territory and of the mind which translate themselves into territorial slavery and mental slavery. Territorial colonialism is the process to capture the decision-choice base of a particular social formation and control it to extract benefit from the colonized people at their own expense. In other words, to change the relative qualitative characteristics defined between the sub-sets of qualitative characteristics of independence and colonialism in favor of colonialism, oppression, colonial exploitation and undemocratic conditions. Territorial colonialism is always accomplished through violence which is then supported and maintained with terror by the colonizer without mercy, morality or shame. The colonial violence and terror are met with force of resistance for freedom and independence. The mental colonialism is a process to indoctrinate the territorially colonized people to individually accept their territorial occupation, loss of sovereignty, lack of democracy and imperial exploitation as normal. Mental colonialism is an attempt to reduce resistance for freedom and the maintenance of the colonial power system. Colonialism and independence exist as duality and translate into conflict between oppression-freedom duality. Colonialism at the negative end of opposites robs the people of their sovereignty and deprives them of the fundamental right to practice their collective will, national interest and social vision in the decision-choice space, where their history becomes an appendage to the history of the colonizer. Independence at the positive end of the opposites installs sovereignty to the people and offers them the possibility for the fundamental right to practice their collective will, national interest and social vision in the decision-choice space to shape the path of their history. This is a distorted structure of colonialism-independence duality by the construct and practice of ideas in the sense that practice is guided by ideas which are placed in the capsule of social thinking. Colonialism, neocolonialism, decolonization, independence, war, peace, imperialism, oppression, slavery, torture, rights, freedom and many others are simply the practice of ideas abstracted from thoughts on the basis of universal polarities with supporting dualities under the principles of relational continuum and unity.

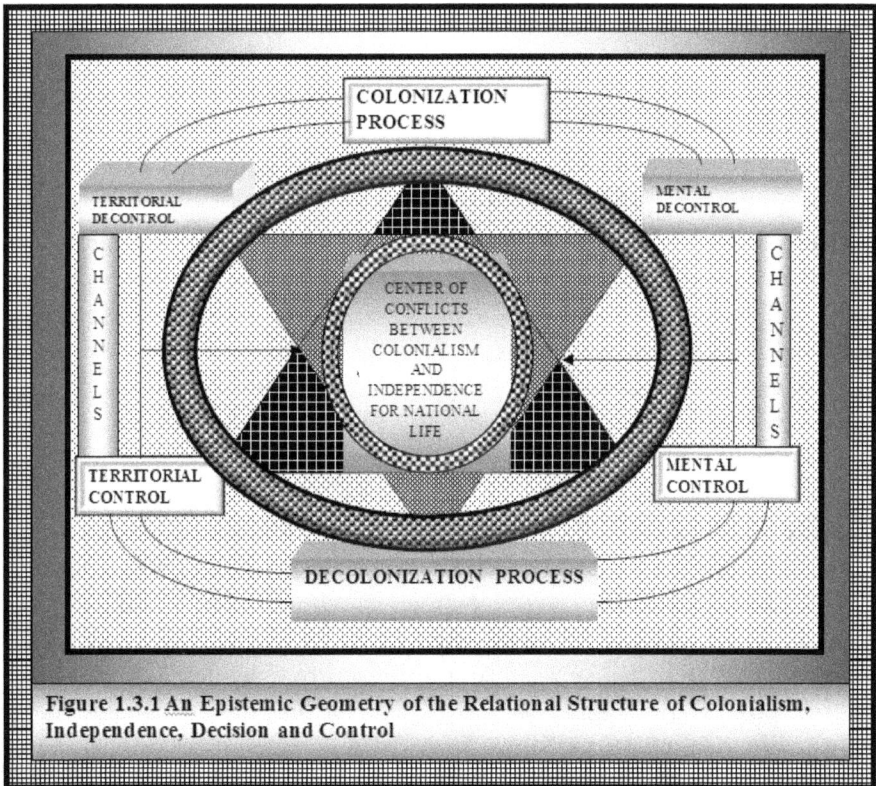

Figure 1.3.1 An Epistemic Geometry of the Relational Structure of Colonialism, Independence, Decision and Control

Just as colonialism proceeds at two levels, decolonization must also proceed at two levels of territorial decolonization and mental decolonization within the colonization-decolonization duality under the principle of relational continuum and unity. To destroy colonialism and bring about national independence and peoples' freedom, social actions must be introduced to negate the process and change the relative terms of conditions that maintain both the territorial and mental colonialism. The process of social actions to change the relative conditions must be guided by opposite ideas which in turn must be encapsulated in a liberation thinking that must overthrow dependency thinking. Let us keep in mind two important conceptual perspectives: 1) when the universe is seen in terms of polarities and dualities, methodology of game theory presents a useful approach to the analytical process which affects the development of ideas and social thoughts in terms of liberation and imperialism, and 2) when the relational information structure is seen in terms of deceptive and defective, the analytical process for understanding and knowledge development is not the classical paradigm of thought but the fuzzy paradigm of thought. The fuzzy paradigm of thought also allow the understanding of conflicts in polarities and dualities under relational continuum and unity where every negative has a positive support and vice versa [R1.90b][R3.7] [R3.8] [R3.13].

Let us keep in mind, the nature of the characteristics structure of both duality and polarity. At the level of the principles of the opposite, every pole of the polarity has a negative duality as well as a positive duality. A negative pole is defined when the negative duality overwhelms the positive duality. Similarly, a positive pole is defined when the positive duality overwhelms the negative duality. In the relational structure, every duality is made up of a complete characteristic set which is divided into two opposing dynamically active subsets of negative and positive characteristics sub-sets. The conditions where the negative characteristic sub-set dominates the positive characteristic sub-set define the negative duality, and hence the negative pole of the polarity. Similarly, the conditions where the positive characteristic sub-set dominates the negative characteristics sub-set define the positive duality, and hence the positive pole of the polarity. Both the negative and positive characteristic sub-sets are always present in differential relative proportions to define the nature of the polarity. Every duality is under mutual struggle and tension in a relational continuum and unity to transmit information and generate energy to bring about a transformation of an actual-potential polarity. For example, when a country is colonized, the process does not completely do away with the characteristics of independence and freedom. In fact, the increasing characteristics of independence and freedom are importantly being manufactured by the internal conflicts which generate powerful energy for either internal-self transformation from a colonial state to an independent state or self-maintenance of the colonial state. Any change comes from the struggle between the negative and the positive sub-sets which first seek to transform the duality into the opposites in the poles to become that which is not such that a colonial state is transformed into a non-colonial state, and the neocolonial state is transformed into an independent state. In other words, the colonial pole with foreign control of the power of the state is transformed into independent pole with domestic political control of the power of the state. The reversal process or the negation requires territorial decolonization for the domestic control of the social power system. The territorial decolonization must then be followed by activities of mental decolonization to restructure the mind for independence, social confidence and national integrity. It is at the point of mental decolonization that liberation philosophy and ideology are called upon to bring relief. The liberation ideology and the supporting philosophy must be combined with the national history and social exigencies from which the philosophy must be drawn.

# CHAPTER TWO

## Restructuring the Mind of Africa and the Oppressed: Defining an Initial Framework for Liberation Thinking in Social Systemicity

Ideology, history and philosophy are relationally linked in unity and intimately connected to society, its organizational structure, behavior and management. When foreign invasion occurs and occupation takes place, the first casualty is truth about the past, present and future. The second casualty is the integrity of the people and the third casualty is the critical thinking of the occupied. It must be kept in mind that the foreign invaders are completely ignorant of the cultural dimensions of the occupied people. When ignorance is elevated to the throne of human social life, knowledge is in critical trouble. Lies become elevated to the deputy of ignorance and knowledge is relentlessly prosecuted for crimes of truth searching. The exposure of truth is punished and the teller of lies is rewarded.

## 2.1 The Art and Science of Creating an Unthinking People for Imperialist and Neocolonialist Domination and Exploitation

The death of truth creates a vacuum of past, present and future history. This vacuum is filled with malicious lies which are then supported with an imperialist ideology. The imperialist ideology is given a justification from the contents of a particular philosophical system, alien to the cultural confines of the oppressed society, to completely alter the history of the occupied from positive to negative achievements. Such an imperialist philosophy is derived from the imperialist experiential information structure and then generalized to create intellectual environment of deception in pursue-avoidance game of hunter-rabbit duality. It should not forget that it is within the basic structure of the imperialist philosophy and ideology that slavery was justified with Devine support the enslavement of Africans. The imperialists are either partially or completely ignorant to the history and culture of the occupied or colonized people who they have subdued with primitive violence and continue to maintain with increasing violence. Here, it is analytically useful to keep in mind that generally, when ignorance is elevated to the throne of human existence, knowledge is in trouble. Lies become elevated to the deputy of ignorance for cover-up, and the first casualty is truth with relentless prosecution of knowledge whose simple crime is truth and continual search for truth.

## 2.1.1 The Modus Operandi of Creating an Unthinking People in Colonial and Neocolonial States

The modus operandi is a process under the principle of deceptive information structure. This process is enforced by discouraging critical thinking and encouraging mimicry and imitations of other achievements except their own. The oppressed people under imperial occupation develop a conceptual system that regards their historical facts as socially insignificant and then embrace the imperialist historical elements as virtues that must be copied. Having cultivated deranged minds and been forced to live in the zone of cognitive imbecility, the oppressed people sing songs of praises for the glory of the occupiers and oppressors. They then condemn everything of their culture and historical importance as insignificant and hence are not of value to emulate. In this way, the history of the oppressed is then written in terms of the views of the occupiers' imperial adventurism which become assumed as good with divine blessings under both ideological manipulations and systematic techniques of all forms of violence against the non-compliance. This cognitive structure happened in Africa and cumulated to the greatest violation of human rights, the enslavement of African people and the holocaust through the doors of no return and the middle passage. These European imperial violence against African culture and people was carried on under the ideological principle of complete reduction of the enslaved and colonized African people to a state of non-humans who are only fit to provide work-machines for the benefit of the European imperial order. This direct cognitive structure has been modified into an indirect structure under neocolonialism and racism in the global politico-legal structure.

The indigenes are then schooled and re-educated to create a new intellectual class whose responsibility is to provide an intellectual support for the oppressive ideology and the supporting philosophy. This class is elevated to the top of the leader of oppression with some material and non-material rewards. It becomes a buffer zone between the oppressed and the oppressors by selling the core elements of the imperialist ideology and supporting philosophy to the masses of the oppressed. The members of the intellectual class of the indigenes are rewarded by the occupiers and promoted to the position of second class oppressors taking instruction from their oppressive masters and perform in the space of imperial and neocolonial cronyism. Armed with this distorted intellectual and ideological system, the members of the intellectual class of the indigenes assume the oppressive responsibility of the oppressors in order to administer the instructions and will of the oppressors for dehumanization and injustice of their kind. In other words, they acquire different qualitative characteristics by categorial conversion from that which they are to that which they are not. They cease to think in their cultural confines and they mimic the thought process of the oppressors whether it fits their conditions of existence or not. The manufacturing and existence of the intellectual class of the indigenes and its impact on the liberation and emancipation of African societies from the shackles of colonialism and imperial oppression were reflected upon and effectively described by Atom Ahuma of the Gold Coast:

> As a people, we have ceased to be a THINKING NATION. Our forbearers, with all their limitations and disadvantages had occasion to originate ideas and to contrive in their own order. They sowed incorruptible thought-seeds, and we are reaping a rich harvest today though, for the most part, we are scarcely conscious of the debt we owe them. Western education or civilization undiluted, unsifted, has more or less enerved our minds and made them passive and catholic. Our national life is semi-paralysed; our mental machinery dislocated, the inevitable consequence being, speaking generally, the resultant production of a Race of men and women who think too little and talk too much. But neither garrulity nor loquacity forms an indispensable element in the constitution of a state or nation [R1.2], [R1.171, p 166, ].

The imperial process is to introduce a categorial moment to alter the positive characteristics of the African personality in order to create a space of distorted personality such that African people's thinking will be anchored in a zone of hatred for themselves and their kind. This process is to enervate the collective mind of the Africans in order to dislocate their cognitive machinery and turn them into perfect imitators of the ways of the occupiers where imitation is elevated to perfection, and thinking is reduced to unessential but simple loquacity and deep preference for trivialities of no national significance where their leaders become mere joker in the global decision-choice space. The occupier's strategy of bringing this to being is through a well-designed program of

89

irrelevant education and active ideological indoctrination where the Africans (indigenes) are taught to mimic the occupiers' tools of analysis without any critical understanding, to adopt the philosophical reasoning that supports the occupiers' ideology of subservience and oppression, and to take this philosophy as the universal system of cognition and social truth. In this way, the existence and the lifestyle of the African become nothing but parody with a complete loss of fruitage of the soul of the African by severing it from its historical and cultural roots. In the process, the African's great intellectual powers are put to deep cognitive sleep creating unconsciousness of their destiny except for the destiny of their oppressors. The African mindset is put in a colonial and neocolonial situation where the African loves the oppressors' destiny and works to promote it more so than he or she loves his or her destiny and works not to promote it. In this colonial and neocolonial mindset, the African sees the African social vision, interest and destiny as simple and irrelevant appendages to the social vision, interest and destiny of the imperial and neocolonial oppressors. The description of this state of negative categorial conversion of the African is efficiently described by Attoh Ahuma:

> The most difficult problem of our times is how to think so that Africa may regain her lost Paradise. How to think the thoughts that galvanize and electrify into life souls that are asleep unconscious of their destiny [R1.171, p168].
>
> After 500 years of continuous intercourse with Europe, we have at least come to the age of Transition – that mental and moral watershed where as a people, we may either stray to the right or to the left, developing into vital forces or else into engines of self-destruction in the national struggle for existence……
>
> We need to think for ourselves to find out the eternal principles that underlie every thought and idea indigenous to the nation. And since the Letter is killing our individuality inch by inch, it becomes our duty, one and all to strive for the spirit that giveth life, and giveth it abundantly.
>
> We have fought valiantly for what we deemed were our Ancestral Rights in the past, and would fight again, if those rights were menaced tomorrow – but the greatest calamity of West Africa that must be combated tooth and nail, we feel, is the imminent Loss of Ourselves. [R1.171, p.434].

Carter G. Woodson reflecting on this problem in relation to the African experience in the United States of America regarding the transformation of the African cognitive force states:

> History shows, then, that as a result of unusual forces in the education of the Negro he easily learns to follow the lines of least resistance rather than battle against odds for what real history has shown to be the right course. A mind that remains in the present atmosphere never undergoes sufficient development to experience what is commonly known as thinking. No Negro [African] thus submerged in the ghetto, then, will have a clear conception of the present status of the race or sufficient foresight to plan for the future; and he drifts so far toward compromise that he loses moral courage. The education of the Negro, then, becomes a perfect device for control from without. Those who purposely promote it have every reason to rejoice, and Negroes themselves exultingly champion the cause of the oppressor. [R1.294, p. 96].
>
> The Negro in this state continues as a child. He is restricted in his sphere to small things, and with these he becomes satisfied. His ambition does not rise any higher than to plunge into the competition with his fellows for these trifles. At the same time those who have given the race such false ideals are busy in the higher spheres from which Negroes by their mis-education and racial guidance have been disbarred. [R1.294, p.111]

When the African mindset has been confined in the colonial and neocolonial space, the colonialist and neocolonialist do not have to do much; the African will make decisions to their own detriment because they are following rules of anti-African reasoning. This colonial and neocolonial mindset is a greatest enemy to African emancipation which is composed of independence from the imperialists, domestic control of sovereignty and nation building in accord with African preferences and will.

## 2.2.2 Conditions for Negation of the Presence of Unthinking Society

To appreciate the decision-choice processes and geopolitical outcomes in the polarities of either colonialist and colonized people or independence and neocolonialism under the principles of imperial aspirations of Western powers and independence aspiration of the African people with diametrically opposing interests, one may associate imperialism with a desert and African states with spring water in the global system of resource distribution, power distribution, production and control. In this spring-water-desert relation, it is also useful for the sleeping leaders of

African decolonized states with their intelligentsia to reflect on the critical and instructive observation of Ayi Kwei Armah who states:

> Spring water flowing to the desert. Where you flow there is no regeneration. The desert knows no giving. To the giving water of your flowing it is not in the nature of the desert to return anything but destruction. Spring water flowing to the desert, your future is extinction.
>
> Hau, people headed after the setting sun, in that direction even the possibility of regeneration is dead. There the devotees of death take life, consume it, exhaust every living thing. Then they move on, forever seeking newer boundaries. Whenever there are living remnants undestroyed, there lies more work for them. Whatever would direct itself after the sun, an ashen death lies in wait for it. Whichever people make the falling fire their aim, a pale extinction awaits them among the destroyers
>
> Woe the headwater needing to give, giving only to floodwater flowing desert ward. Woe the link from spring to stream. Woe the link receiving spring water only to pass it on in a stream flowing to waste, seeking extinction.
>
> You hearers, seers, imaginers, thinkers, rememberers, you prophet called to communicate truths of the living way to a people fascinated unto death, you called to link memory with fore listening, to join the uncountable seasons of our flowing to unknown tomorrows even more numerous, communicators doomed to pass on truths of our origins to a people rushing deathward, grown contemptuous in our ignorance of our source, prejudices against our own survival, how shall your vocation's utterance be head
>
> This is life's race, but how shall we remind a people hypnotized by death? We have been so long following the sun, flowing to the desert, moving to our burial…. It is for the spring to give. It is for spring water to flow. But if the spring would continue to give, the desert is no direction.
>
> The linking of those gone, ourselves here, those coming; our continuation, our flowing not along any meretricious channel but along our living way, the way: it is that remembrance that calls us. The eyes of seers should range far into purposes. The ears of hearers should listen far towards origins. The utterers' voice should make knowledge of the way, of heard sounds and visions seen, the voice of the utterers should make this knowledge inevitable, impossible to lose.
>
> A people losing sight of origins are dead. A people deaf to purposes are lost. Under fertile rain, in scorching sunshine there is no

difference: their bodies are mere corpses, awaiting final burial. [R1.16, pp. xi-xiii].

All the above quotations point to the role of philosophy and ideology in mind control and the manner in which such mind control may restrict or enhance nation building and socioeconomic development of colonized and neocolonial states such as those in Africa. The creation of unthinking society is to maintain a continual flow of spring water to the desert. To reverse the process, it is useful to understand the forces at work, mechanism and conditions of change in general and how to utilize this understanding to bring about a social change. To create the required and relevant understanding, a development of a new African mindset is required to create a new African personality after decolonization. The reversal process of neocolonial and colonial mindset and African personality is the negation of the negation under the principle of categorial conversion whose theory is the subject matter of this monograph in providing the possibility and conditions for reversal. The framework to bring about the negation of negation is taken up in the theory of Philosophical Consciencism which is the subject matter of a companion volume [R1.90a]. The statement of Ayi Kwei Armah presents important elements of struggle in social polarities and dualities with transformation dynamics from the African conceptual system as the way. In fact, the statement which is quoted at length indirectly or directly emphasizes the need for African change of logic of reasoning in the global decision-choice space in which the African leaders and the supporting intelligentsia operate. Such a logical reason cannot be based on the conceptual structure of the imperialist philosophy. It has to be based on the cultural confines of Africa and its experiential information structure with a relational continuum and unity of the past and present which is then projected into the future.

It is through the critical understanding of these elements of the relational structure of philosophy, ideology and education that moved Nkrumah, in retrospect, philosophy to examine in reasonable detail and establish the linkage to oppression and liberation [R1.203]. It is also the conditions of creating parody, irrelevant imitation and lack of development of critical thinking that also led to the discussion of the philosophical and ideological reflections in [R1.92] [R1.159]. [R1.160]. To bring about a change in African conditions of parody and mimicry, requires a development of a philosophy that draws its roots and weapons for its development from the cultural confines of the African people. In other words, the needed philosophy as liberation of the African mind must have a relational continuity and unity from the African conceptual system and the culture that it is to serve. This philosophy, as Nkrumah observed, must support an ideology of liberation to destroy the very philosophical foundation of the imperialist oppressive ideological claims contained in the epicenter of *deceptive information structure* including racial propaganda and crude appropriation of African ancient thought system [R1.24] [R1.25] [R126] [R1.34] [R1.37] [R1.56] [R1.60b] [R1.83b] [R1.243] [R1.249]. This ideology of liberation must introduce into the society a positive action that negates the imperialist and neocolonialist negative action to set liberation forces against the forces of imperial and neocolonial oppression for social transformation. The philosophy and ideology that are in relational continuum and unity of Africa's past and present and projected to the future is to create an African with true African personality. To succeed in the negation process, we must understand the foundational complexity of deceptive information structure and its relationship with philosophy, ideology and education in subduing people to accept their conditions of oppression as normal, unchangeable and perhaps divine. The epistemic geometry of this relational complexity is given in Figure 2.1.1. One pyramid defines deceptive information structure made up of disinformation and misinformation structures, and superimposed on it is another pyramid composed of philosophy, ideology and education to create a complex indoctrination system. It is this problem of deception information structure, ideology and intentionality that it is always useful to practice the principle of doubt.

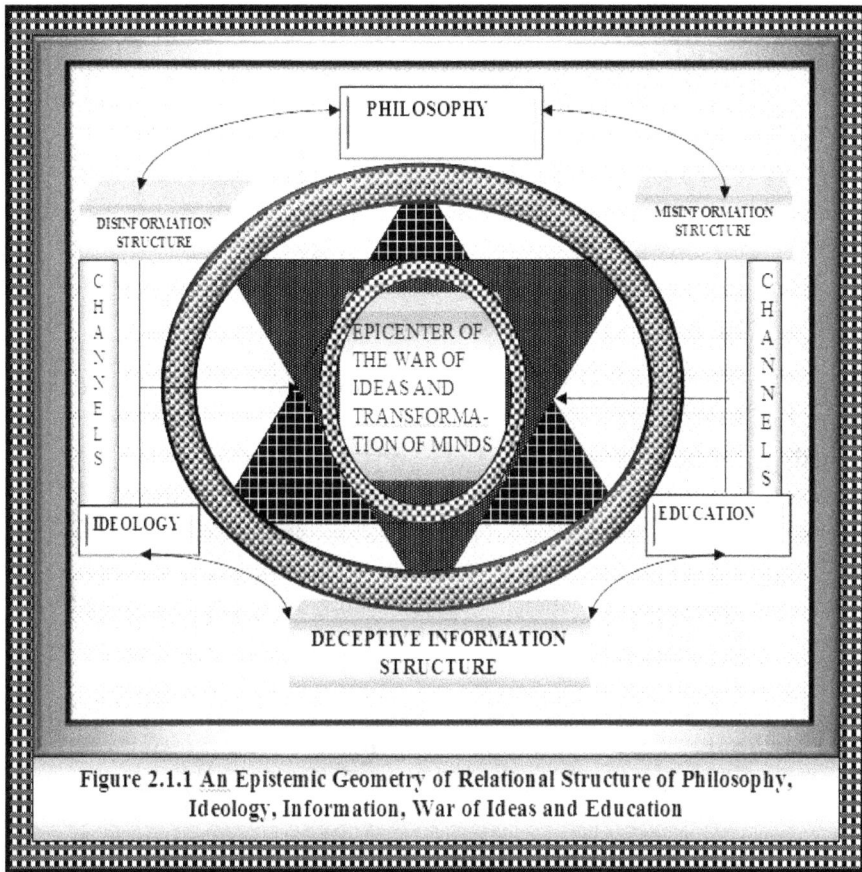

**Figure 2.1.1 An Epistemic Geometry of Relational Structure of Philosophy, Ideology, Information, War of Ideas and Education**

The lessons from known history of ideas as may be abstracted, led Nkrumah to state that African *Philosophy must find its weapons in the environment and living conditions of the African people. It is from these conditions that the intellectual content of our philosophy must be created* [R1.203, p.79]. It was also from the understanding of the same analytical basis that the works in [R1.92] were projected as epistemic and ideological foundations to help the development of a knowledge system that will be relevant for the modern challenges of the current stage of the complete African emancipation. The philosophy needed to support the needed ideology of liberty, freedom and justice is only possible when there is a complete avoidance of the mirage of liberty where freedom has no legs. The development of liberation philosophy and ideology must be complemented with an avoidance of eccentricity of pedants of triflers with trivial scholarship to the African past and her achievements, and the

95

fight to liberate. Additionally, there must be a clear avoidance of a) inanities of simple agitators with lack of sense of freedom, justice and nationhood and ingenuities of sycophants hoping for opportunities of personal gains by filling their minds with the empty ideology and philosophies of the oppressors by living in the comfort *zone of cognitive imbecility*, where the members of the oppressed class are taught to parrot the ideological statements that are created by their oppressors to support their *modus operadi of predation*. In this way and finally, the members of the oppressed crown their personalities with pageantries and triflers that are of little value to their freedom and justice in such a way. In introspect; they create an intellectual and ideological mockery of their own peoples' culture, traditional civil code of conduct and history. Controlled by colonial and neocolonial mindset and working in the space of oppressors' philosophy, they seek entry into the oppressors' class and then participate in the terrorizing zone that has been defined by the oppressors. In other words, the members of the oppressed neocolonial and colonial states act as Amarh's spring water while the imperialists and neocolonialists act as Ayi Kwei Amarh's desert in the decision activities in the social polarities.

### 2.1.3 National Identity, Socio-Cultural Characteristics and Continuity in the Development of Conceptual Systems

Every society may be defined by its characteristic set which is made up of collective and individual personalities. The characteristic set may be partitioned into negative and positive sub-sets. The great achievement of the social set up requires dominance of the positive characteristic set over the negative characteristic set. The conditions of social schizophrenia require dominance of the negative set over the positive set. The negative characteristic set becomes engrained on the collective and individual personalities of a wounded people who are simply begging for relief and survival. The dominant negative characteristic set negates the self-progressiveness of the positive characteristic set of the people in such a way that the people cease to be a thinking people because their cognitive machinery has been dislocated in the sense of Atto Ahuma. In the case of the task facing the African intelligentsia, Attoh Ahuma reminds us that:

> Our forbears, with all their limitations and disadvantages, had occasion to originate ideas and contrive in their order. They sowed incorruptible

thought-seeds, and we are reaping a rich harvest today, though, for the most part, we are scarcely conscious of the debt we owe them [R1.171, p166].

The suggestions of Attoh Ahuma and the reflections by Carter G. Woodson provided an intellectual challenge to any awakened African mind. It is this intellectual challenge that leads us to examine Nkrumah and his response to the African problem, not on the politico-economic space of human action, but on the intellectual space of human thinking in support of the human action. The reason lies in the notion that every social change of great significance is action oriented which must be backed by a well-designed intellectual program in the social decision space. The intellectual program provides guidelines for rational action toward positive changes and social transformations. For the development of the intellectual program to be effective, it must draw its weapons from the cultural confines and experiential space of the society whose transformations are sought. When the collective and individual minds are controlled, the direction of their thinking is also controlled in such a way that their decision-choice activities are effectively controlled by the controller of the minds without direct force which is encapsulated in the thinking process. Here, an advice may be abstracted from the statement of Carter G. Woodson:

## 2.2 The Task of Nkrumah: The Road to the Decolonization, Independence and Emancipation of Africa

We now turn our attention to the description of the task of Nkrumah in relation to the intellectual space of liberation, freedom, justice and emancipation given the colonialist and neocolonialist intellect space. Nkrumah came to understand and accept the advices and reflections of important personalities of African thinkers including the ones we have already mentioned. Among such thinkers we mentioned was Attoh Ahumah who reflected on the problem of African thinking and unnecessary imitations. At the level of thinking, he suggests that *the most difficult problem of our times is how to think so that Africa may regain her lost Paradise. How to think the thoughts that galvanize and electrify into life souls that are asleep unconscious of their destiny* [R1171, p168]? At the level of imitation, he suggests: *imitation reduced to fine Art is much to be deplored through West Africa* [R1.171, p.170]. He further explains: *Thanks to the letter C.O.D, facilities are afforded the youth upstart to gratify his unworthy ambition. What the*

97

*Whiteman eats, he eats, what he drinks and smokes, he drinks and smokes, thereby securing what, in his deluded opinion, is considered the Hall-mark of respectability, civilization and refinement...*[R1.171,p. 170].

At the level of thinking, vision, ideals and Africa's progress, Faduma presents to us that: *There have always been two kinds of men in the world, those who have ideals and those who have not. The men who have ideals, be they political, social, intellectual, moral or spiritual, are often the world's best workers and redeemers. They sometimes fail to reach their ideals but they are better and have the world better for working up to them* [R1.171, p. 197]. At the level of vision for Africa, Casely Hayford suggests that: *The future of West Africa demands that the youth of West Africa should start life with a distinct objective. Of brain power we are assured. Of mechanical skill there is no dearth. What is wanted is the directing hand which will point to the right goal* [R1.171, p.207].

At the level of freedom and liberty of the African people, Williams Ku Kuyé of Gambia suggests to us that: *As I understand it, its meaning is that every people has the right to determine its own destiny by choosing its own institutions and forms of government best suited to its own peculiar circumstances* [R1.171, p. 214]. At the level of regeneration of Africa, Pixley Isaka Seme advises us that:

> Agencies of a social, economic and religious advance tell us a new spirit which, acting as a leavening ferment, shall raise the anxious and aspiring mass to the level of their ancient glory. The ancestral greatness, the unimpaired genius, and the recuperative power of the race, its irrepressibility, which assures its permanence, constitute the Africa's greatest source of inspiration. He has refused to camp for ever on the borders of the industrial world; having learnt that knowledge is power, he is educating his children. [R1.171, p. 265].

At the level of education and thinking, Carter G. Woodson has this reflection:

> When you control a man's thinking you do not have to worry about his actions. You do not have to tell him not to stand here or go yonder. He will find his "proper place" and will stay in it. You do not need to send him to the back door. He will go without being told. In fact, if there is no back door, he will cut one for his special benefit. His education makes it necessary.
>
> The same educational process which inspires and stimulates the oppressor with the thought that he is everything and has accomplished everything worthwhile, depresses and crushes at the same time the

spark of genius in the Negro by making him feel that his race does not amount to much and never will measure up to the standards of other people. The Negro thus educated is a hopeless liability of the race [R1.294, p. xiii]

At the level of economic organization, Du Bois has this instruction: *A body of local private capitalists, even if they are black, can never free Africa; they will simply sell it into new slavery* [R1.96, p. 666]. He further instructs: *Let not the cloak of Christian missionary enterprise be allowed in the future, as so often in the past to hide the ruthless economic exploitation and political downfall of less developed nations whose chief fault has been reliance on the plighted faith of the Christian church.* [R1.96, p.626]. At the level of forces of liberation, Du Bois has this suggestion: ... *the great dilemma which faces Africans today faces one and all: Give up individual rights for the needs of Mother Africa; give up tribal independence for the needs of the nation* [R1.96, p.66]. At the level of inspiration and encouragement Du Bois has this advice:

Awake, Awake, put on the strength, O Zion! Reject the weakness of missionaries who teach neither love nor brotherhood but chiefly the virtues of private profit from capital, stolen from your land and labor. Africa, awake! Put on the beautiful robes of Pan-African socialism. You have nothing to lose but your chains! You have a continent to regain! You have freedom and human dignity to attain [R1.96, p.667].

At the level of history and in relation to deceptive information structure and resource exploitation, Du Bois reminds us with this statement:

Of the debt which the white world owns Africa, there can be no doubt. No black man can recall it without a shudder of disgust and hate. The white followers of the meek and lowly Jesus stole fifteen million men, women, and children from Africa from 1400 to 1900 AD. and made them working cattle in America,; they left eighty five million black corps to mark their trail of blood and tears; then from 1800 to this day their scientists, historians, and ministers of the Gospel preached, wrote, and taught the world that a black man was by the grace of God and law of nature so evil and inferior that slavery, insult and exploitation were too good for him and the virgin purity of white women could only be secured if mulatto bastards were strewn from the Atlantic to the Pacific and from the North to the South pole. Harsh words? But dismal truth. So what? Can bitter revenge erase all this? If Sir John Hawkins could be caught in West Africa today, even I shudder to think what Ghana might do to this blasphemous hypocrite. But he is beyond hurt today. [R1.96, p.668].

These analyses, reflections, words of advice, suggestions, instructions, reminders, historical descriptions and related ones paint an environment and a picture of what problems face an awakened African intelligentsia at the levels of thought and practice. The problems point to the organic problem of the task for Africa's decolonization and emancipation. They require the development of a revolutionary philosophy in support of a revolutionary ideology tailored to a complete African emancipation. The revolutionary ideology supported by a revolutionary philosophy to win the African mind from the colonial and neocolonial intellectual space is intended to create a true African personality and awakened African intelligentsia on the basis of African culture and experiential information structure as well as to abstract social conformity in support of the complete African emancipation.

It is analytically helpful to keep in mind that ideology is simply a system of ideas, beliefs and concepts crafted by a ruling class to reduce complexities in the behaviors of human endeavors as seen in a particular social set-up, where such behaviors are defined in the social decision-choice space for the individual and the social collective. In this way, ideology may be seen as an instrument to abstract social conformity within the domain of individual and collective behavior that will produce social stability. It forms a protective belt of the society for social allegiance and control to protect and defend national interests and social vision as established. At the level of theory, ideology reflects the philosophical concepts and ideas which are accepted as social knowledge. At the level of practice, ideology lays down rules and norms that guide individual and collective behaviors in the general decision-choice space. Ideology, therefore, is a non-violent way of defining the space of elements and processes of decision-choice actions.

The principle of social conformity is an important outcome of any ideological system. The social conformity, however, is social-system specific as well as culturally dependent with supporting justifications provided by a particular social philosophy that supports the national interest social vision relative to a social goal-objective set. In an imperial social system of subjugation, oppression and repression of thought, the ideology is couched in two steps of grand imperial ideology and a supporting set of sub-ideological systems whose objective is to maintain the conditions of social dominance. The sub-system of the sub-ideologies may be divided into colonial ideology, neocolonial ideology and ideology of force. The relational structure may be presented as a cognitive geometry in Figure 2.2.1.

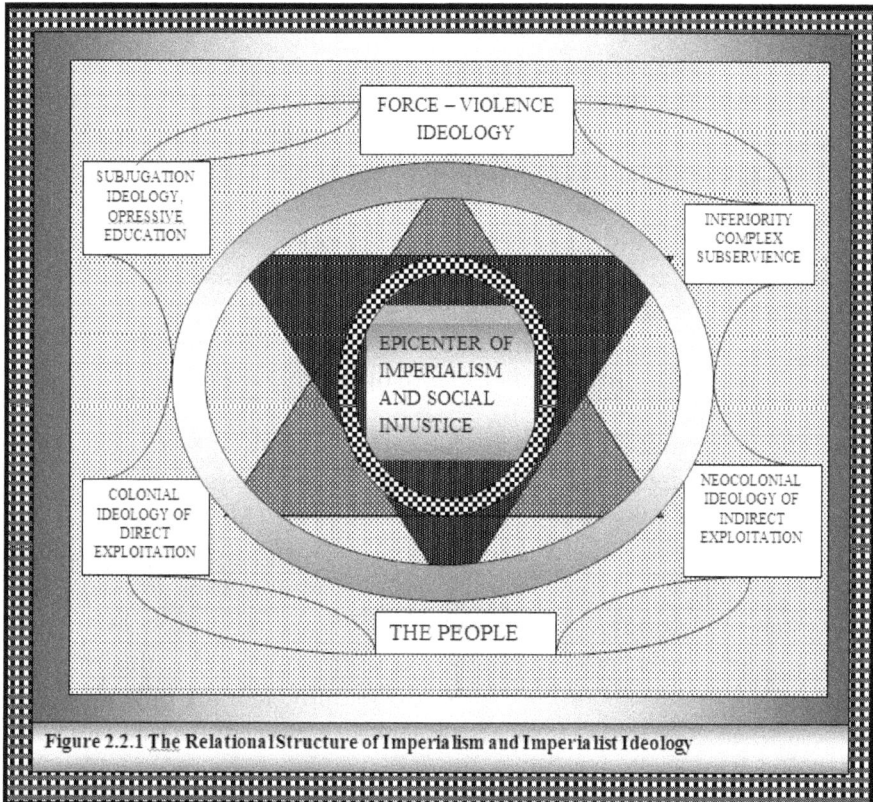

Figure 2.2.1 The Relational Structure of Imperialism and Imperialist Ideology

In these discussions, as they relate to the central conceptual system of Nkrumah, we shall be interested in the morphologies of imperialism and the ideologies of liberation in the global power-resource space. The concept of a liberation ideology implies the existence of an ideology of imperialism, subjugation and oppression. As such, we shall deal with the structure and form of the ideology of imperialism. Given the morphology of imperialist ideology, we shall present the analytical forces leading to Nkrumaism and the development of a philosophical foundation of Nkrumaism which gave rise to what Nkrumah branded as Philosophical Consciencism which relates philosophy to ideology and history of ideas. Here, a critical point is made to relate a need for revolutionary philosophy and ideology to the African social conditions

101

and experiential information structure. The problem of Nkrumah is a search for conceptual tools from within African traditions for the construct of the needed revolutionary philosophy that will give rise to an ideology for Africa's complete emancipation including her diaspora. Such revolutionary philosophy must be dynamic and have the ability to cope with the changing structure of the conflicts within the colonization-decolonization polarity and the supporting dualities with relational continuum and unity. Since such an African philosophy and ideology with their methods of reasoning in social values will in some important areas of human existence stand in opposition to imperialist philosophy and ideology that justified slavery, peonage, colonialism, neocolonialism and war of terror, it will be analytically proper to examine the uses of the imperialist philosophy and ideology to create benefit.

## 2.3 The Uses of the Imperialist Philosophy and Ideology

The imperialist philosophies and ideologies are utilized to construct law and order, and decision-choice sovereignties are claimed to belong to them to control the lives of non-members. The imperialists and colonialists fail to see the dualistic nature of events where propositions and events exist in true-false duality with a continuum in an Africentric conceptual system but not on true-false dualism under excluded middle in the Western logical system. By the use of methodological dualism, as a paradigm of thought, the imperialist philosophy sees an individual and the collective as existing in separable levels of non-interactive relational structures. It perverts the inter-supportive roles of the individual in the society and society as a relational and material collection of the individuals. In the Africentric conceptual system, as seen by Nkrumah, the individual is the society and the society is the individual by organizational construct and its management and administration. In this construct, as it has been extensively explained, the individual has no identity except in the community of individuals, and the community's existence and identity are defined by the individuals [R1.91] [R1.92]. This is the principle of collectivism in the African social set up where every member counts as essential. In the development of social philosophy, which of these entities must constitute the primary category and which one must constitute the derived category? Does the choice of the primary category from which the derived category emerges depend on the context of the socio-cultural milieu in which a particular thought system is born? How does this Africentric view of social set-up sees the societies of the past present and the future?

The caricature of the human thought of individualistic society and individual as a primary category of social existence, as a mode of deception, requires a development of alternative philosophical systems on the basis of collectivism or communalism to emancipate thought from the shackles of deception and imprisonment of justice in the confines of fear. These alternative philosophical systems must strengthen the conceptual and logical forces against imperialist oppression, exploitation, colonialism, neocolonialism and machines of mass destruction against the weak and poor. The important and interesting characteristics of the methodological dualism with excluded middle is the claim of absoluteness of truth and falsity and justified by some *divine law of purity* and relational continuum and unity. The methodological dualism has been the preferred logical tool in the construct of imperialist philosophy and ideology of oppression the existence of absolute truth and falsity leads to arresting the forces of epistemic transformations of ideas, and hence asserts final states which are contrary to social practices and the knowledge production in the epistemological space [R1.92] [R1.159] [R3.8] [R3.10]. It arrests qualitative motion and imposes fixed categories to stop the concept of perpetual creation and transformation without end. It also complicates the understanding of *state-process dua*lity in natural and social transformations. In the imperialist thought system, under logical dualism with excluded middle, final social states are asserted and logically claimed to be permanent with permanent categories.

In this conceptual situation, capitalism in the imperialist thought system is asserted to be the perfect form of a social production system and democracy is the perfect politico-legal state of human social formation and organization under collective decision-choice systems. The perfect states including others are asserted as final and are given divine blessings and couched in immutable walls of divinity, where everything associated is divinized for the oppressed to follow the given divine rules of order for social stability. The reward to the oppressors is to receive the fruits of the oppressed for current enjoyment, and their costs are the creation of institutions to maintain the state of oppression and the oppressed. This was how, for example, the Western brand of Christianity was introduced in Africa. The reward for the oppressed is the future blessings and happiness conceptually established in a divine

103

space of perfection and justice where there is no suffering. The cost of the oppressed, in order to earn the entry passage to this Divine space of perfection is the toiling to create material happiness for the oppressors. Such a divine space of perfection cannot be justified in the Africentric thought system as observed by Nkrumah and as has been discussed at length in [R1.92].

This perfection, the absolute-truth and divinized system of thought without reference to human decision-choice system in the epistemological space, is provided with an abstract thought and justified with a particular logical system of reasoning such as "every statement is either true or false but not both". There are many examples in our known human history, such as the slave economy with a politico-legal structure of social rights given to others to enslave other humans where theological speculations with a particular supporting philosophy are used to divinized these social rights as natural rights inherent in a Devine creation, in the sense that some people are created to be masters and some are created as slaves to serve the masters by Divinity. The initial egalitarianism is dismantled and replaced with a politico-legal cast system of social exploitation and human abuse that are legally entrenched as the final stage, and is maintained by force, violence and fear such as ones observed in imperialist colonialist-colonized polarity, occupier-occupied polarity and other social dualisms under the principle of excluded middle. In this way, a system of natural rights of humans is replaced by a system of social rights crafted and maintained with a particular politico-legal structure which draws its conceptual system from a particular philosophy. In practice, these social dualisms with excluded middle are social dualities that function under the principle of continuum. The social dualism with excluded middle has some basic characteristics. It places permanencies on social states, social categories such as the master-slave relation, oppressor-oppressed relation and other likewise structures, and freezes qualitative motion and categorial conversions. In fact, it freezes social states as permanent. It promotes the creation of logical categories as good-bad-terrorist dualism to justify humanitarian mass killing and the acceptability of just war and mass murdering. The negative and positive duals are cut off from each other and do not logically communicate with one another even though they mutually define themselves and their existence. The opposites are opposites and do not exist in relational unity. The conceptual relationality between either the two poles or between the two duals is dismantled, where self-motion is nullified. In this way, qualitative motion is arrested and the logic of analysis is seen in terms of time-space phenomenon by dimensional reduction of the elemental trinity of quality, quantity and time for any given energy field. Technically, dualism with the principle of excluded middle allows the establishment and analysis of quantitative equation of motion as time-space phenomenon.

Every social polarity with residing dualities under the principle of continuum and unity also has special characteristics. These characteristics include the element that every social state is temporary and is going through a transformation process to the next category. The notion of temporariness rejects permanency of states as well as a final state in all social process. In this respect, social categories such as the master-slave relation, oppressor-oppressed relation, rich-poverty relation and other likewise structures exist in a temporary inter-supportive equilibrating internal forces of the state in transition under tension of the opposites. These tensions must be related to an *energy field* which generates categorial moments and transfer functions required to create forces of qualitative self-motions and self-transformations. The energy field restores the *elemental-relational trinity* of quality, quantity and time as shown in Figure 2.3.1 and then by the principle of continuum links the elemental relational trinity and the *relational-time trinity* of past, present and future in the Akan philosophical concept of *Sankofa Anoma*. In this way and technically, duality with the principle of continuum allows the establishment and analysis of both qualitative and quantitative equations of motion as quality-quantity-time phenomenon. The negative and positive poles exist in inter-pole give-and-take relation in a set of continuum points under mutual determination of their identities. Similarly, the negative and positive duals exist in intra-pole give-and-take relation in a set of continuum points under their mutual determination and their identities [R3.10] [R3.13].

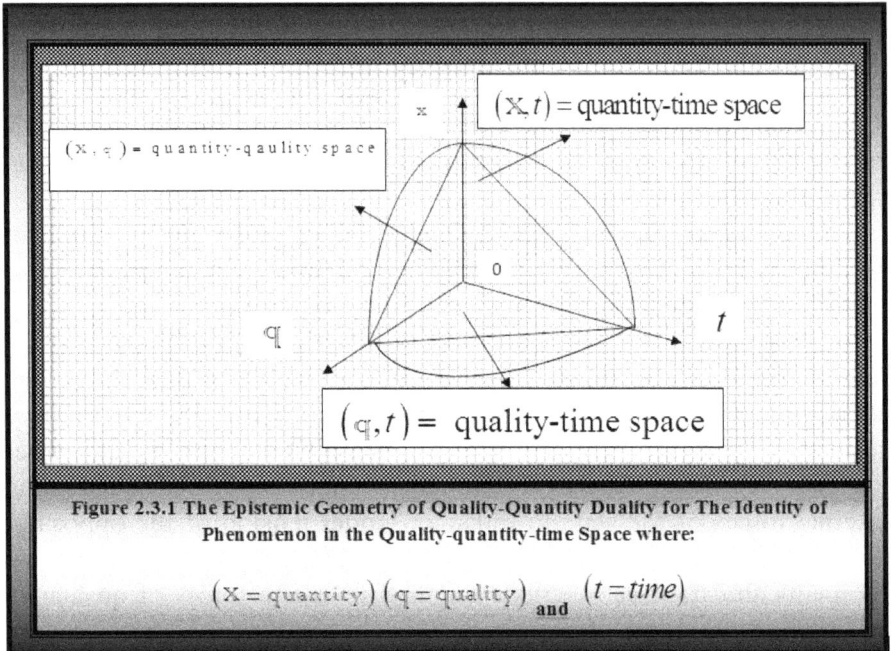

**Figure 2.3.1 The Epistemic Geometry of Quality-Quantity Duality for The Identity of Phenomenon in the Quality-quantity-time Space where:**

$$(X = quantity)\,(q = quality) \quad\text{and}\quad (t = time)$$

Each of these points in the continuum may be viewed as a category under a transformation tension of plenum of forces. This is another way of defining social categories under conversion, tension and moment. Each of these categories is unique relative to the corresponding qualitative disposition. Each of them is in a temporary qualitative equilibrium under categorial conversion through mutual negations where the set of the negative characteristics seeks to transform itself by negating some of the positive characteristics to enlarge the negative characteristic set within a pole to either retain or negate the identity of the pole. At the same time the set of positive characteristics also seeks to transform itself by negating some of the negative characteristics to enlarge the positive characteristic set within the pole to either retain or negate the identity of the pole, where the pole through its internal qualitative dynamics may retain itself or transform itself to its opposite. The internal dynamics within a pole is *intra-polar categorial conversion.* The qualitative dynamics between the poles is *inter-polar categorial conversion.* In this process of continual transformation, every negative-positive position that defines a category exists in a temporary unstable qualitative equilibrium. Each of the temporary unstable categories sits on conditions of potential transformation and is maintained by either a physical force or by instruments of pain deliverance, or by ideological force of mind control or by a combination of them. In all these qualitative dynamics, we must understand the important role that is being played by *matter, energy and information* as the ingredients for categorial conversion with a neutrality of time whether one is dealing with either social or natural phenomenon . The qualitative dynamics is such that intra-polar categorial conversion is associated with inter-dualistic categorial conversion which is associated with intra-negative-positive transformations of characteristics. It may be noted from Figure 2.3.1 that the system of analytical sub-structure for categorial conversion is made up of quality-time analytics, quantity time analytics, quality-quantity relational analytics and simultaneity of quality-quantity-time analytics which bring complexity analytics and synergetic analytics.

For this unstable temporary qualitative equilibrium of the category to be maintained in its state, four transformational actions must create the forces of maintenance. The intra-conditions of both negative and positive characteristic sets must be sustained in their balanced mode to support forces of stability of the category to freeze the *categorial moment* and stop the *categorial conversion*, as long as the forces of stability hold in their equilibrium state. The meaning, here, is that the forces for significant qualitative change are held in check by generating countervailing forces that cease to effect the intra and inter categorial conversions. The hope here is that the position in the negative-positive duality may be maintained with the creation and introduction of countervailing forces. Translated in a social sense, it simply means that the conditions for the existence of master-slave polarity, colonialist-colonized polarity, oppressor-oppressed polarity and categories of similar social polarities with their residing dualities are maintained and considered as immutable social realities. The acceptability of the immutabile social realities will depending on the nature of the transformation dynamics and the instruments of the maintenance of the dualities within categories and their impact on the stability of the corresponding social polarity.

## 2.4 Contending Ideologies, Freedom, and Liberation

In an oppressed society, there are two contending philosophies in supporting two opposing ideologies that are at an intense struggle for negation. One is the dominant philosophy in supporting an oppressive ideology whose thought system gives a rational justification for inhumanity, injustice, war, violence, territorial occupation and terror. The other is a positive liberation philosophy which is planted in the womb of the oppressive philosophy, natured by it and waiting for germination. The germination of the seed of the liberating philosophy is to provide a negation of the system of thought provided by the dominant philosophy as a rational justification of the oppressive ideology. Its germination and growth are then used to support an ideology for freedom, justice, decision-choice sovereignty, peaceful coexistence and egalitarian principles of the common humanity. The liberation ideology and the supporting philosophy become an integrated and consolidated intellectual revolution intended to lay down a program of integrated complex actions in a more or less positive and organic framework. The development and nurturing of this liberating philosophy must be the business and responsibility of the members of the oppressed society who are intellectually liberated from the conceptual system of the oppressive philosophy and its ideology in all their organic structures. The effectiveness of the application of the new ideology and supporting philosophy must be that the members of the oppressed society have an unwavering belief in the new intellectual structure of the liberating philosophy that must support the liberating ideology. It was the need to fulfill the success of the intellectual-revote against the intellectual colonialism that led Nkrumah to state his reflection on the general history of human achievement:

> The history of human achievement illustrates that when an awakened intelligentsia emerges from a subject people it becomes the vanguard of the struggle against alien rule [R1.202, p. 43].

The revolutionary nature of this liberating philosophy was well understood by Nkrumah in all his writings. He understood that philosophy seeks to make sense of human society by analyzing historic events and how the past-present events fit into human experiences, and how they shape the events of tomorrow. In this way, the time trinity is called into human understanding in the true African *Sankofa principle* where the *present* makes sense only in the reflections of the *past,* and the *future* is understood by its seeds in the present. There are thus three dualities in continua that relate to decision-choice actions of cognitive agents in freedom struggle. They are past-present duality, present-future duality and past-future duality. .The epistemic continua and relational continuity involving the three dualities are presented to cognition and thinking in relation to the information-knowledge system in the decision-choice space in a never-ending social transformation of qualitative and quantitative dispositions with neutrality of time. It is here that Nkrumah's statement: *Philosophy, like history, can come to enrich, indeed to define the experience of man* [R1.203, p.i] becomes a useful advice to the people in a liberation movement. The development of such a philosophy, like the development of the content of history, must be rooted in the social environment of the oppressed to acquire the potency of the logical structure needed to fully support the required ideology of liberation. In other words, a liberating philosophy must find its logical instruments in the environment and actual conditions of the social relational structure. It is from these conditions and the social relational structure that the epistemic contents must be developed to liberate thought from the shackles of injustice, subjugation, racism, malicious myths, hypocrisy and many others against the oppressed people. This calls for the restitution of collective freedom and true humanism that are characteristics of African traditions, conceptual system, social formation and governance under the creative principle of opposites with relational continuum and unity as the universal law of understanding states and processes as related to the dynamic behaviors of qualitative and quantitative dispositions of socio-natural elements.

# CHAPTER THREE

## The Intellectual Task for Africa's Decolonization and Emancipation in the Logic of Command-Control Systems

To change the direction of history, we must reconcile the conflict between the actual and the potential. The process involves taking advantage of the dynamics within duality where the potential is actualized by a well calculated action and the actual is potentialized by an opposite action. The place interchange is a task that must be well executed. Colonialism is the actual and decolonization is the potential. Similarly neo-colonialism is the actual and independence is the potential. The organic task for the African in the colonial setting, including its Diaspora, comprises of three sub-tasks: 1) decolonization of the African continent to establish freedom, dignity and self-confidence, 2) emancipation of the African continent from the European imperial occupation and terror in order to establish a path to economic progress, cultural dignity and independence for sovereignty and 3) re-establishing the African traditional greatness and dignity. To tackle this organic task and its sub-tasks, the African must pay attention to the history that has created the current state of being and creating the future history in the making. The African must also pay attention to the advices and instructions from the elders and the wise as was usually the tradition. The implementation of this task requires the African to take some important advice from a number of Pan-African personalities on what they saw and see regarding the structure of the African problem as well as their reflections on the direction of the appropriate solutions.

### 3.1 Reflections by Some Pan-African Personalities on the Task

The African problem of decolonization and emancipation, they seem to suggest, has three dimensions. The three dimensions, physical, mental and spiritual, must be related to conditions of colonialism, occupation and loss of sovereignty, and the African holocaust composed of forceful seizure of resources of labor, land art and history. The interactions of all these problem- solution factors lead to the establishment of the solution center, where the solutions must proceed on all fronts. The relational structure is presented as an epistemic geometry in Figure 3.1.1. The problem-solution duality has a relational continuum and unity which must be related to the global race problem, questions of African ancestry and conditions of African ancient civilization that proves the enviable foundation of evolving human civilization. The solution field must be constructed from the Pan-African logical structure to destroy the conditions of racial myths, oppression, subjugation, imperial armed robbery, Western terrorism, genocide and many other imperial ills that are too many to list. It is from this path of European atrocities and terror perpetuated against Africans that one can understand the emergence of the modern Pan-Africanism and the challenges of its practice as they relate to the works in [R1.95], [R1.96] [R1.97], [R1.120], [R1.286], [R1.287], [R1.290] [R1.227][R1.228], [R1.239] [R1.294]and many others in the African Diaspora and others from the African Continent that I have already mentioned in the previous chapters and also see [R7] [R8].

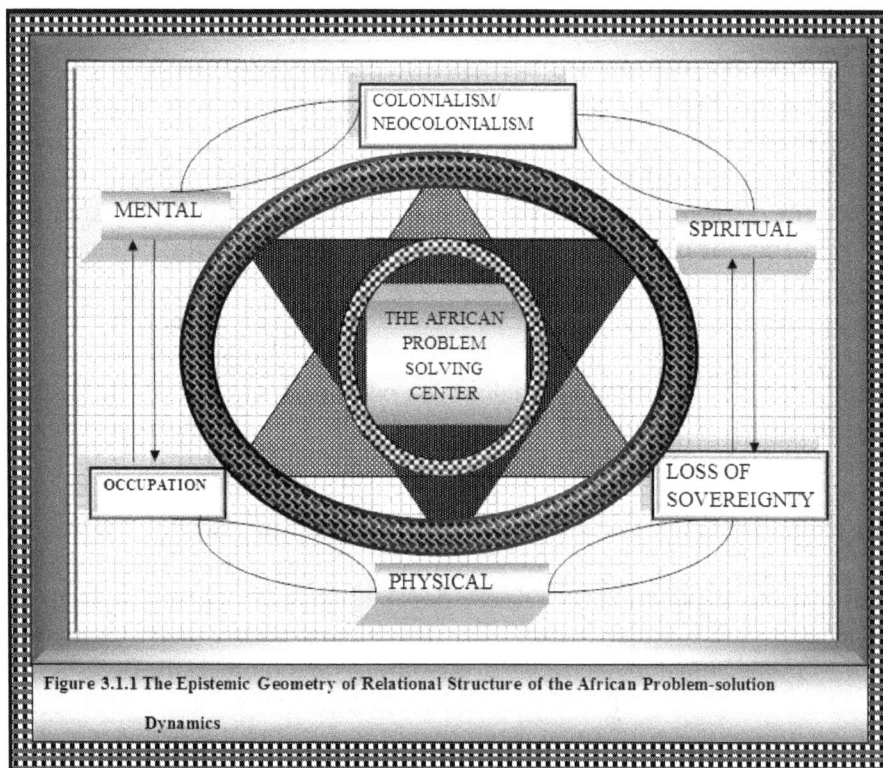

**Figure 3.1.1 The Epistemic Geometry of Relational Structure of the African Problem-solution Dynamics**

The sequential nature of the solution to the African problem is 1) to decolonize the continent by stopping colonial despotism, 2) to emancipate her children from vestiges of colonialism and imperial oppression, 3) to emancipate the African people from spiritual, cultural and material poverty through nation building and development, and 4) to *de-racialize* the global system of thought and its practices against the African and reclaim freedom and justice. Each of these sequential steps while interdependent may be seen in terms of a *social category* that is waiting for transformation by means of *categorial conversion*. The formation of the set of social categories which must be acted upon demands a good and reasonable understanding of the characteristics of these categories and their differences in the epistemological space that distinguish them as well as unite them in understanding the tasks being confronted as Africans. This is the task for the initial understanding to construct the information required for the engineering the strategies to create *categorial moments* to induce categorial conversions through the strategies of the problem-solving sequence for the complete African emancipation. This is part of the implementation of the information structure in the social decision-choice system. The next step is to develope the understanding required to create the strategies to implement the conditions of the categorial moment for categorial conversion of the state categories. The special decision-space is viewed as *control-state space* where appropriate controllers are to be introduced through the actions of the African. Each state is a category with its own set of qualitative characteristics that defines its uniqueness, identity and linkage to other social categories. Each controller is a categorial moment endowed with energy that produces a social force to bring about categorial conversion. Each moment is created within the category through policy constructs and the casting of relevant institutions to bring about self-motion and transformation from one social category to another. The problem-solution process is an information-decision-interactive process under feedback and self-correction principle. The nature of the information-decision-interactive process and its effectiveness directly or indirectly depend on social ideology and the supporting social philosophy.

The creation of a categorial moment requires the understanding of social forces and their utilization. Such an understanding and the application that may be required of it demands the development of an intellectual framework for the practice with the understanding that *Practice without thought is blind; thought without practice is empty* [R1.203., p.78]. The building of the framework for practice requires the creation of a unified African thought system which will guide all different faces of the

113

solution of the African organic problem. This unified African thought system is an African ideology and supporting African philosophy in support of African information-decision-interactive process in the global space. What is the required framework is the appropriateness question that must be answered along the development of the intellectual foundation of revolutionary practices of social movements. The needed thought system cannot simply be derived from the thought system of the oppressors and predators which justified colonization and enslavement of African people under the ideology of Western Christianization and civilization [R1.42] [R1.69b][R1.70b]. It must be derived from the foundations of African conceptual system with the focus on the ideology of liberation, liberty and independence for the affected people. It involves the search for the solution to the problem of relevant thinking on African social vision and the African interest. The development of the required solution is the organic task facing the African world in the sense of creating a relevant thinking system that must guide the movement of the African social system from colonialism to complete emancipation. The development of the relevant thinking system must proceed from Africa's intellectual traditions and experiential information structure.

The solution to the problem forces us to examine the relational structure of ontology and epistemology from the African conditions. The current conditions present the characteristics of the actual which are Western imperial oppressions, subjugation and terror by means of colonialism, neocolonialism and military violence. These are the ontological conditions waiting for transformation to another social category which presents itself as improvement. The exercise for the transformation must be framed from the understanding that philosophy emerges out of real conditions of the people and must be developed from such conditions in the epistemological space. The point here is that social philosophy must be relevant, at the explanatory level, to provide an explanation of social reality which is confronted by the people. It must also be relevant at the prescriptive level of the transformation of social categories. The task for any freedom-loving African demands an uncompromising familiarity with critical studies of the accepted Western philosophies of the past and its contribution to the current African conditions and problems; and with similar critical studies of African traditional thought system and its possible contribution in crafting solutions for negation of negation. These studies are framed in the epistemological space and not in the ontological space. The framing of these studies must involve the consideration of the social milieu, relevant motivational questions, intentionality and the logic or the thinking system that brought about the creation of the philosophies and derived ideologies under studies. These critical studies must point to the direction of the development of needed liberating philosophy that must stand in opposition to the imperialist philosophy of oppression, violence, racism and inhumanity. This is the task facing the African intelligentsia. How should this task be viewed? At the level of thought and movement over the epistemological space, the task of the African intelligentsia is not done and its historical position is not assured until a complete philosophical system which is relevant to the African conditions has been created with its own dynamism. Here, the African intelligentsia must be bold to take the bull by its horns. The older logic of imperialist thought that shuts out new logical possibilities and constrains the African creativity within the impermeable walls of the familiarity of the accepted logic of analysis and synthesis must be avoided at all cost. We must accept those ideas which fit into the African tradition and conditions and reject those that are inconsistent with the African cognition. The advice and reflection on knowledge by British philosophers, Bertrand Russell and Maurice Cornforth are useful for thought:

> It is necessary to practice methodological doubt, like Descartes, in order to loosen the hold of mental habits; and it is necessary to cultivate logical imagination, in order to have a number of hypotheses at command, and not to be the slave of the one which common sense has rendered easy to imagine. These two processes, of doubting the familiar and imagining the unfamiliar, are corrective, and form the chief part of the mental training required for a philosopher. [R12.24, p.24]
>
> At every stage and in all circumstances knowledge is incomplete and provisional, conditioned and limited by the historical circumstances under which it was acquired, including the means and methods used for gaining it and the historically conditioned assumptions and categories used in the formulation of ideas and conclusions [R8.13, p. 179]

The accomplishment of this task requires the art and science of decolonizing the African mind. This decolonization must take place at the epistemological space to free the African mind from epistemic slavery and loss of originality in thinking. It was on this basis that Nkrumah suggested that: *The history of human achievement illustrates that when an awakened intelligentsia emerges from a subject people it becomes the vanguard of the struggle against alien rule* [R1.202, p.43]. It was on this basis of the task to be accomplished that lead Tekyei to state that: *Let there be sane thinking, sane organization, sane methods, and the future is with us to command* [R1.171, p 404]. In support of Tekyei's position Attoh Ahuma added: *The most difficult problem of our times is how to think so that Africa may regain her lost paradise. How to think the thoughts that galvanize and electrify into life souls that are asleep unconscious of their destiny?* [R1.171p.168]. Casely Hayford added by pointing out that; *Brain power we are assured. Of skill there is no dearth. What is wanted is the directing hand which will point to the right goal* [R1.171, p.207]. On the question of obligation directed toward emancipation, Faduma has this advice: *The thoughts, feelings, aspirations must be shared by all and become the common property of all if society is to grow. Wealth must be so distributed by its possessor that the society in which he lives gains by it. In the social organism, the rich and the poor are not and should not be antagonistic forces* [R1.171, p194]. He further instructs us that: *There have always been two kinds of men* [people] *in the world, those who have ideas and those who have not. The men* [people] *who have ideas, be they political, social, intellectual, morals or spiritual, are often the world's best workers and redeemers. They sometimes fail to reach their ideals but they are better and leave the world better for working up to them* [R1.171, p197]. The needed intellectual framework is amplified by the statement that: *every people has the right to determine its own destiny by choosing its own institutions and forms of government best suited to its own peculiar circumstances* [R1.171, p.214]. Let us turn our attention to how Nkrumah took advantage of these suggestions and advice, and how can other Africans learn from Nkrumah.

## 3.2 Ontology, Epistemology and the Prelude to the Logic in Nkrumah's Conceptual System for African Emancipation

To understand the advice and suggestions from some members of the African ancestry and their influence on Nkrumah and the thought that emerged, one needs to examine the relational structure that may be established between ontology and epistemology, and how such a relational structure influences the emergence of different philosophical thought systems which Nkrumah points out in his examination of past European philosophical systems. In general, the output of such a critical understanding will point to the direction of the needed thought system which will allow us to avoid mimicry, cultivate the practice of methodological doubt, develop creative imagination and avoid succumbing to the intellectual slavery in the domain where commonsense has made it possible in cognition. It is here, that philosophy in all forms asserts itself as a general theory of knowledge that requires the development of different tools for specific knowledge development made up of conditions of nominalism, constructionism and reductionism that characterize the methods of thought and the development of epistemic positions and ideology.

In the socio-political set up, the conditions of ontology defining the identity of the ontological actual is that which is taking place, or is what social reality exists to define the actual conditions of the people. The perception of the social actual is cognitively constructed in epistemological space. This is the *epistemic reality* which is constrained in the epistemological space where such perception may generate an epistemic mirage. The constraining elements generating the constrained space are made up of two aggregate structures of the nature of the information structure available for processing and the information-processing capacity structure of the individual and collective cognitive forces. Depending on the effectiveness of the constrained space, the epistemic reality may or may not have an isomorphic relation with the ontological identity. In other words, the characteristics of the social perception may deviate from the characteristics of the ontological reality which is the actual. From the viewpoint of the social decision-choice system, the epistemic reality overrides the ontological reality. In general, there are two sets of organic conditions that we must pay attention to. They are the set of ontological conditions on the basis of ontological information structure and the set of epistemological conditions on the basis of ontological information. These conditions may be viewed in terms of characteristics that present the identities of the ontological elements and the identities of the epistemological elements.

The ontological space presents us with the actual-potential duality in all transformations. The epistemological space presents us with information-knowledge duality that points to cognitive possibilities about the actual-potential dualities in the ontological space. The information on cognitive possibility allows the cognitive agents to expand the ontological elements or to transform them into the *possibility space* or the *possible-world space* with various degrees of uncertainties in the possibility space, and uncertainties and illusions in the possible-world space [R10], [R3.13]. The information-knowledge duality of cognition on the epistemological space presents itself as a limited, vague and deceptive one in degrees that define the epistemological information structure which must be abstracted by some of our senses to constitute the combination of sense and non-sense data by acquaintances. The information structure from the combined data, constrained from the epistemological space, comes to us as differential information structures from the ontological space containing the ontological elements of relevance to the social decision-choice actions. The information-knowledge duality involves the processing activities of learning, knowing and decision-choice actions of social change. The logical system connecting them in theory and conception is driven by social intentionality that is discussed in Amo [R1.11 ]. Such intentionality finds different expressions in the epistemological space and may be amplified into illusions in the possible-world space that may be directed to social set-ups to bring about change, maintaining the existing social structure or social transformation. The most important logical thing for knowing and change of social categories is the set of initial conditions that is established by cognitive perceptions of or acquaintances with the ontological elements in the social formation.

The first important task in the process of thinking within the information-knowledge duality is the recognition that there is the existence of the *ontological space* independent of cognitive agents, and there is the existence of the *epistemological space* in which all cognitive agent conduct thinking and knowing activities to construct knowledge systems that connect the two spaces for social transformations on the basis of doubting the old and imagining the new. The thinking and knowing process must lead to the development of an appropriate thinking system which will bring into being a liberating philosophy and the corresponding liberating ideology that are opposed to the imperialist philosophy and the corresponding ideology of oppression, exploitation and slavery for profit and national interests. The ontological space is completely defined by the ontological characteristic set that constitutes the primary organic category for knowing. The primary organic category is an organic ontological family of ontological families defined by sub-sets of the ontological characteristics. These ontological families are called ontological categories where each category contains the same ontological elements which define the distribution of the micro-ontological realities over the categories. The epistemological space, on the other hand, is also completely defined by the epistemological characteristic set that constitutes a derived organic category.

The derived organic category is an organic epistemological family of epistemological families also defined by sub-sets of epistemological characteristics. Like the ontological space, these epistemological families are called epistemological categories, where each epistemological category contains the same epistemological elements that also define the distribution of the micro-epistemological realities. Each epistemic reality constitutes a claim to a knowledge item which is a derivative from the ontological space. The collection of all epistemic realities constitutes the *social-knowledge space* that may function as an input into social decision-choice actions given the distribution of decision-choice sovereignties. The methodological movement from the ontological space to the epistemological space for knowledge claims is called *methodological constructionism*. The test of any knowledge claim is a methodological movement from the epistemological space to the ontological space. Such a methodological movement is called *methodological reductionism*. The knowledge development process takes place within the construction-reduction duality with relational continuum and unity.

## 3.3 The Ontological-Epistemological Relationship in the Development of either Imperialist or Liberation Philosophy and Ideology

In this framework of knowing, every ontological reality, defined by an ontological element, belongs to a primary ontological category which exists independently of knowing by other ontological elements including humans. The ontological characteristic set of any ontological element is its ontological information that reveals the identity of the category to which it belongs in the ontological space. The ontological space contains infinite ontological elements. The ontological information of each ontological element is projected to the epistemological space through signals of the characteristic sets for cognition. The set of the signals are received by cognitive agents who consolidate them into an epistemological characteristic set which is a perceptive interpretation of a particular element in the ontological information. The epistemological information characteristics set becomes an input into the process of knowing by first defining an epistemic element, and then processing the information characteristic set with some logical system to derive an epistemic reality. The epistemic reality is then tested against ontological realities for claims to knowledge.

The claim to knowledge is always conditional, where such conditionality reflects the structure of the distance between ontological and epistemic realities. It is called *epistemic conditionality*. The distance is called the *epistemic distance* that measures how much is not known or

119

known of an ontological element. The epistemic distance is mapped onto the decision-choice space as an *epistemic risk* which may be separated into stochastic and fuzzy ricks [R3.7] [R3.8] [R3.9] [R3.13]. The defining characteristics, the size of the epistemic conditionality, epistemic distance and epistemic risk depend on the epistemological information structure which is characterized as incomplete and vague (called defective information structure composed of are *fuzzy information structure* and *stochastic information structure*). The fuzzy information structure and stochastic information structure combine to generate *fuzzy-stochastic uncertainty* and *risk* in all knowledge phenomena. In the social systems and knowledge concerning societies, which were of interest to Nkrumah, the fuzzy-stochastic uncertainty and risk are amplified by a *deceptive information structure* composed of *disinformation* and *misinformation* to change preference ordering of social categories. It is here that ideology operates in the phantom possible-world space. The processing of knowing is made up first by methods of information representation after an acquaintance to initialize the knowledge set. The methods may be empirical or axiomatic or both to create the epistemological information set. The epistemological information set may come to the cognitive agents as either an empirical information set or an axiomatic information set or hybrid of the two for the epistemic transformation of an ontological category to an epistemic category to create a knowledge item. The process paths of ontology and epistemology for the creation of the knowledge claims are illustrated in Figure 3.2.1.

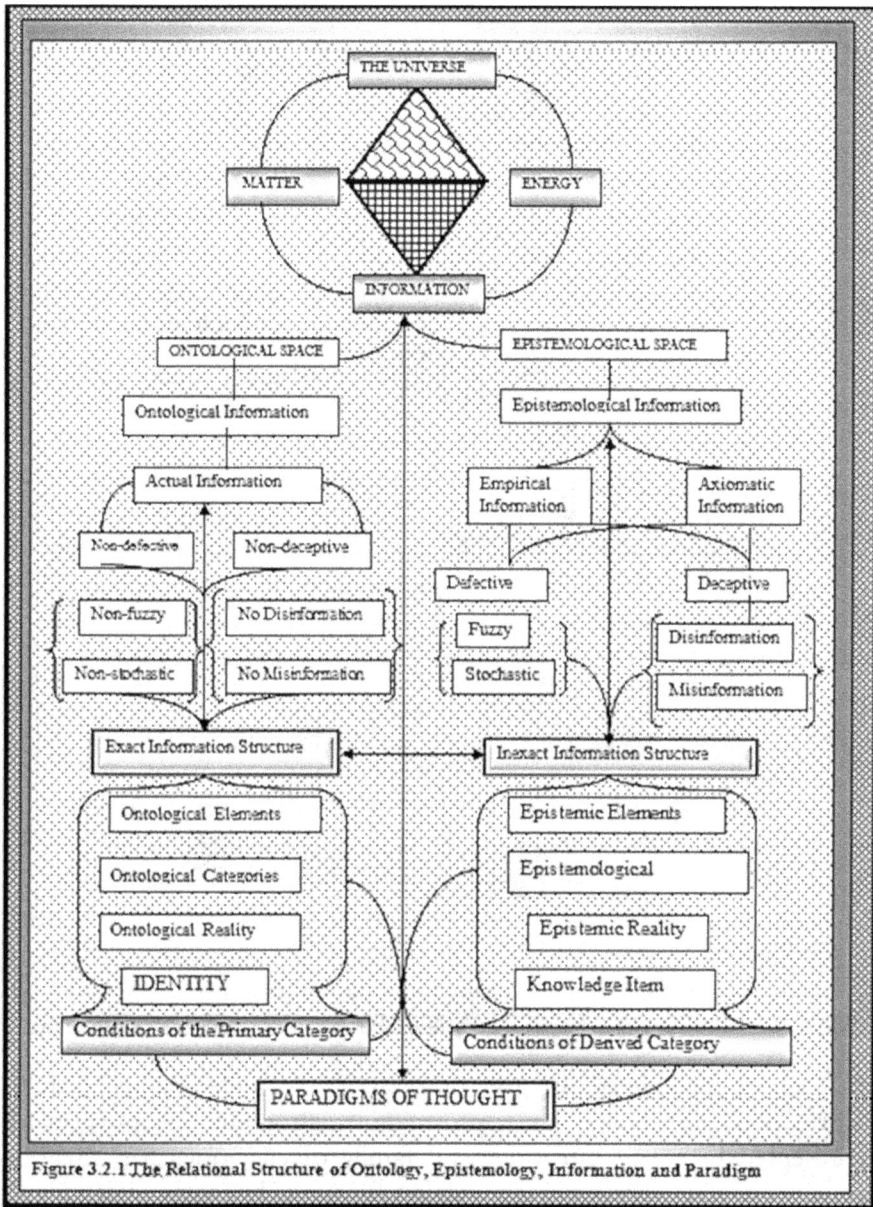

Figure 3.2.1 The Relational Structure of Ontology, Epistemology, Information and Paradigm

The information structures become inputs into epistemic processing machines which are generated by paradigms of thought with corresponding information-representation forms, laws of thought and mathematics. The structure of the current paradigms of thought is provided as a continuity of Figure 3.2.1 in Figure 3.2.2.

121

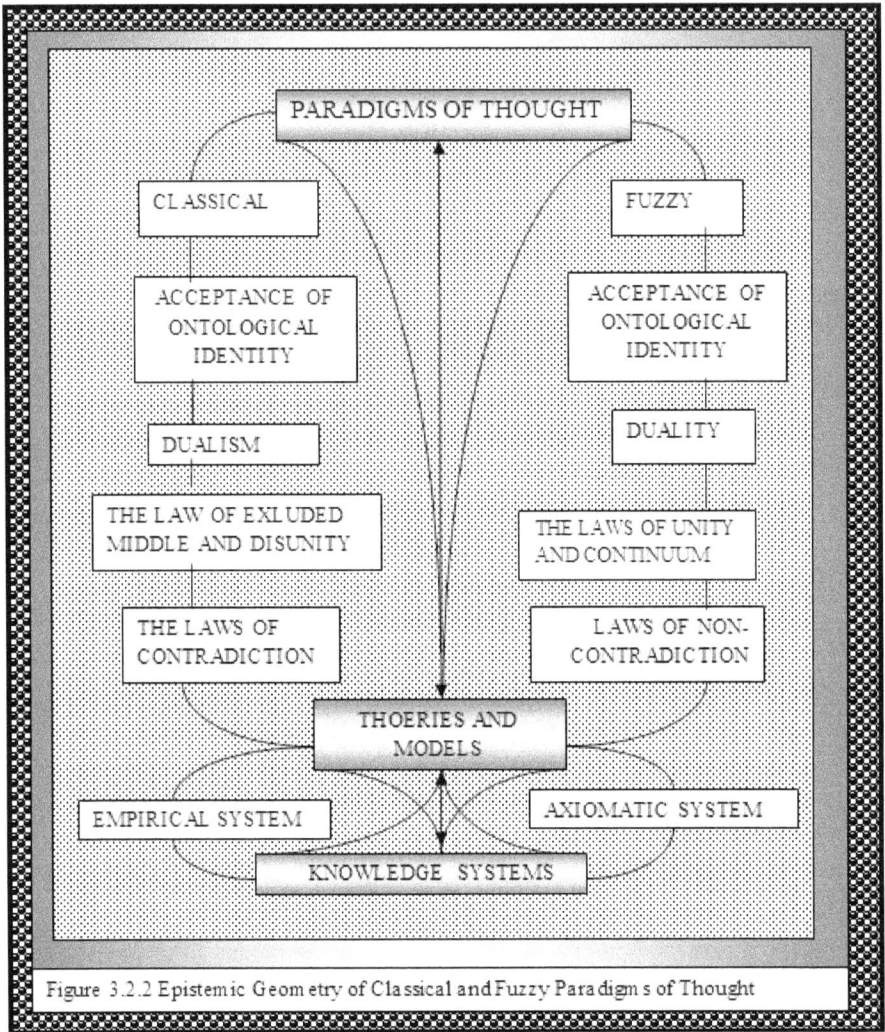

Figure 3.2.2 Epistemic Geometry of Classical and Fuzzy Paradigms of Thought

The transformation of the ontological elements begins with an information structure. The information structure is categorically represented to create categories and family of categories for categorial conversion. The categorial conversion is done by some methods of information processing. These methods are induced by some acceptable laws of thought which are defined by some socially acceptable paradigms of knowledge development. The methods of information processing and the paradigm of thought with its logic and mathematics of reasoning are sensitive to the assumptions imposed on the general information structure. For example, the assumption of an exact information structure, whether complete or incomplete, will demand a particular methodological structure and a paradigm of reasoning. Similarly, the acceptance of a non-exact information structure, whether complete or incomplete, will demand different methods of information processing and a paradigm of reasoning. The information inexactness in the knowledge production of natural phenomena is due to general vagueness while the information inexactness in social phenomena is due to both vagueness and deceptiveness. The deceptiveness amplifies the degree of intentionality associated with conclusions attached to a particular social knowledge item in order to manipulate decision and choice.

The analysis and understanding of any epistemological item require information that is relevant to the item. The analysis and understanding provide relevant conditions for initiating categorial conversion into a desired category. The first task of the African intelligentsia is to define an appropriate structure of information input for the development of the understanding of the African social conditions. The second task is to create from these social conditions, appropriate intellectual instruments for epistemic transformation in the epistemological space. These tasks from the explanation and understanding in the epistemological space break down into the following: 1) construction of a correct information input structure relevant to the African conditions at any social category; 2) an appropriate representation of the information structure for effective processing; and 3) the design of an epistemic paradigm with its laws of thought to process the information into social knowledge items with supporting ideology that is appropriate to the African conditions. The knowledge items become inputs into the social decision-making in support of the African social vision of emancipation and interests with the view of conversion of social states. Knowledge and ideology perform two inter-supporting roles in all social set-ups. Knowledge provides credibility for decision-choice action while ideology provides the motivating force to action on the basis of the perceptive interpretation of knowledge.

The ideology provides an environment for harmonizing the behavior of the individual to the collective existence for stability and compliance. In all cases, a liberating ideology becomes an instrument to fight the oppressive ideology that imperialism, colonialism and neo-colonialism have placed in the minds of the peoples of the colonized territories or territories under occupation and negative action. In these imperialistically and colonially subjected territories, the elements of imperialist ideology have become important inveteracy in the Africa's cognitive structure placing the Africans in a zone of cognitive imbecility that has, on the aggregate reduced Africa's collective vision to a simple mimicry and demand of trivialities of what the imperialist ideology has epistemologically mapped for them. This is an important social process for bringing into being the African global vision and interests. It is also very important in creating the required conditions for nation building, and to work out the needed strategies to solve the race, colonial, neo-colonial and the transformation problems that have become endemic to Africa's social thinking. These problems show themselves at the level of thought and social practice where thought must be created to guide practice. It is at the level of unity of theory and practice that Nkrumah's following advice acquires transformational potency in decision-choice systems: *Practice without thought is blind; thought without practice is empty* [R1.203, p.78]. The failure to acknowledge this unity in the thought-practice duality with relational continuum will lead to an African intellectual destruction that is generated by an acute malignant epistemic schizophrenia that will rack the useful and important development of needed social policies and applications for Africa's social transformation. These simple ten words were central to Nkrumah's approach in finding solutions to African organic problem and sub-problems. We have examined in some detail Africa's problem at the level of thought. Let us now turn our attention to the level of practice.

At the level of practice of the desired African knowledge system and ideology for bringing progress and emancipation, the problems are revealed in: 1) designing social policies that relate to the African global interest and social vision; and 2) casting relevant institutions through which the policies may be transmitted to tame the antagonistic elements and overcome hostile environments. It is at the level of practice to achieve emancipation, nation building and African progress that Nkrumah's statement is instructive to Africans and their leadership at all times:

> Progress does not come by itself, neither desire nor time can alone ensure progress. Progress is not a gift, but a victory. To make progress, man has to work, strive and toil, tame elements, combat environment, recast institutions, subdue circumstances, and at all times be ideologically alert and awake [R1.207, p. 113].

This ideological alertness must A) reflect the practice of a liberation ideology and supporting knowledge structure as well as B) be aware and awake to the imperialist's ideological maneuvers, camouflages and chameleon behavior. Every ideological framework contains a propaganda locked in a deceptive information structure with implicit intentionality. Propaganda may be a means of liberation by taking a positive instrument of clarification of the path of the existing and evolving information-knowledge structure. The intentionality of liberation propaganda is to overcome the propaganda of an oppressive ideology that holds the people's collective mind in bondage, decimates their freedoms, destroys their collective decision-choice sovereignty, strips off their national independence, steals their vision, enslaves their labor and freely abstracts their national resources and wealth for the benefit of others. Every liberation ideology is violently opposed to oppressive propaganda that denies freedom, justice and democracy. The statements reflect fundamental propositions in the Nkrumah's conceptual system. They are also part of the essential points in the development of Nkrumah's approach to the solution of Africa's organic problem and the extensions that are being developed in this monograph. The relational interactions among the essential components for effective transformational dynamics through decision-choice actions and the social institutions are provided as an epistemic geometry in Figure 3.2.3. There are two pyramidal interactions. There is the ideology-thought-practice pyramid and superimposed on it is the society-institution-policy logical pyramid with the social decision-choice center that generates a transformation force.

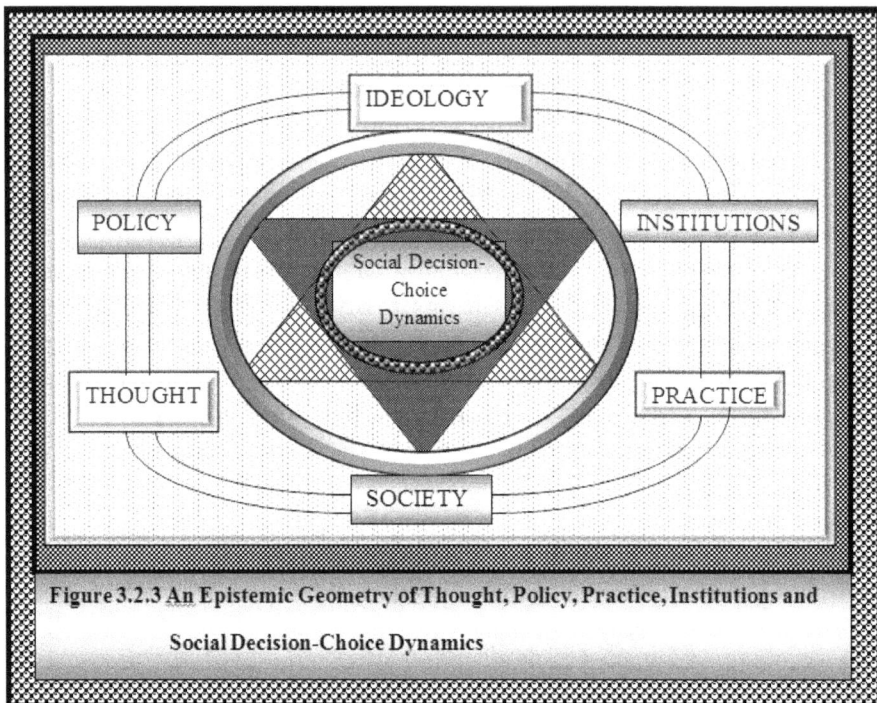

**Figure 3.2.3 An Epistemic Geometry of Thought, Policy, Practice, Institutions and Social Decision-Choice Dynamics**

The liberation ideology and the oppressive ideology exist in duality with relational continuum, defined with attributes of intense conflict, negation and negation which define temporally social equilibrium-disequilibrium states with a transitional stability-instability process. Every temporally social equilibrium state is under double negation for either negative (reversal) transformation or positive (forward) transformation. The conditions of the oppressed people in general and Africa in specific were observed by Nkrumah, in relation to the phenomenon of transformation reversal who pointed out that: *The achievement of freedom, sovereignty and independence is the product of matter and spirit of our people* [R1.207 p. 77]. In this respect of the people's effort, he also observed that: *It would be a great mistake to imagine that the achievements of political independence by certain areas in Africa will automatically mean the end of the struggle. It is merely the beginning of the end of struggle* [R1.207, p78]. Complementing this observation on ideology and emancipation, he added that: *It is only in conditions of total freedom and independence from foreign rule and interference that the aspirations of our people will see real fulfillment and the African genius find its best expression* [R1.207 p.5]. The application of African ideology is to reintegrate and enhance the original African personality before the

126

imperialist destruction of African social set-ups and their internal fabrics with their organizational philosophy of collectivism, where the moral principle guiding the social relations is that *each for all and all for each* under a production-consumption system of collectivism. This organizational philosophy is opposed to the imperialist organizational philosophy of individualism, where the moral principle of decision-choice action is *each for himself and God for us all* under a production-consumption system of capitalism.

This philosophy of individualism, Nkrumah observed, brought about the Western imperialist terror defined in terms of different forms of slavery, executions, and exile of leaders, limb cutting and de-humanitarian principle that created the greatest global human suffering on the people of Africa from South to North and from the West to East of Africa. The anti-humanist principles were promoted and administered through distorted and twisted types of religion where the ideology of anti-paganism was levied on the Africans to convert them into submission to the zones of imperialist oppression and torture. This is the African tragedy under the Western imperial design on the basis of which current African problems are sown, watered and continuously being managed by the imperialists through the methods of neocolonialism. It is here that the revolutionary theory of social knowledge in support of liberation and revolutionary ideology is called into action. As Nkrumah observed, this revolutionary ideology must be crafted by the people who think as people of action and act like people of thought. They must clearly understand the enemies of the African people, the modus operandi that they use, the complex system of enticements and their ideological system to create confusion in the African mind for thought and practice. They should practice methodological doubt in order to rid themselves from colonial mental habits and develop imagination to solve the Attoh Ahuma's stated problem that *the most difficult problem of our times is how to think so that Africa may regain her lost Paradise* [R1.171, p.168]. They should refrain from allowing themselves to be simply enslaved by what the colonial mentality has rendered for the African commonsense to imagine and work with. The methodological doubt and Ahuma's conditions of thinking are fundamental requirements for using Nkrumah's thought-practice duality in relational continuum and unity to work out the strategies for Africa's emancipation.

The relational structures among thought and practice, doubting the present and imagining the future in reference to the past in the tradition of the *Sankofa Anoma* place on the African intellectuals a special demand for which Nkrumah points out to us that:

> Africa needs a new type of citizen, a dedicated, modest, honest, informed man. A man who submerges self in the service to the nation and mankind, a man who abhors greed and detests vanity. A new type of man whose humility is his strength and whose integrity is his greatness [R1.202, p.130].

The new African must be equipped with the *African personality* as specified by Nkrumah within the space of the African tradition and extensively discussed in the book [R1.74][R1.75] [R1.92, pp. 109-138] [R1.263]. An important question arises in this frame of thinking. The question is simply how to create this new African from the colonial conditions and make it a mass phenomenon. In the search for relevance, we must keep in mind that there is nothing more dangerous for African emancipation and social transformation than the old intellectual delusion and existence of cognitive imbecility which imprison Africa's creative thinking within the dark walls of colonial mentality and the familiarity of irrelevant commonsense. To Nkrumah, the tasks of creating the thought and putting it into practice are inseparable. The framework of needed revolutionary ideology and the supporting philosophy as well as a methodological approach to solving the problem of the relational unity of thought and practice lead Nkrumah to the development of Consciencism [R1.203] and other related works for which an attention will be turned to in Chapter 4.

The work in Consciencism may be seen in two important inter-supportive conceptual and logical parts. The first part deals with the justification and philosophical foundation of internal self-motion which Nkrumah calls *categorial conversion*. Here, a justification is made for Africa to self-transform herself from within from the understanding of natural transformations. The second part which Nkrumah entitles philosophical *Consciencism* deals with the analytical foundation for the construct of the required tactical and strategic conditions that will make the categorial conversion applicable to social transformations. This monograph, entitled *The Theory of Categorial Conversion: Analytical foundations of Nkrumaism,* is devoted to the former. The analytical foundations are made up of philosophy, logic, mathematics and computable systems to construct the needed theory of general mechanism of transformations. The theory helps to derive the convertibility conditions of categories as well as explanatory and prescriptive conditions. The part two which involves philosophical Consciencism will be dealt with in a separate monograph as the theory of Philosophical Consciencism. The theory will involve the derivation of general and specific tactical and strategic mechanisms for implementing the convertibility conditions in social transformations. The theory, therefore, will present explanatory and prescriptive conditions for human actions in the social decision-choice space.

# CHAPTER FOUR

## The Theory of Categorial Conversion
## Philosophical Foundations and Extensions of
## Nkrumaism in System Dynamics

To solve the problem of alternative logical system that will allow the construction of a thinking frame to support the corresponding African ideology needed to develop a thought system and the corresponding African ideology, Nkrumah observes that this thinking frame must draw its weapons from Africa itself. He states that *our philosophy must find its weapons in the environment and living conditions of the African people* [R1.203, p.78]. The nature and usefulness of Nkrumah's statement and the required weapons are discussed in some details in the book [R1.92, pp.33-81 and 182-225]. In the current discussion, it will be pointed out, contrary to what other writers' think, that Nkrumah kept his eyes on the African conceptual system which has the conceptual building blocks of polarity, duality, continuum, time-trinity, continual change, relationality and unity under the principles of opposites. The roles of these conceptual building blocks in the philosophical and mathematical foundations with Nkrumaism will be made explicit in this monograph.

### 4.1 Searching for an Alternative Paradigm

In the lure of his own intellectual stand, Nkrumah exits from the classical paradigm based on the Aristotelian logic and corresponding mathematics with its *methodological dualism* and the laws of thought that are based on the principles of excluded middle and non-acceptance of contradictions. In the classical epistemic system, the problems of subjectivity and quality as characteristics of human reasoning are hard to deal with. In dealing with the African problem and social transformations from colonialism to different stages of emancipation, Nkrumah realizes that he must confront the difficult problem of *relationality* between quality and quantity within the *time trinity* of the past-present-future phenomenon as philosophically represented by the *Sankofa-anoma* [R1.74] [R1.175] [R1.124][R.125][R1.219]. The phenomena of qualitative disposition and quantitative disposition and their relational problem must be related and solved in the logical plane. The relevance of the solution the solution must then be shown at the level of practice of Africa's complete emancipation. Changes in qualitative disposition relate to transformational phenomena within the subjective-objective duality under continuum where the transfer function is seen in terms of qualitative motion but not quantitative motion. Nkrumah realizes that he must analytically deal with and understand the qualitative, quantitative and qualitative-quantitative motions that must individually or collectively bring about an explanatory theory of change or prescriptive theory of change which will guide social actions to move one social category to another, where such movements involve intra-state and inter-state changes. The conditions of the African logical form are such that the intra-state and inter-state changes are in relational unity with necessary and sufficient conditions for social transformations.

At the level of thought, the motions require the conceptual definitions and specifications of quality and quantity under what Nkrumah discussed as qualitative and quantitative dispositions which must be related to something. It is the solution to the need for a relational something that leads to the introduction of the concept of the *primary category* as initializing the conditions of qualitative motions that provide links to the chain of *derived categories*. At the level of both philosophy and mathematics, the concept of category requires a definition and explication. To develop a theory of dynamics within quality-quantity duality, a force must be identified and indicated. For the dynamics to be internal to any category, the force must be internally generated and linked to conflicts in the category. These conflicts must be the result of the activities of the positive and negative characteristics that reside in the same unit. The intra-category and inter-category changes must be related to qualitative and quantitative transfer functions. The analytical approach needed by Nkrumah for solving the transformation problems embedded in qualitative changes cannot be drawn from the conditions of the classical paradigm that constitutes the imperialist pillar of thought and practice where exactness defines information and relational points. It must be drawn from a non-classical paradigm of knowledge production which must relate to the existing African conditions and future possibilities in relation to the past realities, present conditions and future possibilities. The logical approach must see events, outcomes and relations in gray areas under the conditions of inexactness.

The comparative discussions on the classical and non-classical paradigms with the implied laws of thought are discussed in the works

[R3.7] [R3.10] [R3.13]. The initial analytical problem to be solved in the search of a *non-classical paradigm* is whether quality and quantity constitute a duality with a continuum and unity or dualism with excluded middle and separation. Given that we have quality-quantity duality with a continuum and unity, the problem to be solved is whether quality emerges from quantity, quantity emerges from quality or quality and quantity mutually define themselves in the quality-quantity duality by transformation in dealing with motions in nature and society.

Can quality be separated from quantity, and what is the relationship between them in understanding changes in nature and society? Does our theory of measurement that presents quantity to us not proceed from quality to quantity? What are the relationships between matter and quality-quantity phenomena? What role does *energy* play in these changes? Importantly, what roles do both quantity and quality play in the distinction and classification of elements of nature and society as seen in states and processes? How should matter, information and energy be related to qualitative and quantitative motions? The search for answers to these questions points to the discussions on self-motion and non-self-motion. It further points to the discussions on information requirements on the ontological identity in terms of *what there is* as representing the ontological elements. The information requirements allow us to solve the problem of ontological identity which must be related to cognitive actions in the epistemological space. The cognitive actions in the epistemological space raise the question of knowability and the methods of knowing. They also raise the question as to *what does the knower know and how does the knower know what he or she knows.* How does *what is known* relate to the *possibility* and *possible-world spaces* of social and natural transformations? Simply stated, the knowability question relates to the questions: Can 1) cognitive elements operating in the epistemological space know the elements in the social and natural ontological space, and 2) what are the methods and process of knowing? There are also a few related questions. How should we believe that what the knower claims to know is an ontological element and not an epistemic mirage in the desert? What is the intentionality of knowing? Can such intentionality alter the methods and process of knowing and interpretational support of the known? Can such knowing become an important input into the human collective decision-choice actions to change natural and social conditions? These questions and the corresponding answers are intimately connected to thought and practice or alternatively, theory and application are the underlying conditions of social ideology.

These questions are answered in the epistemic space of which Nkrumah was aware. The ontological space is taken as given in the form of ontological elements from which epistemic elements will be created by reasoning with information acting as a linkage between the ontological and epistemological spaces. The process of reasoning connects the epistemological elements to the ontological elements for knowledge discovery. In this cognitive frame, Nkrumah tackled the problem of the information requirement by accepting the empirical information structure as established by sense data on the basis of which knowledge is abstracted by epistemic categorial conversion. The epistemic conversion must meet the conditions of convertibility for any conversion. Here, we must deal with the notion of conversion from what to what. Nkrumah's approach is to accept the existence of ontological identity which is taken as independent of cognitive awareness in the sense that it exists as independent of the minds of cognitive agents. In a broad general existence, this ontological identity is to be matter and not spirit or mind, where matter, spirit and mind are taken to be organic categories that will be shown to emerge by the means of their internal forces from the category of matter which is selected to be the prime.

The question to be answered is this: Can matter emerge from spirit or mind through their internal dynamics, or can the spirit and mind emerge out of matter by the dynamics of internal conversion of matter? The first categorial emergence relates to the question of whether spirit or mind is the primary category and whether by its internal dynamics can spirit or mind give birth to matter as its derived category. The second categorial emergence relates to the question of whether matter is the primary category, and that by its internal dynamics can it give birth to either spirit or mind as its derived category. Similarly, one may examine the internal relations in the matter-energy duality. The acceptance of matter as an ontological identity implies that matter is taken as the primary category and spirit or mind is taken as the derived category over the epistemological space. The selection of any one of the three as representing the primary category depends on one's philosophical view of the universe. The selection initializes the epistemic search on the epistemological space. It is then the job of the knowledge seeker to show by some convincing methods of thinking, how by the internal dynamics of the primary category, other categories emerge as derivatives thereof. It is because of the need for methods of reasoning that the comparative understanding of 1) dualism with the principle of excluded middle and relational separation, and 2) duality with the principle of relational continuum and unity becomes useful in the search of a useful thought system. It is important to note that both dualism and duality work under the polarity. The difference is found in the relationality of the opposites. The dualism maintains logical excluded middle with a separation while the duality maintains a logical continuum with unity in epistemic constructs. This analytical distinction is important to understand the differences and similarities between the classical laws of thought and fuzzy laws of thought and their applications to qualitative dynamics

At this point, we are confronted with the problem of *categorial distinction* between either matter or mind as a category or between matter and spirit as a category. To deal with this problem of categorial distinction, Nkrumah appealed to the African conceptual duality with a relational continuum and not to Aristotelian conceptual dualism with excluded middle. The duality involves the fundamental existence of quality-quantity duality which is linked to matter-energy duality with infinitely relational continuum under the principle of information sharing. There is an important extension that must be made to connect the dualities to polarities. Such an extension and its role in the problem of categorial conversion will be philosophically discussed, analytically worked out and mathematically presented in the chapters that follow. The categorial distinction demands an epistemic classification from the information structure which relates to the elements on the basis of which categories are formed and established for distinction and identity. The information is generated by the *epistemic characteristics* from sense data. In the epistemological space, the conceptual building blocks of knowing are matter (M), information,(I) and spirit (S, energy) (MIS) or quality, information and quantity (QIQ). The relational structure may be illustrated by superimposition to generate a cognitive geometry of a pyramidal logical system of interactions. This is shown in Figure 4.1.0.1.

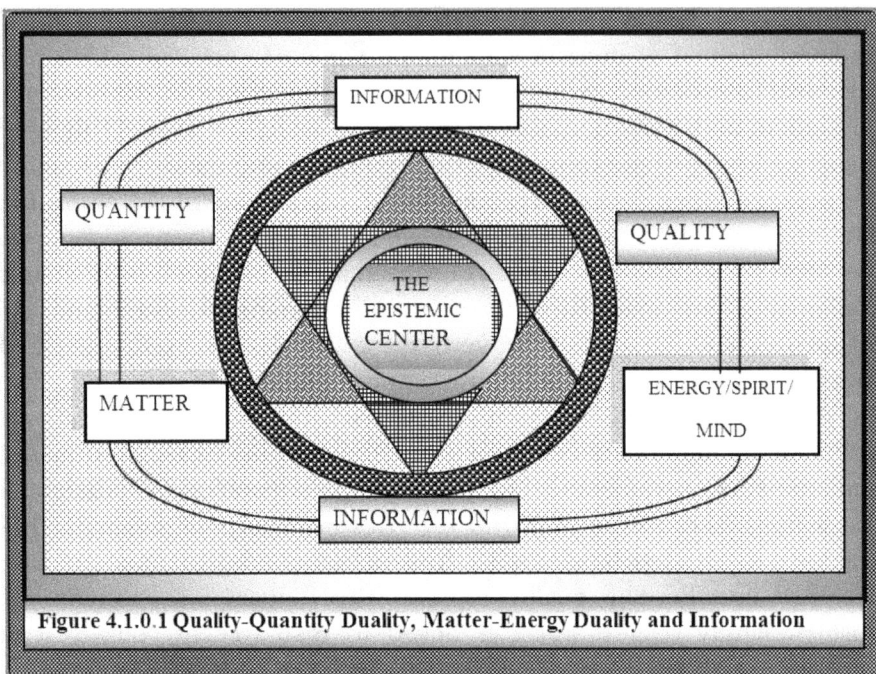

**Figure 4.1.0.1 Quality-Quantity Duality, Matter-Energy Duality and Information**

In the epistemic process of knowing, when one accepts matter as the primary category then the primary organic category must be related to quality and quantity as specified in terms of disposition. The categorial difference and similarities of matter and energy are linked by information and distinguished by their qualitative dispositions. These qualitative dispositions are captured by the characteristics of elements as abstracted by acquaintances, broadly defined, from the ontological elements. The analytical structure of *categorial conversions*, *categorial moment* and the *conditions of categorial convertibility* as used by Nkrumah is firmly grounded on the meaning of a category and how categories are formed. The conditions of categorial convertibility define the conditions of the existence of categorial moments needed to create a *qualitative transfer function* that moves one element from one category to the other to complete the categorial conversion journey. Let us keep in mind that the qualitative transfer function is defined in terms of qualitative motion and must be constructed on the basis of qualitative information or characteristics (it is useful to examine the discussion in [R3.8][3.9] [R3.13]).

## 4.1.1 The Reconciling of Elements in the Ontological and Epistemological Spaces

In this respect, there are two organic sets of initial interest. They are the set of ontological categories in the ontological space, and the set of epistemological categories in the epistemological space. The family of the ontological categories serves as an identity for the family of the epistemological categories. The epistemological categories are established by epistemological information that serves as an input into the knowledge production process to obtain knowledge items over the epistemological space. The epistemological information is always defective irrespective of whether natural or social phenomena are under investigation. The defectiveness is made up of *information limitation* which is associated with incomplete information and captured by probability, and then *information vagueness* which is associated with fuzzy information and captured by possibility. In dealing with social phenomena, the defective information structure may be amplified by a *deceptive information structure* that may lead to a different social knowledge item after the information processing. As it has been pointed out in the previous chapter, the deceptive information structure is made up of disinformation and misinformation sub-structures. The ontological identities are established by perfect information structures in the sense of exactness and completeness as specified in terms of characteristic sets which define the concept of *what there is*, which the ontological reality is. The epistemological identities are established by an imperfect information structure, the processing of which leads to cognitive uncertainties and *epistemic risks* [R3.7] [R3.9].

Let us understand the problem of the dynamics of general transformations where one ontological item in a given category loses its qualitative properties and emerges from another ontological item into yet another ontological category, or one ontological item in a given category loses its qualitative properties and fades into a seemingly different category by the internal dynamics to induce changes. This is the *categorial-conversion problem* which may also be viewed as the *transformation problem* of qualitative states defined in terms of categories with conditions of similarities and differences. The problems of transformational dynamics in the ontological space and the solutions that must be constructed were central to Nkrumah's philosophical and analytical works. The most important thing to observe is the intentionality of Nkrumah's framework

137

of the philosophical and analytical structures of the problem and solution. The intentionality is related to how the framework can be applied to the understanding of social change composed of decolonization and resistance to neocolonialism as transformation phenomena, as well as applied to nation building composed of institution casting and socio-economic development. This is the central core of Nkrumah's philosophical and analytical works that also provide us with methods and techniques for scientific inquiry. To see the structure of the problem, let us separate the matter-energy-information structure from the quality-quantity-information structure and bring them into unity. The relational structure is presented as an epistemic geometry in Figure 4.1.1.1. Nkrumah identifies the transformation as the *categorial conversion,* and hence we may refer to the transformational dynamics as the categorial-conversion dynamics. He defines the categorial conversion in his statement as*: By categorial conversion, I mean such a thing as the emergence of mind from matter, quality from quantity* [R1.203, p. 20].

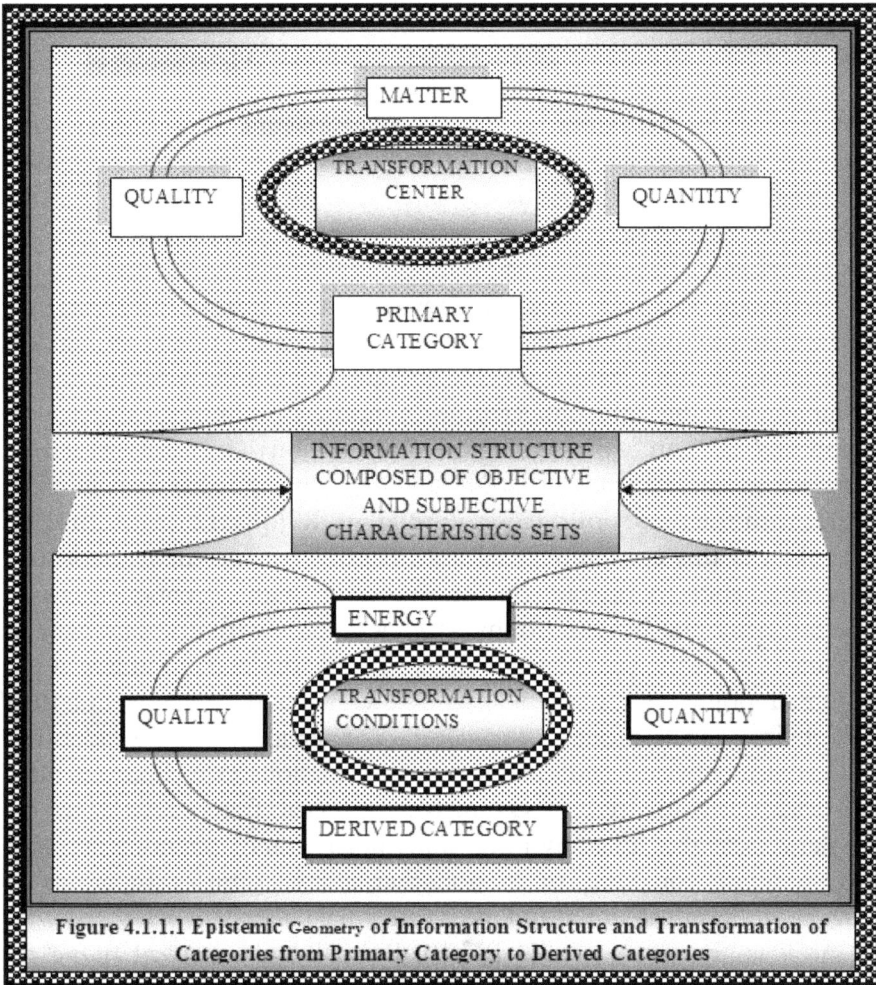

**Figure 4.1.1.1 Epistemic Geometry of Information Structure and Transformation of Categories from Primary Category to Derived Categories**

In this respect, there are two categorial conversions that must be dealt with within Nkrumaism. They are ontological categorial conversion due to natural activities, and epistemological conversion due to cognitive activities under a rationality and a paradigm of thought. The ontological categorial conversion is the identity and constitutes the primary categorial conversion. The epistemological categorial conversion is the derived categorial conversion that seeks to explain the ontological process of the emergence of the new from the old to become the actual and the disappearance of the old from the actual to become the potential. This is the dynamics of the actual-potential duality with a relational continuum under construction-destruction process. From the viewpoint of knowing,

139

the primary categorial conversion is taken as given. The job of the epistemological conversion is to solve the problem of understanding the ontological categorial conversion and its enveloping path. In solving the problem of an epistemic understanding, Nkrumah appeals to the toolboxes of logic and science as he states: *The conceptual tools which philosophy has fashioned in logic, and by means of which it can cope with the formal problems of categorial conversion, are contained in nominalism, constructionism and reductionism* [R1.203, p.21].

Nkrumah suggests that to advance the model of categorial conversion, philosophy turns to science. He considers matter and energy to exist as distinct categories but in an interconnected mode. The distinction and the interconnectedness over the epistemological space are established by an epistemological information structure which is discussed as an extension of an Nkrumaistic conceptual system in order to deal with the theory of category formation [R2] [R2.12] [R2.18][R2.23] [R3.13]. It is this epistemological information structure that philosophy calls on science to operate on, or process it with acceptable laws of thought fashioned by philosophy as logical tools within a paradigm of thinking. Given the paradigm of thinking with the corresponding tools, Nkrumah points out that the matter-energy connectedness and their reducibility provide models of categorial conversions. These models can be related to physical and chemical transformations as well as extended to social transformations. Nkrumah's analytical frame relates to the problems of the quantity-time and quality-time phenomena with corresponding equations of motion that induce transfer functions to bring about intra-categorial and inter-categorial changes.

As it is argued in [R1.92] [R3.10] [R3.13], the analytical frame for calling upon science to help the solution of the problem of categorial conversion encompasses the theoretical blocks of categories, polarity and duality in relation to matter, energy and information. Here, the polarity and duality under the principles of continuum help to establish distinction and connectedness of categories that may be seen in actual-potential duality under continuum. The conceptual framework of polarity and duality with the principle of relational continuum and unity is based on the African system of thought such as the Dogon or the Adinkra and the Pharonic conceptual system where everything is seen in opposites, and as relational continuum to its opposites, such as truth and falsity, good and bad, individual and community or live and death and other similar examples all under the principle of relational continua and unity

[R1.8][R1.12][R1.37][R1.60b][R1.86][R1.92] [R1.218a] [R1.218b[R1.218c] [R1.243] This is the conceptual frame that Nkrumah was working from, and not the Marxian system with the Aristotelian classical laws of thought as it is sometimes argued by others even though the two have epistemic similarities under the principle of opposite. It is analytically important that a distinction be made between methodological Nkrumaism and methodological Marxism in understanding social change. It is within the African conceptual system in relation to the African social formation from the individual-community duality, that one must interpret the concept of Nkrumah's scientific socialism. Nkrumah's analytical links among constructionism, reductionism and nominalism is itself logically powerful and this logical power is enhanced by further linkage to the concepts of category, primary category and derived category. The development of the conceptual and analytical power of this frame to deal with the general transformation problem is the ultimate concern of these two monographs. To solve the analytical problem of categorial conversion, non-classical laws of thought must be developed and must be developed from the African cognitive traditions.

### 4.1.2 The Framework for the Development of Non-Classical Laws of Thought

In terms of the development of the laws of thought, the relational structure of nominalism, constructionism and reductionism must be clearly understood, and in terms of how Nkrumah uses them to energize his conceptual system on categorial conversion within the development of Consciencism [R1.203]. It is also useful to understand how Consciencism is linked to the development of thought which must then be associated with the understanding of events and how such an understanding can be used to guide social practice and revolutionary actions for social change. Nominalism is epistemologically linked to language which is used to specify the epistemological elements and categories on the basis of *epistemological characteristics sets* which provide epistemological information. The primary category formation is the foundation of linguistic primitives where the derived categorial formation is associated with linguistic definables as has been discussed and explained in [R3.7] [R3.10] [R3.13]. The primary and derived linguistic categories constitute the foundation of the *vocabulary of thought*. The initialization of the thought process in the epistemological space is such that one takes an ontological category as the primary category of reality (that is the identity) from which other categories are derived. The conditions and principles of nominalism allow us to establish distinctions between the primary category and the derived categories and among different derived categories. Given the initial primary category, every derived category has a successor to which it acts as its primary category by methodological constructionism in the categorial conversion. Every successor has a preceding category that serves as a primary category by a methodological reductionism in the categorial conversion [R3.13]. The analytical system of thought involving categorial formation, categorial conversion and categorial moments, categorial transfer functions and categorial transversality conditions that is being constructed is extremely powerful. It allows the development of explanatory and prescriptive theories of socio-natural transformations within the general theory of exact and inexact scientific knowledge under structural dynamics. In fact, the system of thinking allows us to develop a unified theory of engineering sciences in all areas of human endeavor. The vocabulary, grammar and language constitute one pyramid of logical reasoning while nominalism, constructionism and reductionism constitute another pyramid of logical reasoning. The superimposition of the two produces a relational foundation of methodology and language for the development of thought through a particular paradigm. This is shown in Figure 4.1.2.1 as a cognitive geometry of methodology and language with a paradigm as a center of reasoning in thought production. The system of reasoning under duality with relational continuum and unity is such that life and death , good and bad, cost and benefit and similar opposites are inter-supportive and inseparable in identity and existence.

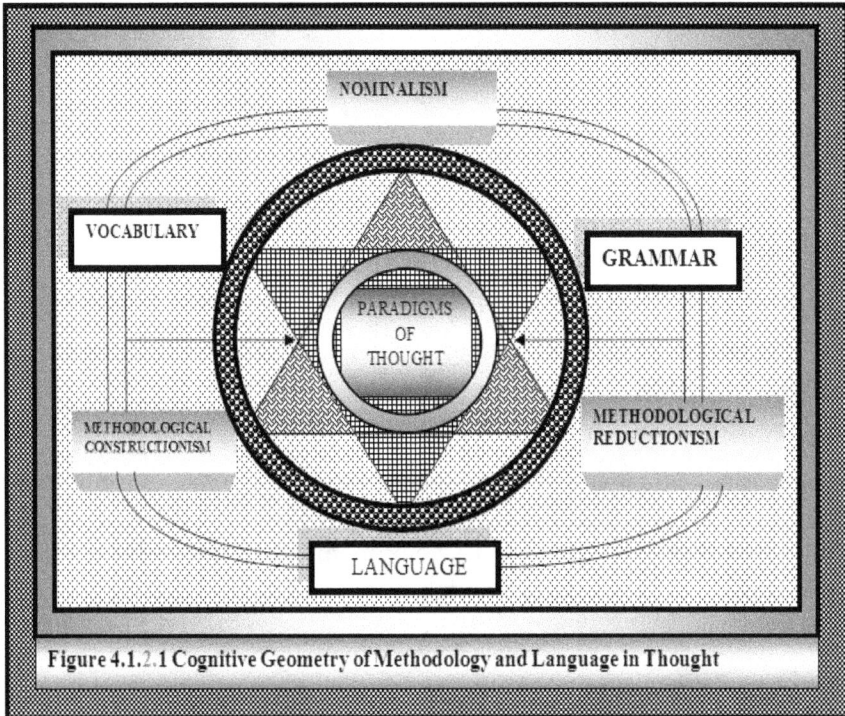

**Figure 4.1.2.1 Cognitive Geometry of Methodology and Language in Thought**

Given the primary category, every derived category serves as a primary category by distinction through the relational structure for defining the separating characteristics sets of the categories over the enveloping of the categorial conversion in the domain of vocabulary, grammar and language of a particular knowledge development. For example, a complete freedom must, by process, emerge from freedom which serves as its primary category while freedom emerges by a process from non-freedom that serves as its primary category over the categorial-conversion enveloping. In other words, the propositions contained in the characteristics set about complete freedom are true, and make sense only if the propositions contained in the characteristics set about freedom and non-freedom are true and make sense also. As applied to societies, and in relation to social theory and transformational dynamics, if we postulate a development, then we must consider its ancestry of underdevelopment which may be taken as the primary category from which development must belong to as a derived category, both of which must be distinguished by their corresponding sets of qualitative characteristics. In this case propositions describing the characteristic set of development are true and meaningful if and only if the propositions describing the characteristic set of underdevelopment are true and meaningful. This framework is very important in understanding planned and unplanned categorial-conversion dynamics over the enveloping of the social development. It is also important in understanding natural and social transformation and the use of information-knowledge structure to engineering natural and social transformation including medical treatments and the creation and management of industrial systems and innovations.

We can consider social systems $\mathbb{S}$ with characteristic measures of a state of complete underdevelopment, $\mathbb{U}$ and the state of complete development $\mathbb{D}$ such that $\mathbb{S} = \mathbb{U} \cup \mathbb{A}$, and $(\mathbb{U} \cap \mathbb{A}) \neq \varnothing$. Both of them are opposites which constitute a duality with a relational continuum. The relational continuum defines the categorial-conversion enveloping curve with various degrees of underdevelopment and development. By the method of fuzzy logic [R3], through membership characteristic function, $\mu_{\mathbb{U}}(s) = \alpha$ the degrees of underdevelopment and development may be assigned *dualistic weights* of $(1-\alpha, \alpha)$ respectively such that $\alpha \in [0,1]$, and $(\alpha + (1-\alpha) = 1)$ where $\mu_{\mathbb{U}}(s) = \alpha$, and $\mu_{\mathbb{D}}(s) = (1-\alpha)$ such that $(s \in \mathbb{S})$ We have complete underdevelopment when $(\alpha = 0)$ and complete development when $(\alpha = 1)$. As the categorial conversion moves a social system $(s \in \mathbb{S})$ through increasing phases of development, the value $\alpha$ increases and $(1-\alpha)$ decreases to maintain the relational continuum and unity. Each state $(s \in \mathbb{S})$ is an actual-potential polarity waiting to be transformed where the value $\mu_{\mathbb{U}}(s) = \alpha$ defines the degree of development. Over the enveloping path of the categorial conversion of social actual-potential polarity, every point is characteristically specified and identified by a duality of the form $((1-\alpha)\mathbb{U}, \alpha\mathbb{D})$ in each pole of the polarity. In the logic of categorial conversion, every degree of development has a supporting underdevelopment that is reflected in the potential which is development. The condition of increasing development may be specified with a set of increasing weights of successful categorial conversion of the form $\{\alpha_1 < \alpha_2 < \cdots < \alpha_i < \cdots \leq 1\}$.

Every categorial conversion defines a point that belongs to a category which is different from the previous one. Each point of the set of categorial conversions in the development-transformation process may be specified with a corresponding set of weights in terms of duality as $\{(1-\alpha_1, \alpha_1) < (1-\alpha_2, \alpha_2) < \cdots < (1-\alpha_i, \alpha_i) < \cdots \leq (0,1)\}$ in a relational continuum and unity The path of conversion and enveloping is a set of the form $\{(\mathbb{U},0) \leq ((1-\alpha_1)\mathbb{U},\alpha_1\mathbb{D}) < ((1-\alpha_2)\mathbb{U},\alpha_2\mathbb{D}) < \cdots < ((1-\alpha_i)\mathbb{U},\alpha_i\mathbb{D}) < \cdots \leq (0,\mathbb{D})\}$ where each time position of a social system $(s \in \mathbb{S})$ is a polarity subject to categorial conversion, and $\alpha_i$ asymptotically approaches $(1)$ and $(1-\alpha)$ asymptotically approaches $(0)$ in an infinite time horizon. This conceptual weight distributional specification is extremely important in understanding the analytical strength and the logical boundaries of the *theory of categorial conversion* in solving the qualitative transformation problems in nature and society. The question that arises is how does one know that conversion has occurred? In other words, what are the conditions of *categorial convertibility*, the point of *categorial transition and categorial transfer function*? The construct of these conditions demands the discussions of the *categorial transversality conditions* of the optimal conversion. This question is important in natural and social transformations and acquires an increasing complexity when a social system is under decision-choice actions.

Let us keep in mind the essential logical structure of the conceptual system that is being developed here. Every element in the universe is defined by its characteristic set that places it in a category. The total characteristic set of each element exists as negative and positive sub-sets to define it as a duality with a relational continuum and unity. There is a negative duality and a positive duality whose interactive structure defines a polarity with a negative pole and a positive pole. From the viewpoint of the concepts of negative and positive, one may view complete underdevelopment as a negative pole in which negative and positive dualities reside with the negative duality overwhelming the positive duality. The complete development may, alternatively, be viewed as a positive pole in which negative and positive dualities reside with the positive duality overwhelming the negative duality. The negative duality contains negative and positive characteristics subsets and is identified by the condition that the negative characteristics subset dominates the positive characteristics subset. Similarly, the positive duality contains negative and positive characteristic sub-sets and is identified by the condition that the positive characteristic sub-set dominates the negative characteristic sub-set. This logical structure is such that $\mathbb{S}$ may represent a general set of economic systems, $\mathbb{U}$ may represent a set of underdeveloped economies and $\mathbb{D}$ may represent a set of developed economies in terms of specific definition and measure. The economy $(s \in \mathbb{S})$ containing the poles of a social polarity is such that $\mathbb{S} = \mathbb{U} \bigcup \mathbb{A}$, and $(\mathbb{U} \bigcap \mathbb{A}) \neq \emptyset$ with implicitly defined negative positive dualities and negative and positive characteristics sub-sets. There are three organic systems (spaces) that must remain in our cognition. They are the ontological system (space), the epistemological system (space) and the engineering system (space). The engineering system (space) is a creation of cognitive agents from the information-knowledge construct of the epistemological system (space) whose elements are abstracted from the ontological system (space). The engineering system contains all human artistry including socioeconomic formation, its transformations and management. The engineering system is the social categorial conversion. The natural categorial conversion is the creation of natural forces from the decision-information interactive processes in the ontological system whose inputs and outputs belong to the ontological space. For a complexity analysis, a clear distinction must be made between *social categorial conversion* and *natural categorial conversion* to partition the *socio-natural categorial conversion.*

The definitional building blocks of nominalism, constructionism, reductionism and categorial conversion as related to Nkrumah's conceptual system were presented in his book *Consciencism.* The qualitative dynamics of derived categories from the primary category, irrespective of what is, Nkrumah calls it categorial conversion and he states: *By categorial conversion, I mean such a thing as the emergence of self-conscious from that which is not self-conscious; such a thing as the emergence of mind from matter, of quality from quantity* [R1.203, p.20]. This Nkrumah's definition will conditionally belong to events in both the ontological and epistemological spaces. This categorial conversion is a transformational dynamics where an element of one category is shown to be converted to another category by a process. In dealing with the mechanism of the categorial conversion and the analytical explanation of the process,

Nkrumah appeals to methodological constructionism and methodological reductionism. The relational structure to the theory of knowledge, symbolic representation and test for knowledge is discussed in full detains in [R3.7] [R3.10 [R3.13]. This relational structure must be linked to social and natural transformations and then connected to a unified theory of engineering sciences at all areas of human action.

It is here that category formation becomes an essential concept in understanding transformations of qualitative states and social development that involves categorial inter-reducibility. Nkrumah points out at this point that:

> The inter-reducibility of matter and energy offers a model for categorial conversion. And another model is given in the distinction between physical change and chemical change, for in chemical change physical quantities give rise to emergent qualities [R1.203, p. 21].

The manner in which the sequences of categories are treated in qualitative transformations is specified as:

> In nominalism, constructionism and materialist reductionism, one holds some category to be a primary category of reality, and holds other real things to become real only in so far as they are ultimately derived from the primary category of reality. The derivation is such that for every true proposition about an item which falls under a derivative category, there are provided true propositions about items falling under the primary category, such that the former proposition could not be true unless the latter propositions were true; and, further, such that the former propositions could not even make sense unless there were items falling under the primary category [R1.203, p. 21-22]

Given 1) nominalism, constructionism and reductionism, 2) that every individual element exists as opposites in relations to polarity, duality, under relational continuum unity, then every condition of an element, by the methodological constructionism, has a supporting condition of methodological reductionism. The methodological distinctions among nominalism, constructionism and reductionism are explained by Nkrumah in the following statement:

> In constructionism, one has a picture how those concepts which are proper to derivative categories might be formed, using as raw materials concepts which are proper to the primary category. In reductionism one sees how concepts proper to a derivative category can be reduced

147

completely to concepts which are proper to a primary category. When a certain reductionism holds matter to be primary, such a reduction has, for its product, concepts which are directly applicable only to matter. In nominalism, only concrete existences are held to be primary and real, all other existence being, as it were, surrogate of concrete existences on a higher logical plane [R1.203. p. 22]

The conceptual meaning of category is implicit in Nkrumah's use of nominalism but it is given an explicit definition in [R3.7] [R3.10]. For analytical convenience, the conceptual meanings of these terms will be defined here. It is important to understand the roles that category, nominalism, constructionism and reductionism play in Nkrumah's system of thought. Nominalism is used to develop and enlist the vocabulary of thought while constructionism and reductionism are used to develop the grammar and system of thought. The relational structure is extensively discussed **in** [R3.10] [R3.13].

### 4.1.3 Analytical Definitions of Some Important Concepts

Let us now turn our attention to presenting some important definitions. These definitions are critically important in understanding Nkrumah's conceptual system and its philosophical foundations and extensions that are being developed here. It is through the understanding of this conceptual system with philosophical foundations and extensions that one can appreciate how Nkrumah's thought and actions are integrated for positive socio-political action and the lessons that may be learnt for Africa's social transformation and nation building.

### Definition 4.1.3.1: Nominalism

Nominalism is a methodological process for the development and enlistment of the vocabulary of a language system on the basis of category formation which may be formal, natural or theoretical. The vocabulary is a set of symbols either in a written and spoken language used to represent information needed to identify elements and categories of belonging where such information forms input to be processed into knowledge claims. It may be divided into linguistic primitives and linguistic derivatives.

# Note 4.1.3.1

In terms of Nkrumah's conceptual system and its extension, the linguistic primitive belongs to a primary category whose exactness is cognitively imposed. Every symbol in a vocabulary, whether written or non-written, carries with it exact and inexact characteristics. The meaning of every symbol is thus established within a context. The conditions on which a symbolic representation may be accorded the property of exactness are discussed in [R3.13]. Further discussion on the general theory of definitions, linguistic primitives and the development of paradigms of thought and their uses may also be found in [R3.7] [R3.10] [R9.2] [R9.13] [R9.18] [R9.24]. It is the dualistic presence of exact and inexact characteristics in vocabulary and concepts that defines the motivation of the debate between intuitionist and non-intuitionist information representation and the appropriate paradigm of reasoning [R12.5] [R12.6] [R15.14] [R15.15] [R12.16].

## Definition 4.1.3.2: Methodological Constructionism

Constructionism is a methodological process of reasoning where one holds an element in a system of knowledge production as belonging to a primary category. From a set of concepts about the primary category a set of material concepts, using the acceptable grammar, is used to create a derive category from the element which belongs to the primary category.

## Note 4.1.3.2

In methodological constructionism, the development of the system of thought is initialized with a primary category where a system of derived categories is shown to exist by a logical process. Methodological constructionism is called classical if its information representation is assumed to be exact, irrespective of whether it is complete or incomplete and the information processing machine used is that of the classical paradigm and Aristotelian laws of thought. It is a forward logical process to establish logical claims. Methodological constructionism is also called fuzzy if its information representation is assumed to be inexact, irrespective of whether it is complete or incomplete and the information processing machine is that of the fuzzy paradigm. We may also keep in mind that a hybrid of the two may be used under either a fuzzy-stochastic or stochastic-fuzzy processing machine when the information is both inexact and incomplete (see the discussions in [R3.13]. These discussions are consistent with nominalism, a theory of definition, and explication and thought development where exactness is claimed and maintained by decision-choice action.

## Definition 4.1.3.3: Methodological Reductionism

Reductionism is a methodological process of reasoning where one holds an element in a system of knowledge production as belonging to a derived category, and then reduces the set of concepts about the derived category, by using the acceptable grammar and paradigm of thought, to the set of concepts that reveal the primary category from which the entity is derived.

## Note 4.1.3.3

Methodological reductionism is simply the opposite of the methodological constructionism. It is a backward knowledge development process. It may be related to falsification, justification, corroboration and verification principles in truth-confirmation by developing comparability and acceptability principles. Methodological reductionism is such that the development of the system of thought is initialized with a derived category where a primary category is shown to exist as a parent by a logical process. It is a backward logical process as a check on the claims in methodological constructionism. Methodological reductionism is called *classical* if its information representation is assumed to be exact, irrespective of whether it is complete or incomplete and the information processing machine used is that of the classical paradigm with Aristotelian laws of thought. It is also called *fuzzy* if its information representation is assumed to be inexact, irrespective of whether it is complete or incomplete, deceptive or non-deceptive, and whether the information processing machine is that of the fuzzy paradigm with fuzzy laws of thought. We may also keep in mind that a hybrid of the two may be used under either a fuzzy-stochastic or stochastic-fuzzy processing machine when the information is both inexact and incomplete (see the discussions in [R3.7] [R3.10] [R3.13]).

## A General Note 4.1.3.1

Within the Nkrumah's conceptual framework, four analytical paths are identified to help the understanding of the forces at work in transformations which may be related to the behavior of elements in the qualitative space. There are the classical constructionism and reductionism paths with exact information structure and exact information representation under the classical paradigm with its laws of thought. Additionally, there are the fuzzy constructionism and reductionism paths with inexact information structure and fuzzy information representation under the fuzzy paradigm with its laws of thought. As it will be made clear, the fuzzy paradigm allows us to deal with the problem of quality, subjectivity and inexact reasoning in addition to dealing with the problems of the classical paradigm and elements in the classical thought space in transformation and non-transformation processes. A relational structure of the two paradigms of thought is presented as a cognitive geometry where the paradigms of thought form the center of knowledge development that will become an input into the decision-choice process. The exact information structure, classical constructionism and classical reductionism form a logical pyramid that creates the foundation of the classical logic and corresponding mathematics to constitute the classical paradigm for knowledge production. The inexact information structure, fuzzy constructionism and fuzzy reductionism form a logical pyramid that creates the foundation of the fuzzy logic and corresponding mathematics that constitute the essential core of the fuzzy paradigm. These are represented in Figures 4.1.3.1 as an epistemic relational geometry.

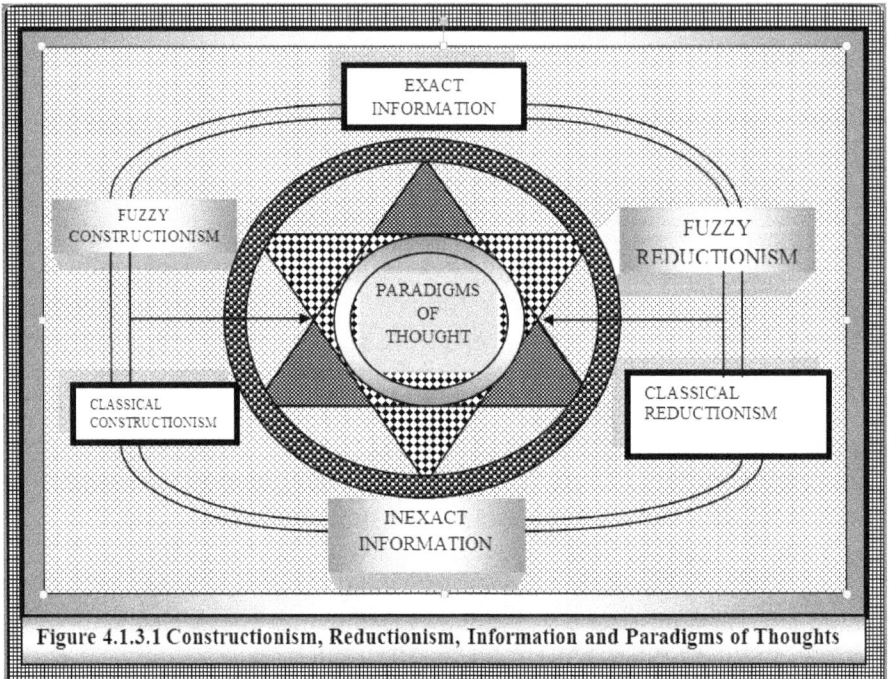

**Figure 4.1.3.1 Constructionism, Reductionism, Information and Paradigms of Thoughts**

## 4.1.4 A Comparison of the Classical Paradigm and the Fuzzy Paradigm in Cognition

It is analytically useful to keep in mind that a paradigm presents a logical system of information processing to create a knowledge output which may enter as an input into a decision-choice system. The nature of a paradigm, therefore, defines a limitation on the information-procession capacity of decision-agents who use the paradigm. As defined, dualism and duality are different with distinguishing characteristics that specify conditions of mutual exclusivity such as false and true characteristic sets. In the knowledge-production process, dualism is associated with the classical paradigm and its laws of thought with *excluded middle, relational separation* and the presence of *non-contradiction*. Duality is associated with the fuzzy paradigm and its laws of thought with a *relational continuum, unity* and acceptance of the presence of *contradiction*. In the classical paradigm, true and false are not allowed to simultaneously exist in the same proposition. In the fuzzy paradigm, true and false characteristics are allowed to be present in the same proposition where true or false is discounted by judgment with a fuzzy conditionality. Both paradigms have their corresponding mathematics and systems of logical symbolism

152

with different rules of mathematical and symbolic operators that may be used to generate and accept propositions and conclusions over the epistemological space. The similarity of the two paradigms is found in the logical polarity for all knowledge constructions and the study of dynamics of qualitative and quantitative dispositions of actual-potential polarity. Keep in mind that the paradigm with its laws of thought, logic of reasoning, and mathematics to create computable frame is simply an information processing module to guide the decision-choice activities that connect matter and energy.

The classical mathematical and symbolic operations function under the conditions of dualism and exact symbolism with exact laws of thought. The fuzzy mathematics and symbolic operations function under conditions of duality and non-exact (vague) symbolism with fuzzy laws of thought. For a comparative understanding, let us state side by side in Table 4.1.4.1, the two laws of thought that guide symbolic and mathematical operations. In the classical paradigm, words, propositions and conclusions are seen as points on the meaning and true-false lines where two opposing statements are seen in dualism and in separation with no connections, and where every statement is either true or false but not both. Here lies the principle of logical separation in the classical laws of thought on the basis of which propositions are accepted or rejected. In the fuzzy paradigm, words, propositions and conclusions are seen as sets on the meaning and true-false line where two opposing statements are seen in duality and in unity, each of which has sets of true and false characteristics that connect them in a relational continuum and in *relative true-false proportions*. Any statement is claimed to be either true or false by a decision-choice action with optimal fuzzy conditionality. Here, lies the logical unity of the fuzzy laws of thought in dealing with exactness and inexactness in the knowledge-production process on the basis of which propositions are accepted or rejected. The fuzzy paradigm applies to all decisions where every benefit has cost support and every cost has a benefit support in such a way that one cannot take a benefit and leave the cost. This is the *asantrofi-anoma* problem in decision-choice rationality and cost-benefit analysis that has been discussed in [R4.17] [R4.18]. In the unified system of thoughts, analyses and decision-choice activities, the truth value is constrained by the false value and similarly, the benefit value is constrained by the cost value.

| THE CLASSICAL PARADIGM | THE FUZZY PARADIGM |
|---|---|
| 1.The law of identity (ontological condition) What there is, is, what there is. | 1. The law of identity (ontological condition) What there is, is, what there is. |
| 2. The law of contradiction (epistemological condition) Nothing can be and not be at the same time and space. | 2. The law of relative commonness: or contradiction (epistemological condition) Everything is both what is and what is not in degrees. |
| 3. The law of excluded middle (epistemological condition) Everything is either *there is* or *not there is* but not both. | 3. The law of continuum (epistemological condition) Everything is in the process of being and not being. Alternatively, everything is both *what there is* and *what there is not* in degrees. |
| 4. The Principle of Separation (epistemological condition) All opposites exist as separate entities. | 4. The Principle of unity (epistemological condition) All elements exist in relational unity. |

**Table 4.1.4.1.** The Essential Foundational Differences between the Laws of Thoughts in the Classical and Fuzzy Paradigms.

The classical laws of thought, as applied to the acceptance of propositions and representations, satisfying the laws of identity, contradiction and excluded middle with dualism reduce to the statement : *all propositions are either true or false* (with an excluded middle) [R14.100]. Comparatively, the fuzzy laws of thought, as applied to reasoning and acceptance of propositions and representations reduce to the statement that *every proposition is a set of true-false characteristics in duality and continuum with varying relative proportional distribution and hence the acceptance of a proposition to be either true or false is obtained on the basis of subjective decision-choice action by cognitive agents in reconciling the conflicts in the true-false proportionality distribution in the continuum of true-false duality* [R3.7]. The reconciliation comes as a constrained cognitive optimization problem where the functional description over the true characteristic set is optimized with the constraint of the functional description over the false characteristic set. As applied to the transformations an element is simultaneously *is and is not* because every element exists in a relational continuum of changing-not-changing duality where the attainment of degree of change is constrained by the degree of not changing.

The classical paradigm admits of absolutism in truth and falsity in the knowledge-production process. As applied to qualitative dynamics, the classical paradigm of reasoning accepts the condition that there is a change or no change, but not both in the same element. In other words, the laws of thought of the classical paradigm do not allow us to examine gradual qualitative transformation. They deny a relational continuum and unity of opposites by relying on discrete processes. As applied to information process, the fuzzy paradigm accepts the relativism of truth and falsity in the knowledge-production process. As applied to qualitative dynamics, the fuzzy paradigm of reasoning accepts the condition that there simultaneously exists a change and no change, in the same element. In other words, the laws of thought do allow us to analyze gradual qualitative transformation in the same element. The fuzzy paradigm accepts relational unity of opposites by relying on the principle of continuum. The epistemic importance and analytical interest of these two sets of laws of thought are that the classical laws of thought taken as an aggregate statement cannot satisfy its own required condition of the statement: *all propositions are either true or false but not both.* This presents an epistemic dilemma in the classical system of thought. The set of fuzzy laws of thought can be subjected to its internal requirement, in that its statement for guiding true-false acceptance is also defined with true-false duality in degrees and resolved through decision-choice actions by cognitive agents, and by introducing epistemic conditions and conditionality of partial truth.

There are other differential dimensions of the classical and fuzzy paradigms. The classical paradigm does not allow the acceptance of transitional values where a simultaneous existence of truth and falsity can be claimed, in particular, when a qualitative transformation is gradual and decision-choice indeterminate. The fuzzy paradigm allows for relational inter-connectedness of true-false proposition as a valid claim which is revealed in the use of conceptual duality under the principle of opposites, tension and struggle as is the case in the life-death duality of living-beings where both of them reside in relational continuum and unity in ontological transformation. At the level of universal understanding, the classical paradigm is compatible with discrete and digital reasoning where a point is a point. The fuzzy paradigm sees digital and discrete as approximation in the continuum and analog, which is useful in the study of gradual transformation where the universal system is seen as a collection of polarities with residing dualities under relational continuum and unity. From the application of the fuzzy paradigm to polarity with residing dualities under relational continuum and unity, a possibility is opened to us to say that death is derived from life and life is derived from death, where death is a category and life is another category both of which are defined by their respective characteristic sets. Let us keep in mind that every analog (continuum) is an enveloping of discrete (digital) points which belongs to the analog (continuum). It is the understanding of this analog-digital (continuum-discrete) process that one can develop a theory of general human creation such as various elements of engineering that allows characteristics of elements to be transformed in intra and inter-categorial existence. It is through the same understanding that one can develop a general theory of mechanism of socio-natural transformations through the logics of category formation and categorial conversion. The discussion on the classical and fuzzy paradigms and the corresponding laws of thought is to establish the appropriate paradigm and corresponding laws of thought with its information representation, mathematics and laws of thought for the construction of the theories of category formation and categorial conversion. The appropriate paradigm is the fuzzy paradigm.

## 4.2 Category Formation and the Logic of Categorial Conversion

The central core of the development of the theory of categorial conversion is the acceptance of the simultaneous existence and interaction of qualitative and quantitative dispositions in a relational continuum and unity for any given element in either ontological or epistemological space. The logical structure is that in any element, every qualitative disposition has a supporting quantitative disposition and vice versa. The qualitative disposition is associated with the distinction of ontological and epistemological elements to establish categories [R2].

Categorial conversion is logically meaningless without the existence of categories and how categories are formed.

## 4.2.1 Category Formation in the Logic of Qualitative Transformations

The distinction of elements reveals them in categories. Variations in the qualitative disposition give rise to category formation leading to the construct of the family of ontological and epistemological categories. Every variation in qualitative disposition has a supporting variation in quantitative disposition. The study of the transformation of elements is a study of inter-categorial movements where such movements are governed by qualitative transfer functions. In this respect, we must understand the nature of categories and how categories may be formed. We must then tend our attention to the definitional structure of the concept of a category.

**Definition 4.2.1.1**

A category $\mathbb{C}$ is a collection of objects with the same *characteristics set* $\mathbb{X}_{\mathbb{C}}$ that is specified by the same qualitative disposition. Given the characteristics set $\mathbb{X}$ the set $\mathbb{X}_{\mathbb{C}} \subset \mathbb{X}$ is such that $\left( \bigcup_{i \in \mathbb{I}^{\infty}} \mathbb{X}_{\mathbb{C}_i} = \mathbb{X} \right)$.

Furthermore, $\left( \bigcup_{i \in \mathbb{I}^{\infty}} \mathbb{C}_i = \mathfrak{U} \right)$ where $\mathfrak{U}$ is the total space which is infinite and $i \in \mathbb{I}^{\infty}$ is a generic element where $\mathbb{I}^{\infty}$ is an infinite index set of categories.

**Note 4.2.1.1**

As specified here, a category is formed with a qualitative disposition in terms of its qualitative characteristics. Every element in a category is defined by two opposite sub-sets of negative characteristics set $\mathbb{X}_{\mathbb{C}}^{N}$ and positive characteristics set $\mathbb{X}_{\mathbb{C}}^{P}$ such that $\mathbb{X}_{\mathbb{C}} = \left( \mathbb{X}_{\mathbb{C}}^{P} \bigcup \mathbb{X}_{\mathbb{C}}^{N} \right)$ and $\left( \mathbb{X}_{\mathbb{C}}^{P} \bigcap \mathbb{X}_{\mathbb{C}}^{N} \right) \neq \varnothing$ with $\# \mathbb{X}_{\mathbb{C}}^{P} \gtreqless \# \mathbb{X}_{\mathbb{C}}^{N}$ in order to establish its identity and the category to which it belongs. Sub-categories may be formed within the qualitative categories on the basis of quantitative disposition. Both the ontological space and the epistemological space are infinite in existence and are partitioned by the categories of being. The epistemological categories are inexactly known to cognitive agents while a small set of the ontological categories are accessible to them. The category formation is essential in the development of language with the corresponding vocabulary and grammar of thought through the principle of *nominalism* as defined above. In Nkrumah's conceptual system and the extensions of its philosophical foundations, the categories are divided into the primary category and the derived categories. The primary category constitutes the identity of being from which the derived categories may be shown to emerge by some epistemic process. In simple terms, the ontological space is filled with objects, things, states and processes which may be grouped into categories by means of their qualitative characteristics. These characteristics present themselves as information to cognitive agents operating in the epistemological space. The cognitive agents must process this information with *epistemic tools* to try to find an isomorphism between the ontological elements in the ontological space and epistemological elements in the epistemological space. The epistemic tools are couched in methodological constructionism and reductionism as defined above. The foundational discussions on these critical concepts and their roles in reason may be found in [R3.7] [R3.10][R3.13]. Every name and every concept in the vocabulary must meet the qualitative characteristic conditions of a category where every category is distinguished from others by its qualitative disposition. Alternatively viewed, a vocabulary is a set of qualitative dispositions in nominalism that creates foundation for the application of methodological constructionism and reductionism. The concept of category defined in terms of common characteristic set is based on the implicit theory of *characteristic-theory of categorial analytics* which requires the characteristics of each object to be identified and then placed in an appropriate category.

## 4.2.2 Categorial Conversion in the Logic of Qualitative Transformations

The study of transformation of elements is a study of inter-categorial movements where such movements are governed by qualitative transfer functions. In other words, we must study how an element in one category loses its old categorial characteristics, acquires new characteristics, and migrates from its parent category into a new parent category which is a derived category. The question then becomes what is the meaning of categorial conversion? This requires a conceptual definition.

## Definition 4.2.2.1: Categorial Conversion

Categorial conversion is a process where derived categories emerge from a primary category where the derived category has a direct or indirect continuum with the primary category for a reversal process. At the level of ontology, categorial conversion is a natural process called *ontological categorial conversion*. At the level of epistemology, categorial conversion is an epistemic process called *epistemological categorial conversion*. The distinction between the *primary and derived categories* is established by *categorial differences* as revealed by their qualitative characteristic sets.

## Note 4.2.2.1

As defined, the ontological categorial conversion is the identity as well as the primary category for epistemological categorial conversion which constitutes the derivative in the knowledge production process. For any given phenomenon, the epistemological categorial conversion leads to a knowledge discovery if it is equal to the ontological categorial conversion under epistemic conditionality [R3.10] [R3.13]. The work in the epistemological space is to discover the events, states and processes taking place in the ontological space where these events, states and processes are independent of the cognitive and non-cognitive elements contained in the ontological space. To illustrate the categorial conversion process, let the elements under the qualitative process in the epistemological space be a set of the form $\{\mathbb{C}_0, \mathbb{C}_1, \mathbb{C}_2, \cdots \mathbb{C}_i \cdots \mathbb{C}_\infty\}$ and corresponding to it a time set $\{T_0, T_1, T_2, \cdots T_i \cdots T_\infty\}$. Let $\mathbb{C}_0$ be the primary category at the initial time $T_0$ to be transformed in an infinite time $T_\infty$ where we consider methodological constructionism in a forward moving process.

159

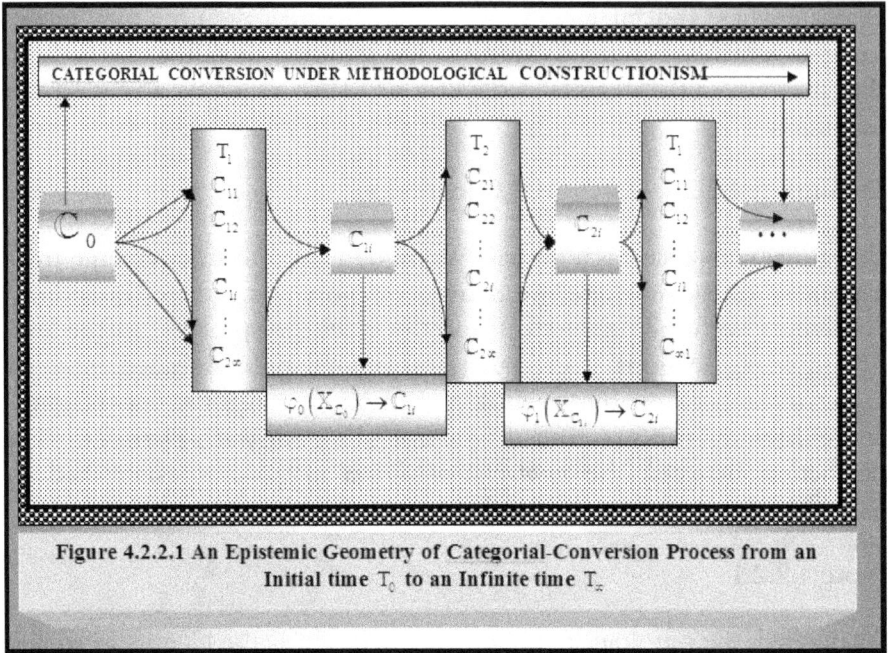

CATEGORIAL CONVERSION UNDER METHODOLOGICAL CONSTRUCTIONISM

$$\varphi_0\left(X_{C_0}\right) \rightarrow C_{1i}$$

$$\varphi_1\left(X_{C_{1i}}\right) \rightarrow C_{2i}$$

**Figure 4.2.2.1 An Epistemic Geometry of Categorial-Conversion Process from an Initial time $T_0$ to an Infinite time $T_\infty$**

The functions $\varphi(\Box)$ are categorial-conversion functions that alter the relative relationship between the negative set and the positive set of the element in a given category and then become qualitative transfer functions through the creation of a categorial moment to move the element in its category into a new category which is derived. The forward arrows represent a methodological process of constructionism as shown in Figure 4.2.2.1 while the backward arrows represent a methodological process of reductionism as shown in Figure 4.2.2.2.

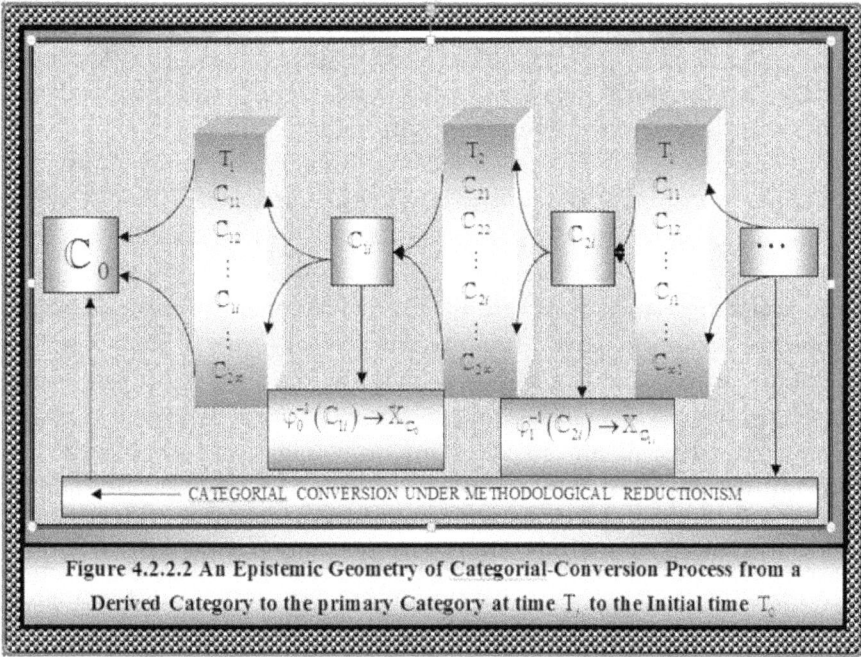

**Figure 4.2.2.2 An Epistemic Geometry of Categorial-Conversion Process from a Derived Category to the primary Category at time $T_i$ to the Initial time $T_0$**

## 4.3 The Epistemic Structure of Categorial Conversion

The concept of categorial conversion and the logic behind it are critical attempts to understand transformation in both the ontological and epistemological spaces. The ontological categorial conversion is taken as the identity in the sense of, *what there is.* This, *what there is,* defines the ontological reality. The epistemological categorial conversion, however, is a derivative in the sense that it presents a way of understanding qualitative motions in natural and social processes by cognitive agents for any given quantitative disposition as they present themselves in the ontological space. It helps us to understand, for example, how matter and mind exist in a relational continuum with unity, how mind emerges from matter and how death emerges from life by processes. Any result of the epistemological categorial conversion is an epistemic reality that may deviate from the ontological reality. If we take the ontological categorial conversion as an identity of transformations, then how do cognitive agents come to know, how do they know that they know, and how do they justify what they know as consistent with any element in the ontological space? The cognitive essence of this justification, verification

161

or corroboration is through methodological reductionism which may involve reasoning in the fuzzy-stochastic topological space due to defective information structure composed of a vague information substructure and a limited volume of information structure.

The process of knowing begins by developing an *epistemological information structure* from the *ontological information structure* through a direct or indirect process such as acquaintances through sense data and enhanced sense data. On the question of information for knowing, the ontological information structure is taken as the identity and constitutes the *primary information category*. It is the ontological reality in the sense that it is complete and exact for any ontological element in the further sense that *what there is*, is what is there without vagueness and without incompleteness. Each ontological element is specified and defined by a set of ontological characteristics that reveals its identity and ontological category to which it belongs. When cognitive agents encounter these ontological characteristics by acquaintance, they bookmark them, transform them into epistemological characteristics by differentiating elements, assigning names and epistemological categories and place them into the epistemological space for epistemic activity. The collection of all ontological categories constitutes the ontological space as has been explained in [R3.7] [R3.10]. From the process of knowing, the ontological information structure constitutes the primary information structure for knowledge production. The epistemological information structure constitutes its derived category.

The epistemological information structure in the process of knowing is a derived information category from the ontological information category. The question then, is how the derived information category is obtained by cognitive agents. Each epistemological element is specified and defined by a set of epistemological characteristics on the basis of which the epistemic element and its epistemic category are identified as a potential knowledge item. The collection of all the epistemological categories constitutes the epistemological space and the collection of all the claimed knowledge items constitutes the *space of epistemic reality* that is obtained by an *epistemic process*.

## Definition 4.3.1: An Epistemic Process

An epistemic process is a sequence of cognitive decision-choice activities which takes an epistemological information structure about a category and by methodological constructionism shows its successor (child) or by methodological reductionism shows its precursor (parent) for a comparison to an ontological reality. The epistemic process is thus to construct an isomorphism between an epistemic reality and an aspect of ontological reality. It is a relational operator between the epistemological and ontological spaces.

**Note 4.3.1:**

In an epistemic process, the information about the phenomenon of any ontological element under knowing will constitute a primary category. The empirical and axiomatic abstractions of information to be processed into knowing will constitute a derived category and the corresponding element will constitute an epistemological element. The results of epistemic process will lead to knowledge claims with *epistemic conditionality* while the distance between the ontological reality and epistemic reality which is called the *epistemic distance* and is under continual minimization and asymptotically approaches an *irreducible epistemic ignorance* in a repeated construction-reduction process in knowing. [R3.10].

The epistemic process is simply the cognitive machine acting as an input-output device which transforms an epistemological information input into knowledge claims that constitute the epistemic reality. It is useful to remember that while ontological information is identical to ontological knowledge, epistemological information is not epistemic knowledge, and epistemic knowledge is not necessarily ontological knowledge, otherwise, the concept of the epistemic distance is not necessary and methodological reductionism will be superfluous. All cognitive agents work in the epistemological space with the ontological space given as the absolute in the sense of collection of all *what there is*, satisfying the principle of identity required for the construction of a paradigm with laws of thought and corresponding mathematics as it has been explained in the general note 4.1.3.1 [R3.7] [R3.8] [R3.9][12.5] [R12.14] [R12.15][R12.85] [R12.88]. Over the epistemological space, the initial information about one of the epistemological items is taken to be a primary epistemological item from which cognitive categorial conversions may occur to obtain a set of derived epistemic categories as an epistemic world picture in the sense of Planck [R12.75] [R12.76].

The epistemological information structure for any phenomenon that enters as an input may be initialized as an empirical information structure or as an axiomatic information structure or combination of both (for extensive discussion see [R3.13]). Both the empirical and axiomatic information structures may be assumed to be derived from an experiential information process. For example, matter may be taken as the primary epistemic category from the ontological space and transformed by acquaintance into the epistemological space for processing. Corresponding to matter we have energy which may be taken as an ontological derivative from matter by *ontological categorial conversion.* The matter and energy are linked in the ontological information structure that allows information sharing, and distinguishes them by their ontological characteristics which define their categories. The ontological information structure revealed by the ontological characteristics is then mapped by some *acquaintance process* into the epistemological space as two relationally distinct epistemic items as an epistemological matter and epistemological energy, one of which must be initialized as the epistemic primary category and the other as an epistemic derivative that is linked by an epistemological information structure.

Matter and energy exist in the ontological space as *what there is* independent of whether cognitive agents know them or not. They exist on the epistemological space as a result of awareness on behalf of cognitive agents by a process of an epistemic mapping of the ontological characteristics set onto the epistemological space as epistemological characteristics, and then as the epistemological information structure which allows epistemic distinctions for the establishment of the derived epistemic categories. The cognitive agents have no preview to the ontological process of whether energy is a derivate of matter or matter is a derivative of energy. The processes of establishing epistemological categories and logical application of categorial conversion allow one to demonstrate at the logical plane whether energy is a derivative of matter as a primary category or matter a derivative of energy as a primary category given the connecting epistemological information structure. In other words, there are always epistemic uncertainties defined by an epistemic distance that gives rise to epistemic conditionality in terms of claims of validity and truth. It is this epistemic uncertainty, that the construction-reduction process finds its greatest epistemic value.

The process is to minimize the epistemic distance by choice of methodological constructionism subject to methodological reductionism and methodological nominalism. It is here, that tools of classical and fuzzy statistics act as helping epistemic tools over the epistemological space relative to the ontological space. Firstly they may help to establish the epistemic information structure as input into a selected paradigm for methodological constructionism. Secondly, they may contribute to methodological reductionism for truth-falsity verification for the selected paradigm. In this construction-reduction process in minimizing the epistemic distance to reduce the epistemic uncertainty, every epistemic reality in the epistemological space is always provisional with epistemic conditionality; the final reality resides in the ontological space. The fuzzy paradigm is a complete covering for the classical paradigm and more. It also allows us to examine gradual and slow changes such as climate due to human action and social transformation due to human action. The analyses of gradual and slow changes are not open to us with the use of the classical paradigm. We shall examine how this paradigm helps in the development of the theory of categorial conversion.

## 4.4 The Fuzzy Paradigm and the Explication of Qualitative Disposition in Category Formation and the Theory of Categorial Conversion

It has been argued that the universe is a collection of dualities with relational continuum. Every element in the universe exists as a duality whose existence is defined by negative and positive characteristic sets under a relational continuum and unity. The relational structure of the negative and positive characteristics sets provides a particular qualitative disposition and category of belonging. There are a number of things that are conceptually and analytically important here. The qualitative disposition which helps to define the category of belonging depends on the relative size of the positive and negative characteristics sets, the negative and positive relational structure and the point of reference in the duality. There is a set of actual-potential polarities where is pole of a polarity is defined by a corresponding duality. Every element exists as an actual-potential polarity where every residing duality is under a relational tension that generates energy and force of qualitative motion under the principles of relational continuum and unity. Every element is under a plenum of forces under tension and hence belongs to an actual-potential polarity in the universal system. The simultaneous presence of actual and potential in the qualitative dynamics presents us with the same symbolic, analytical, logical and mathematical difficulties that must be resolved in the epistemic process within any given formal language and paradigm of thought. The search for resolutions to these difficulties proceeds from the nature of the information sending-receiving mode, methods of information representation, reasoning, problem formulation and analytical systems, all of which constitute a paradigm of thought in understanding universal transformations where one category emerges from another as a universal creation. Any selected paradigm must provide not only how to resolve these analytical difficulties, but must provide us with a decision-choice rationality to help alter categories through human creation. It must also offer us a way of analyzing qualitative risks of transformations. We have discussed the epistemic problems of the classical paradigm, composed of its logic, mathematics and decision-choice rationality, in dealing with the explanation of qualitative motion and transformations in terms of intra-categorial and inter-categorial movements with transfer functions. We shall now discuss the relevant aspects of the fuzzy paradigm, composed of its laws of thought, logic of reasoning, mathematics and fuzzy rationality, in order to examine the possibility of solutions to the epistemic problems in categorial conversion at the level of information deficiency.

## 4.4.1 The Fuzzy Paradigm, Its Nature and Relational Structure to Categorial Conversion

Categorial conversion is about dynamics of actual-potential polarity. The explanatory problem of categorial conversion, therefore, is to explain the process by means of which an actual pole becomes an element in the potential space and an element in the potential space becomes an actual pole. This is the *qualitative transformation problem* or simply the transformation problem or dynamics of qualitative disposition in nature and society whose solution is being sought under an appropriate paradigm of thought. The essential characteristic of the classical paradigm that creates epistemic difficulties in dealing with conditions of qualitative disposition that gives rise to fuzziness which involves vagueness, and ambiguities in transformations, the knowledge-production system, is its laws of thought with relational excluded middle and non-acceptance of contradiction in existence. This basic characteristic presents a logical dualism in epistemology. The classical paradigm, functioning with logical dualism and non-acceptance of contradictions, allows the exactness of science, classical mathematics and a discrete process to be justified, where the principle of non-contradiction provides decision-choice rationality for the acceptance of simultaneous existence of negative and positive characteristics in the same element in the transformation processes. The interpretations of the concepts of negative and positive are content dependent in either duality with continuum or dualism with excluded middle. These conditions of the classical paradigm require us to do away with quality, subjectivity, contradictions and the role of cognitive agents as internal parts of natural and social processes. The cognitive agents are thus externalized, with the effect that the qualitative motions in the system of universal dualities are stripped of the basic fact that cognitive agents are internally essential and integral parts of the socio-natural system. They are both being transformed as well as being transforming objects, thus paying for the cost and receiving the benefits of the socio-natural transformations.

The epistemic nature of the fuzzy paradigm and its development are a serious attempt to incorporate fuzziness, composed of qualitative disposition of things, linguistic quality, vagueness and ambiguities, whose understanding requires subjective decision-choice actions in the reasoning process to assess whether a contradiction is relevant or not in affecting the nature of the qualitative motions and the outcomes of transformations. In this way, the subjective actions of cognitive agents as well as their very existence as elements in the universal system are internalized into the enterprise of the universal transformations as it should be. The cognitive agents are not simply machines that follow rules that are external to them, but they make the rules and execute judgments in an error-correcting process. Living and non-living things are subjects of change as well as objects of change. They bring about change which then changes them. In this way, every element in the universal system is being transformed no matter how slowly in the sense that no-change is also a change. This is in line with the position that the transformation process requires coordination of matter, energy and information that allow a categorial traverse over the penumbral regions of socio-natural actions which involve inexact inductive and deductive thought processes in all areas of the transformation processes. In fact, it is useful understanding to keep in mind that the universal transformation system is self-correcting and self-organizing within destructive-constructive duality under the principle of continual substitution in a continuum. Every transformation is also a substitution which is also a substitution. In analytical reasoning, the fuzzy paradigm accepts the classical law of identity in the sense that ontological objects are what they are, but rejects the classical position of dualism and the law of excluded middle. It accepts contradictions as part of elemental existence and human thought. These contradictions exist in elements and thought. At the level of thought, the contradictions are resolved through reconciling the degrees of contradiction and non-contradiction in the logical unity through the decision-choice actions of cognitive agents under the principle of continuum. At the level of transformation and qualitative motion, the contradictions are resolved through reconciling the relative strength of negative and positive characteristics set with an element under the principles of substitution and continuum.

To explain the transformation process through which contradictions are accepted in reasoning, resolved in thought, and projected to categorial conversion, the *classical logical dualism* is replaced by *fuzzy logical duality*. This is an important fundamental change of the methodology in understanding categorial conversion that will allow the inclusion of subjectivity, approximate reasoning and cognitive computing of gradual qualitative changes as internal processes of epistemology in relation to elements in ontology. In this respect, an acknowledgement is made regarding limits of language in the representation of information, derivation of thought, communication of ideas, interpretations of meanings and the use of the understanding to engineer the categorial conversion that is desired. Even though the fuzzy paradigm is a revolution against the classical paradigm and also a contestant of the future paradigms, it actually enhances the subjective nature of thought in symbolic reasoning and widens the areas of mathematical and logical application in social, natural and other sciences. Under the principle of universal polarity and duality, the fuzzy paradigm provides an explanatory process of the general principles of engineering when it is combined with category formation and categorial conversion. Our new scientific age, with information dominance, demands of us a new path of epistemic processes, and a new methodology of logical inquiry that will enhance our current methodological regimes in dealing with the understanding of complexities, synergetics and energetics in phenomena, engineering of all forms and human creation. Here, it is interesting to consider new areas of informatics and energetics in our understanding of quantum phenomena at the level of nature and human behavior in relation to the mutual flows and conversions of information and energies into forces of change in categorial conversions under the relational structure of primary and derived categories. In these discussions, one shall concentrate on the essential characteristics of the acceptance of duality and the rejection of dualism and their implied meanings in category formation and fuzzy paradigm within the theory of categorial conversion. The central conceptual building blocks of the theory of categorial conversion are characteristic sets, categories, opposites, negativity, positivity, polarity, duality, continuum, relational unity, tension, qualitative disposition, quantitative disposition, energy and force. Most of these concepts have been introduced in previous discussions. We shall turn our attention to duality relative to dualism. We have discussed the role of duality in the construct of the fuzzy paradigm and its laws of thought; we have also discussed the role of dualism in the construct of the classical paradigm and its laws of thought. Let us now tighten the discussions on duality and dualism and their relationship to categorial conversion of actual-potential polarity.

## 4.4.2 Dualism, Duality and Relational Unity in Categorial Conversion

In understanding transformations that take place through intra-categorial conversions that bring about inter-categorial conversions, the question as to why classical dualism with relational excluded middle and separation is being replaced with duality with relational continuum and unity arises. Furthermore, is dualism not the same as duality, and if not, what are the distinguishing characteristics? In general, the concepts of duality and dualism are related to the principle of opposites and polarity. They have many different meanings involving morality, deity, ontology, mind-matter existence and many others where both of them represent a state of two opposite parts. These opposites may be assumed to exist in relational unity or in separate existence without organic ties to one another. The concepts of dualism and duality as they relate to the paradigm of thought and theory of knowledge have been discussed in [R1.92] [R3.7] [R3.10] [R3.13]. In this discussion, however, we are interested in dualism and duality in relation to the theory of categorial conversion and internal transformations relative to socio-natural qualitative changes that are brought about by the internal forces. As we have stated previously, duality accepts the existence of contradiction under the same element. These contradictions are under relational tension. For example food is simultaneously good and bad, an individual is simultaneously good and evil, and a living thing is both living and dying from within. In other words, positivity and negativity always characterize the same element from within. Let us offer some working definitions by first considering the total characteristic set $\mathbb{X}$ on the basis of which the identity of an element is established in the universal object set, with two opposite sets

169

of the negative characteristic set, $\mathbb{N}$ and the positive characteristic set $\mathbb{P}$, such that the relation $\mathbb{X} = (\mathbb{N} \cup \mathbb{P})$ always holds. We shall sometimes write $\mathbb{N}$ as $\mathbb{X}^{\mathbb{N}}$ and $\mathbb{P}$ as $\mathbb{X}^{\mathbb{P}}$ and hence $\mathbb{X} = (\mathbb{X}^{\mathbb{N}} \cup \mathbb{X}^{\mathbb{P}}) = (\mathbb{N} \cup \mathbb{P})$.

Duality and dualism will be explicitly defined and specified in terms of characteristic sets that will allow the construct of the logical relation to the development of the theory of categorial conversion as a general framework for transformations.

## Definition 4.4.2.1: Dualism

Dualism represents a conceptual state of two mutually exclusive and collectively exhaustive opposite characteristic sets, called the negative and positive characteristic sets, that exist in the definition of the aggregate qualitative disposition of a unit element without which the element has no defining identity in the universal object set. The negative and positive characteristics sets have no relationality. In this case, $\mathbb{X} = (\mathbb{N} \cup \mathbb{P}) = (\mathbb{X}^{\mathbb{N}} \cup \mathbb{X}^{\mathbb{P}})$ and $(\mathbb{N} \cap \mathbb{P}) = (\mathbb{X}^{\mathbb{N}} \cap \mathbb{X}^{\mathbb{P}}) = \varnothing$ where overlapping categories are not defined.

In terms of a Venn diagram this may be represented as in Figure 4.4.2.1.

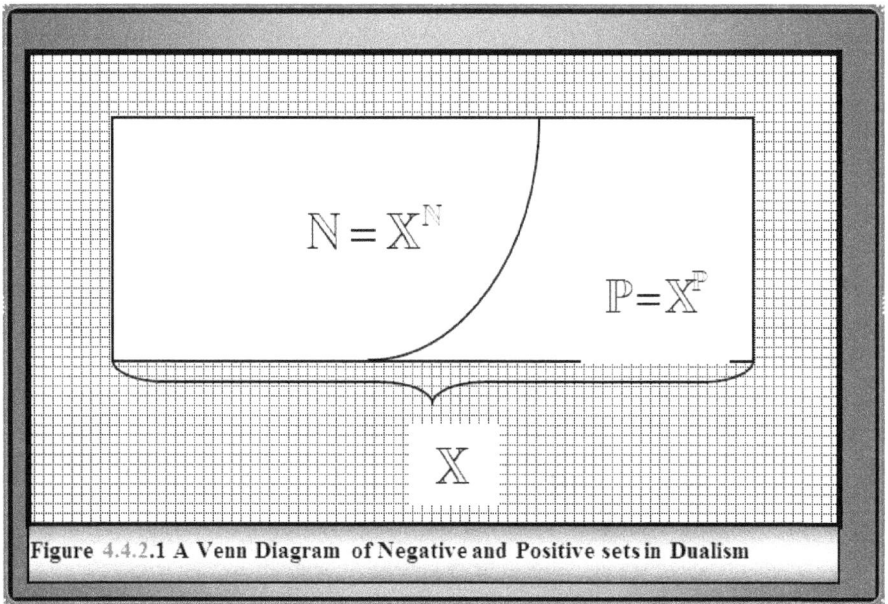

**Figure 4.4.2.1 A Venn Diagram of Negative and Positive sets in Dualism**

## DEFINITION 4.4.2.2: Duality

Duality represents a conceptual state of two opposite non-mutually exclusive and collectively exhaustive characteristic sets called negative and positive characteristics sets that exist in the definition of the aggregate qualitative disposition of a unit element without which the element has no defining identity in the universal object set. The negative and positive characteristics sets have defined relationality. In this case $\mathbb{X} = (\mathbb{N} \cup \mathbb{P}) = (\mathbb{X}^{\mathbb{N}} \cup \mathbb{X}^{\mathbb{P}})$ and $(\mathbb{N} \cap \mathbb{P}) = (\mathbb{X}^{\mathbb{N}} \cap \mathbb{X}^{\mathbb{P}}) \neq \emptyset$ where overlapping categories are allowed and defined.

In terms of a Venn diagram this is represented in Figure 4.4.2.2.

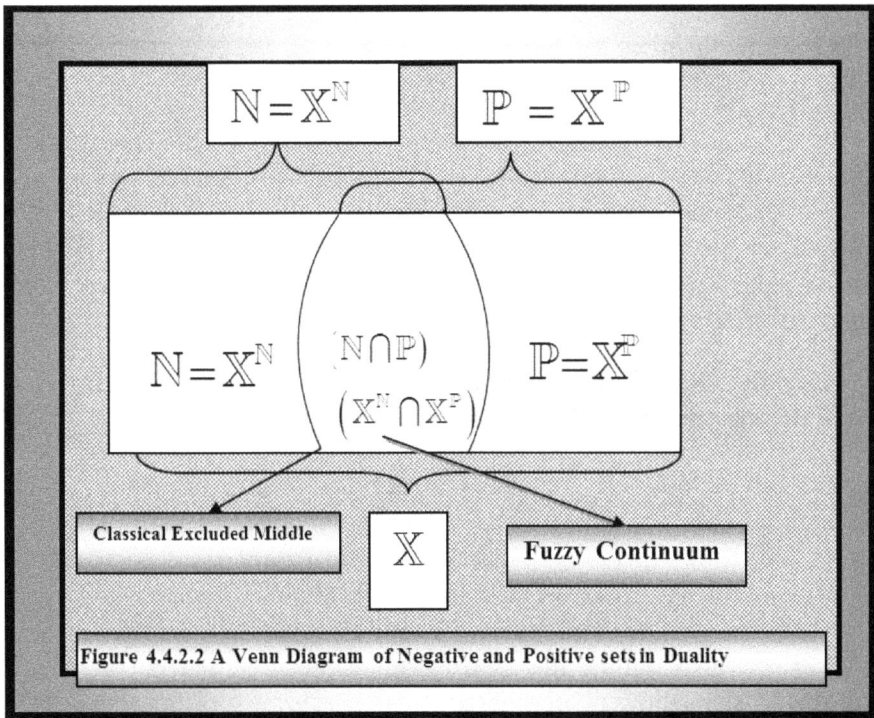

Figure 4.4.2.2 A Venn Diagram of Negative and Positive sets in Duality

Throughout the categorial conversion process and in examining the relational interactions of opposites, we shall associate $\mathbb{N} = \mathbb{X}^{\mathbb{N}}$ with the set that defines the negative qualitative disposition and $\mathbb{P} = \mathbb{X}^{\mathbb{P}}$ with the set that defines the positive qualitative disposition of the element of interest. For the benefit of the discussion on socio-political transformation of colonialism we could allow $\mathbb{P} = \mathbb{X}^{\mathbb{P}}$ to represent the characteristic set of *progressive elements* of the culture that rejects colonialism, and $\mathbb{N} = \mathbb{X}^{\mathbb{N}}$ to represent the characteristics set of the *reactionary elements* of culture that is passive to colonialism. The relational structure of the characteristics sets of the progressive and reactionary elements will define the qualitative disposition of the colonized people. The structure of these negative and positive characteristic sets defining either duality or dualism must then be related to colonization that represents a negative pole and decolonization that represents a positive pole in a political actual-potential polarity. In general, we can specify duality and dualism in set representations as in equations (4.4.2.1) and (4.4.2.2) respectively.

Duality implies that: $\mathbb{X} = \left\{ x \mid x \in \mathbb{N} \text{ or } x \in \mathbb{P} \text{ or } x \in (\mathbb{N} \cap \mathbb{P}) \neq \varnothing \right\}$ (4.4.2.1)

Dualism implies that: $\qquad \mathbb{X} = \left\{ x \mid x \in \mathbb{N} \text{ or } x \in \mathbb{P} \text{ and } x \notin (\mathbb{N} \cap \mathbb{P}) = \varnothing \right\}$ (4.4.2.2)

Where $x$ is a distinguishing characteristic relative to an element in the universal object set. The negative and positive characteristic sets as well as the qualitative disposition will acquire different conceptual and material representations as different elements are considered. They, further, engender differential cost-benefit configurations and rationality as categorial conversions are intentionally induced.

## 4.5 Duality, Relationl Continuum and Unity in the Categorial Conversion under Fuzzy Laws of Thought

In the discussions on the fuzzy paradigm relative to the classical paradigm as they may relate to the reasoning on categorial conversion in nature and society including the construction of thought, the process of knowing and the possible use of knowing to engineer transformation, we mentioned the concepts of duality, continuum and relational unity. It will now be useful to outline the essential characteristics of the concept of duality and show how it presents a useful analytical tool for the general understanding of the conversion of socio-natural categories and the design of a powerful cognitive framework for reasoning in simple and complex systems under static and dynamic conditions in various stages of qualitative disposition. Complementing the concept of duality are the concepts of polarity, opposites, relationality, continuum and relational unity, given the universal object set and the fuzzy paradigm of reasoning.

## 4.5.1 The Concepts of Duality, Continuum and Relational Unity in Category Formation

In section 4.4, we examined the similarities and differences between the concepts of dualism and duality relative to relational unity negative and positive characteristics sets. Their respective definitions were provided. We now want to examine in some analytical detail the reasoning foundations of the concept of duality and its use in understanding variations in qualitative disposition that bring about intra and inter categorial conversions. The use of the concept of duality conceives that every element in the universal object set exists in duality composed of two opposites which help to define its quality and identity. Every element in the ontological space is seen to be characterized by a set of characteristics in negative and positive standings, where such characteristics are mapped onto the energy space by a *conflict function* that generates forces of qualitative change. It is this characteristic that allows the claim that every element is a plenum of forces under tension. The energy space for qualitative transformation must be constructed and its understanding must proceed through the construction of cognitive operators. The nature of the qualitative disposition of things is seen in terms of ontological elements which must be related to epistemological elements such as propositions, statements and conclusion for the understanding of the categorial-conversion process by cognitive agents. For the identity of each element to be defined, the negative and positive characteristics sets must exist in relational unity in both ontological and epistemological spaces in which qualitative transformations must take place through the internal forces. The opposites are the extremes, defined by sets of the elemental characteristics that exist in a continuum without excluded middle in their residential duality, where the points of indifference are cognitively computed in the epistemological space by decision-choice actions under a constructed rationality and judgment to claim categorial conversion. Every element that is defined by duality under relational continuum and unity, also exists as actual-potential polarity under past-current-future time process.

To illustrate the point of discussion, let us take the negative-positive duality which characterizes all elements in the universal object set. The negative-positive duality is composed of negative and positive characteristic sets whose varying relative proportions define the qualitative dispositions of all points in the continuum. Every point in the continuum has a positive characteristics set that is supported by a negative characteristic set. The negative characteristics set produces internal negative action while the positive characteristic set produces internal positive action in such a way that both exist under a contradiction to generate tension which then creates force for internal categorial conversion. The direction of the categorial conversion will depend on the relative degrees of effectiveness of the actions by the negative and positive characteristic sets. In the fuzzy paradigm, the degree of the effectiveness of the action of the negative characteristic set may be specified as a fuzzy set with a membership characteristic function. Similarly, the degree of the effectiveness of the action of the positive characteristic set may be specified as a fuzzy set with a membership characteristic function. The direction of the internal categorial conversion and the nature of the conversion is the output which may be seen as the degree of conversion. The transfer function generated by the positive characteristics set and the (impulse) resistance function generated by the negative characteristics set will constitute the input function. The degree of success of the categorial conversion may then be defined as the ratio of the membership characteristic function of the positive effectiveness to membership characteristics function of the negative effectiveness. We must keep in mind that the negative and positive actions are continuous-time input signals. As specified, positive effectiveness and negative effectiveness are fuzzy sets. Their intersection defines the point of indifference around which we can specify the needed categorial transition from one category to the other. For any particular element, we define a distribution of degrees of negative effectiveness of the negative characteristics set and the distribution of the degree of positive effectiveness of the positive characteristics set associated to the intra-categorial conversion in the same observational space in a spectrum. The distributions of the degrees of actions are in the fuzzy sets and membership characteristic function. The geometric structure of their interactions and mutual support are expressed as cognitive geometry in Figure 4.5.1.1.

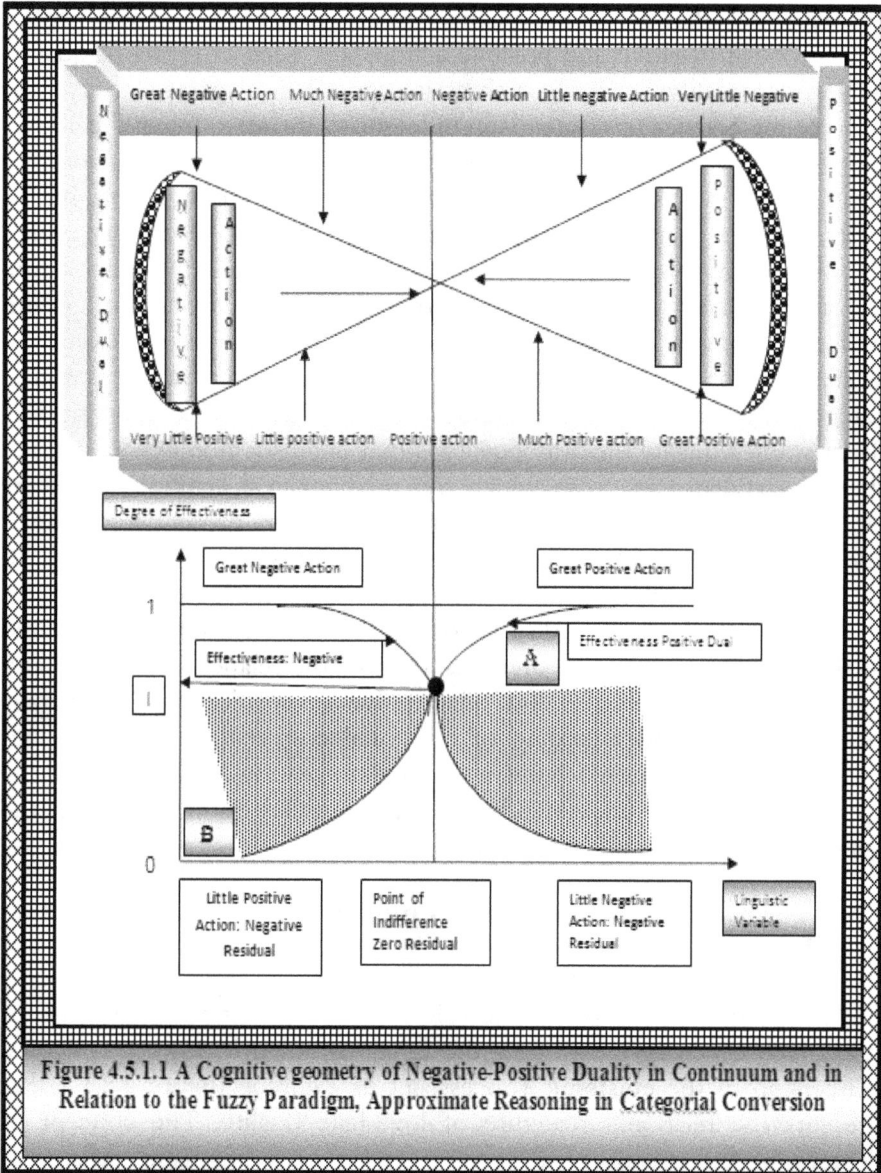

**Figure 4.5.1.1 A Cognitive geometry of Negative-Positive Duality in Continuum and in Relation to the Fuzzy Paradigm, Approximate Reasoning in Categorial Conversion**

In all analyses of categorial transformation in dealing with intra-categorial and inter-categorial movements, we must always keep track of the relationships between qualitative disposition and quantitative disposition with neutrality of time. Every quality can be expressed as a linguistic quantity whose comparative magnitude is expressed in terms of linguistic numbers. A quality requires the specification of the

175

characteristics that define the quality. For example, a set of characteristics of degrees of negative and positive actions are seen in terms of quality, while their qualifying words such as low, medium or great effectiveness provide the essence of comparison in linguistic quantity. From the above, it is easy to understand the quality-quantity relationship in the logic of approximate reasoning and internal logic of categorial conversion. Just as every negative disposition has a positive disposition, so also every quality has quantitative disposition and every quantity has qualitative disposition. In an analytical works and epistemic processes, the meaning and understandings of qualitative and quantitative dispositions are expressed as sets but not points, and where every point of meaning in relation to a category has a fuzzy set covering. Another way of looking at this discussion of quality-quantity duality is that linguistic numbers constitute the primary category from which the mathematical symbolic numbers emerged as derivatives, where linguistic numbers are qualitatively defined and the mathematical symbolic numbers are quantitatively defined. In fact, symbolic numbers may be viewed as explicated linguistic numbers and hence quantitative disposition may be viewed as explicated qualitative disposition. The understanding of the relational structure and unity of both qualitative and quantitative dispositions are extremely essential in the understanding of categorial conversions in nature and society as a general mechanism of transformation as well as in the application of the understanding to the unified foundations of the general theory of engineering science.

## 4.5.2 The Categorial Conversion Problem in the Structure Fuzzy-Decision Analytics

For a further clarification of the role of duality, continuum and use of fuzzy laws of thought and how the fuzzy paradigm connects categorial conversion in the epistemic process, let us examine Figure 4.5.1.1 that presents the relational structure of love-hate duality. The concept of degrees of effectiveness of positive action as a linguistic variable is represented as a fuzzy set $P$ with a generic element $\ell \in P$, with hate, as a linguistic variable, also represented as a fuzzy set $N$ with a generic element $h \in N$, while the success of categorial conversion may be considered as a set of characteristics $C$ with a generic element $f \in C$, where $C = (P \cup N)$ and $C = (P \cap N) \neq \varnothing$. To specify the fuzzy set for any degree of successful action variable or linguistic variable that connects negative disposition to positive disposition, we define the *membership characteristic function* that specifies the degree to which an element belongs to the sets of effectiveness of both negative and positive actions as $\mu_P(\square)$ and $\mu_N(\square)$ respectively. The membership characteristic functions are more or less mappings from the qualitative input space to the quantitative output space while their inverses are mappings from the quantitative input space to the qualitative output space as we examine the categorial conversion process. The geometric structures present the following fuzzy sets:

$$\tilde{N} = \left\{ \left(h, \mu_N(h)\right) \mid h \in N, \mu_N(h) \in [0,1] \text{ and } \frac{d\mu}{dh} \leq 0 \right\} \quad (4.5.1.1)$$

$$\tilde{P} = \left\{ \left(\ell, \mu_P(\ell)\right) \mid \ell \in P, \mu_P(\ell) \in [0,1] \text{ and } \frac{d\mu}{d\ell} \geq 0 \right\} \quad (4.5.1.2)$$

The two equations combine to form a fuzzy union $\tilde{\mathbb{F}}$ that defines a linguistic love-hate duality in the form:

$$\tilde{C} = \left\{ \left(f, \mu_C(f)\right) \mid f \in P, \text{ or } f \in N \text{ or } f \in N \cap P, \ni \mu_C(f) = \left(\mu_P(\ell) \vee \mu_N(h)\right) \in [0,1] \right\} (4.5.1.3)$$

Equation 4.5.1.2 defines conditions of a positive categorial moment that will create an intra-categorial conversion which will induce a positive transfer function needed for inter-categorial conversion, where an element emerges from one category as a derivative and enters another category. The positive transfer function may be seen as defining benefit characteristics in the categorial conversion process. Equation 4.5.1.1 defines a resistance function that creates conditions of a negative categorial moment that will restrict the intra-categorial conversion in the process of inducing a positive transfer function needed for inter-categorial conversion, to ensure the success of the overall categorial conversion. The negative resistance function may be seen as defining cost characteristics in the categorial conversion process. The benefit and cost characteristics must be seen in terms of *creative destruction* in construction-destruction duality in socio-natural transformation-substitution processes. The cost of categorial conversion is defined by the conditions of the resistance function through the set of negative actions while the benefit of the categorial conversion is specified by the conditions of the transfer function through the set of positive actions.

The concepts of negative and positive actions, and characteristic sets are general in the duality with continuum under the principle of relational unity. In Figure 4.5.1.1, the negative-action zone in the categorial conversion is where $\left(\mu_N(h) > \mu_P(\ell)\right)$ and the relative positive-action zone is where $\left(\mu_P(\ell) > \mu_N(h)\right)$, and $\left(\mu_P(\ell) - \mu_N(h)\right)$ is the *fuzzy residual* in the positive-negative space. The fuzzy residual is zero at the point of indifference with $\ell, h, f \in \left(P \cup N\right)$ and $\ell, h, f \in \left(P \cap N\right)$.

We may also define the *fuzzy conditionality* around the point of indifference where $\left(\mu_P(\ell) - \mu_N(h)\right) = 0$ and $\left(\mu_P(\ell) = \mu_N(h)\right)$ in the form:

POSITIVE ACTION SET $= \left\{\ell \mid \ell \in P \cup N \text{ and } \mu_P(\ell) \in \left(\mu_C\left(f^*\right), 1\right]\right\}$

Fuzzy conditionality for positive (4.5.1.4)

NEGATIVE ACTION SET $= \left\{h \mid h \in P \cup N \text{ and } \mu_N(\ell) \in \left[0, \mu_C\left(f^*\right)\right]\right\}$

Fuzzy conditionality for negative (4.5.1.5)

Fuzzy conditionality must be seen in terms of mutual negation in duality and polarity under the principles of continuum and unity. There are two points of transition that must be specified. They are the positive categorial point of transition of categorial conversion and the negative point of categorial transition of categorial conversion in the reversal process. The positive point of transition is defined in the positive action zone where $\left(\mu_P(\ell) > \mu_N(h)\right)$. The negative point of transition is defined in the negative action zone where $\left(\mu_N(h) > \mu_P(\ell)\right)$

Fuzzy conditionality will be constructed by the method of fuzzy decomposition and will be used to specify the *categorial transversality* conditions for conversion.

The set $A = \left\{\ell \mid \ell \in P \cup N \text{ and } \mu_P(\ell) \in \left(\mu_C\left(f^*\right), 1\right]\right\}$ defines a set of net positive actions for intra-categorial conversion required to create conditions necessary for positive transfer function to move an element from one category to another. The degrees of effective positive action for the categorial conversion may also be seen in terms of real benefit characteristics. Similarly, the set

$B = \left\{h \mid h \in \left(P \cup N\right) \text{ and } \mu_N(\ell) \in \left[0, \mu_C\left(f^*\right)\right]\right\}$ defines a set of net negative actions for intra-categorial conversion required to create conditions necessary for negative resistance function to stop the movement of an element from its home category to a different category. The degrees of effective negative action for the categorial conversion may also be seen in terms of real cost characteristics examined in construction-destruction duality with a continuum under relational unity. The conditions of fuzzy conditionality for categorial conversion will be used to develop categorial transversality conditions for the optimal control of the qualitative motion of the categorial conversion. These discussions form part of the grammar of fuzzy reasoning in understanding categorial conversion as logic for explaining the relational structure between the primary category and the derived categories in nature and society. As we have stated before, every element in the ontological and epistemological space is characterized by a positive-negative duality in continuum under a relational unity. The categorial conversion process is a composite of intra-categorial conversion and inter-categorial conversion. The process conceives of polarity where every pole contains either negative or positive duality in continuum under the principle of relational unity. Every duality is also composed of both negative and positive characteristics sets.

The socio-natural transformations are qualitative in nature and the process is explained by the theory of categorial conversion that keeps in focus creative destruction with the construction-destruction duality under the general principle of the transformation-substitution process. In this process, the construction is benefit and the destruction is cost in a relational unity. Similarly, transformation is destruction and substitution is construction in the relational unity. All the conceptual terms and analytical structures form part of the grammar of the *theory of categorial conversion* and the fuzzy paradigm with its laws of thought and the corresponding mathematics. The theory of categorial conversion is both explanatory theory and prescriptive theory. At the level of explanatory science, its intent is to explain how by the internal forces of a category (called a primary category), a new category emerges as a derivative. At the level of prescriptive science, its intent is to develop a set of prescriptive rules which when used will allow a new category to emerge as a derivative from a category (called the primary category) by the activities of its internal forces. The theory of categorial conversion in collaboration with the theory of Philosophical Consciencism provides a general mechanism to the solutions of the transformation problems. The theory of categorial Consciencism establishes the *necessary conditions* for internal self-transformation while the theory of Philosophical Consciencism establishes the *sufficient conditions* for internal self-transformation. It is suggested as a framework for the development of the theory of socio-

179

economic development as well as a framework to develop a general theory of engineering sciences. This framework has been used in the development of the theory of knowledge production under the complex interactions of decision-choice actions in methodological constructionism-reductionism duality with relational continuum and unity given nominalism [R3.10] [R3.13].

### 4.5.3 The Characteristics of Duality in the Theory of Categorial Conversion

Let us expand the analytical richness of the concept of duality and its distinction from dualism in relation to the theory of categorial conversion. The grammar of the theory of categorial conversion involves polarity with negative and positive poles. Each pole has a residing duality. The pole is said to be negative if it has a residing negative duality. The pole is said to be positive if it has a positive residing duality under the principles of continuum and relational unity. The positive and negative poles are in relational tension as they function in a contradictory mode. Every duality is composed of negative and positive characteristics sets that reside in conflict, continuum unity and defined relationality. The negative characteristics set constitutes the negative dual while the positive characteristics set constitutes the positive dual. The negative duality is identified by the conditions that the negative characteristic set with the production of a set of negative actions is greater than the positive characteristics set with the production of a set of positive actions in the categorial-conversion process. All the negative and positive attributes are in a mutual negation under contradiction. It is here that the study of the theory of contradiction acquires analytical significance in the knowing process and social change under the principles of continuum and unity. The categorial conversion process is initiated in duality by the relational conditions of the negative and positive characteristic sets. This approach will allow us to identify the matter, energy and information structure. In nature, categorial conversions are created by the internal decisions of nature on the basis of matter, energy and information. In society, categorial conversions follow the natural process but by decision-choice actions of cognitive agents to create the social categorial conversion on the basis of matter, energy and information structure. The essential defining characteristics of duality and polarity that allow the categorial-conversion action are:

1. The mutually interdependent opposites that are defined by characteristic sets which may be called positive and negative characteristic sets that establish the positive and negative duals and poles.

2. The opposites exist in a complex complementarity in that they mutually define the existence of each other in categories and categorial conversions without which each other's existence is conceptually non-definable.

3. The mutual complementarity induces supplementarity, interdependence, reciprocity in terms of give and take, and other attributes defining qualitative and quantitative statics and dynamics of change in relations such as independence, interdependence, identity and others.

4. The internal tension generated by the dynamics of internal conflicts between the sets of positive and negative characteristics in the process of accomplishing the basic relational characteristics specified in (1, 2 and 3) above.

5. The mutual negation that tends to change sameness into a difference and a difference into sameness, and the relative proportionality of the characteristic sets in a defined continuum of elements in terms of categorial conversion in the dynamics of the substitution-transformation processes in a category. For further discussion see [R1.92].

6. The mutual negation is through the internal qualitative and quantitative motions which are produced by internal conflicts through the contradictions in the internal unity of the negative and positive attributes and actions which in turn are generated by the elements of the residing characteristics sets.

7. Every element in the universe belongs to a category and every category is defined by qualitative-quantitative dispositions in a relational unity and continuum; and every element in any category is an actual-potential polarity with relational tension from the corresponding dualities, where each pole has a residing duality under the principle of continuum and unity with the defined duality in the other pole.

These seven statements are essential attributes in defining the concept and meaning of duality as it relates to polarity. The postulate of relational continuum may be related to the structure of quality and quantity of four points of qualitative and quantitative extremes. On the qualitative side, we have the extreme negative and the extreme positive. On the quantitative side of things, we have negative big and positive big. These extreme negative and positive qualitative characteristics and extreme negative and positive big define the identity of an element in unity under an epistemic tension and cognitive resolution. They are presented as an epistemic geometry in Figure 4.5.2.1 where there are movements from extreme positives to extreme negatives in the quality-quantity space. At the level of the primary category of the knowledge-construction process, the concepts of quality, quantity and time are epistemic fundamentals. These epistemic fundamentals are violated by the acceptance of linguistically quantitative purity with exact measurements and mathematical representation that fail to account for the role of quality and subjective judgment in the quantitative positions. The qualitative characteristics in human experiential manifolds must guide us in the development of mathematical and logical reflections in representation of reasoning and geometry of thinking in reducing complexities into simpler forms towards irreducibility.

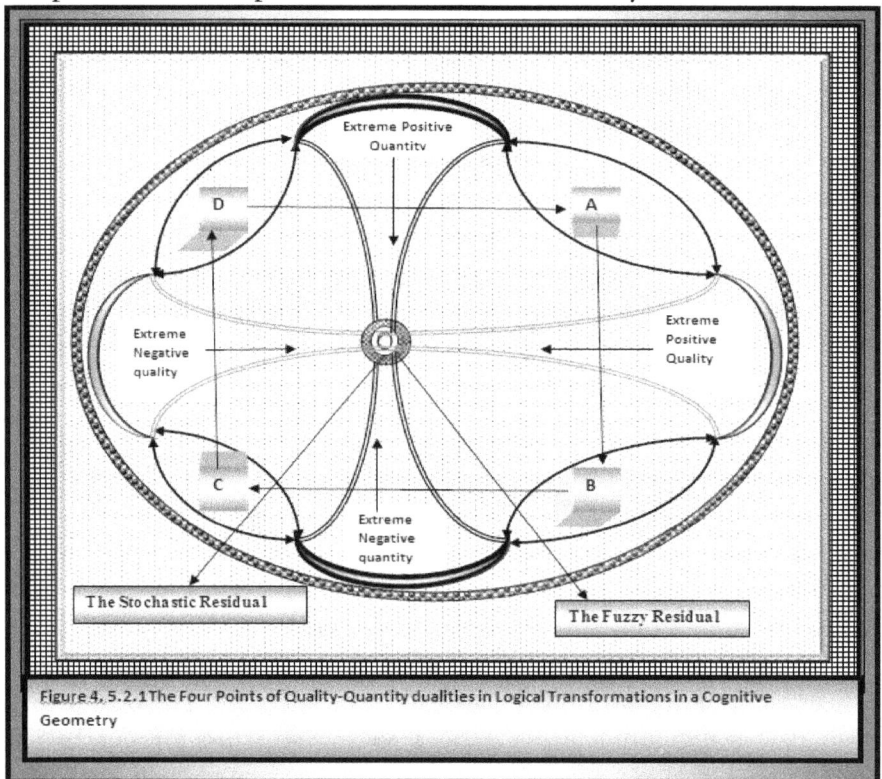

Figure 4, 5.2.1 The Four Points of Quality-Quantity dualities in Logical Transformations in a Cognitive Geometry

The extreme positive and negative qualities may be associated with complete exactness and complete inexactness respectively as the poles, and with the logic of excluded middle they are transformed into dualism where they do not share elements of their transformations. The exactness and inexactness may be applied to quantitative characteristics as what we know to be true. This position is not consistent with the actual world of operations where the concept of exactness can only exist and be understood with the existence of the concept of inexactness and vice versa. The complete inexactness and complete exactness are conceptual poles of quality and quantity. The acceptance of exactness and inexactness comes with supporting subjective belief values which the fuzzy laws of thought advocate. In the classical laws of thought, opposites are recognized, but they are viewed as mutually exclusive and collectively exhaustive with nothing in common, where the membership characteristic function assumes the value of either 0 or 1 with nothing in between.

The connection of positive to negative through a give-and-take process and mutual negation as expressions of the path linking the complete exactness and complete inexactness must be done in such a way that we abandon the law of excluded middle and the non-acceptance of contradiction that form the conditions of the true-false acceptance in the classical paradigm. The classical laws of thought are also incompatible with substitution-transformation processes of mutual negations that allow the study of the dynamics of quality-quantity changes in a specified time domain. It is here that the fuzzy laws of thought that may be derived from the conditions of fuzzy paradigm, as presented in Table 4.1.4.1, enter.

First, we observe that the law of identity establishes the ontological elements in the universal object set where the elements are free from the conceptual notions of exactness and inexactness because they are what they are and would be what they would be. Given the ontological elements, *what there is*, is brought to the epistemological sphere for examination, study and understanding. The laws of duality of commonness, unity and continuum (non-excluded middle) advance rules of reasoning and criteria of acceptance of contradictions where every element is and is not, and every element is always on the trajectory of being or becoming and not being or not becoming. The fuzzy laws of thought establish epistemic conditions for the acceptance and non-acceptance of the knowledge propositions where each acceptance has a fuzzy conditionality. Thus, it helps to establish approximate-reasoning tools to navigate through the penumbral regions of decision-choice actions between the positive and negative extremes such as complete exactness and complete inexactness or absolute truth and absolute falsity in thinking and in the knowledge production process, as well as in any epistemic duality with defective information structure.

The conditions of the epistemic duality with the law of continuum in reasoning suggest that there is a process of interaction between sets of true and false characteristics that produces tensions which is resolved by defined decision-choice rationality as a utility tool for the cognitive agents. Since tension involves two tendencies of falsity and truth, it also involves two epistemic directions where acceptance of truth of proposition suggests that the truth is accepted to contain more truth characteristics than false characteristics. Another way of saying this is that the membership characteristic function of the truth characteristic line lies above that of the false characteristic line as exemplified in Figure 5.4.1.1.

The fuzzy laws of thought, given the fuzzy paradigm that encompasses ontological and epistemic conditions, may simply be stated again as:

> Every proposition contains true and false characteristics in varying proportions in such a way that true and false statements are determined by decision-choice action in resolving the conflict between truth and falsity, where the truth characteristic set is optimized subject to the false characteristic set, to provide a measure of fuzzy conditionality that is used to decompose the set into true and false statements for choice [R3.7, p. XIII]

This is an acknowledgment of the role that judgment plays as an important tool that is used by knowledge agents in the knowledge-production process and that inductive and deductive reasoning are constrained by defective information and cognitive limitations. In the fuzzy system of the theory of knowledge, every claimed knowledge item, therefore, cannot be definitive. Any claim of knowledge can be accorded varying degrees of validity for specified applicable domain with both stochastic and fuzzy risks defining the relevant domain. The stochastic risk in the knowledge production and human action is generated by the stochastic residual whose measure provides the stochastic conditionality, while the fuzzy risk is generated by the fuzzy residual whose measure provides the fuzzy conditionality for every claimed knowledge item. The sum of fuzzy and stochastic risks presents us with a total risk of inexactness contained in the region of acceptance of truth or falsity and in the space of human action. All these operate in the cost-benefit space where every benefit has a cost support and ever cost has a benefit-support.

# CHAPTER FIVE

## The Theory of Categorial Conversion: Axiomatic Foundations

The essential philosophical structure of Nkrumaism and its extensions has been discussed in the previous chapters, where the idea is to find an intellectual framework in support of the decolonization and complete emancipation of Africa based on the African conditions as seen by Nkrumah. The task is defined in the quality-time space where the qualitative characteristics of the African societies under imperial domination must be politically and economically transformed. Nkrumah knew that the task could not be accomplished with the intellectual framework of Africa's oppressors. So a new intellectual framework must be constructed. This new intellectual framework must be equipped with analytical tools to deal with both the qualitative and quantitative dispositions of the African societies. The most important thing to note is that making quantitative changes alone in the quantity-time space without changing the qualitative disposition in the quality-time space does not essentially alter social conditions. The skeleton of the new intellectual framework was advanced by Nkrumah in his Consciencism, as a toolbox of logical guides to practice in the African politico-economic transformations. This intellectual framework is based on the African conceptual system as has been argued in this monograph and in [R1.91] [R1.92] with further generalized extensions in [R3.7] [R3.10] [R3.13]. In the previous chapter, the philosophical foundations and extensions were advanced. The intension in this chapter is to crystallize and tighten the foundation and show how the categorial conversion works in Nkrumaism. As has been advanced in the studies on general knowledge development, logic and mathematics, the *theory of categorial conversion* presents a logical framework in the study of qualitative motion in general and social transformations in particular, where subjective phenomena are an essential part of change. It is also applicable to the underlying logic of all engineering phenomena, such as biomedical, electrical, genetic, and social and all forms of human creation. It offers an important framework in the study of economic development, social transformation, revolutionary change and the process of gradual transformation in society and nature. The understanding of the theory of categorial conversion is essential in understanding the conceptual system of

Nkrumah and foundations of Nkrumaism and its philosophical and analytical extensions. It is also essential in the understanding of the African traditions in thought where all things are relationally connected and exist in dualities under relational continuum and unity which together define the oneness of existence under the principle of never-ending creation.

## 5.1 The African Intellectual Framework, Fuzzy Paradigm and the Development of the Theory of Categorial Conversion

The construct of evolving African intellectual needs is the work of active and critical African thinkers whose main concern is the understanding of the working mechanism of natural forces. These thinkers must belong to the awakened African intellectual class whose members act as people of thought and people of action. The intent is directed towards the complete emancipation of the African and towards the improvements of African human welfare through knowledge acquisition, effective reasoning and efficient utilization of that which is known. Generally an accumulated knowledge defines the level and the state of human ignorance while the process of knowing leads to the growth of knowledge accumulation and the reduction of human ignorance and fear. For the members of the awakened Africa intelligentsia, this knowledge accumulation and reduction of ignorance must be relevant to the African conditions and the changes that they engender. The active process of knowing leads to the growth of knowledge accumulation and the reduction of ignorance and fear, to provide foundations for the practice of freedom and justice as may be abstracted from African traditions. Here, an important *duality* emerges between knowledge accumulation and the process of knowledge production. Knowledge accumulation owes its growth to the process of knowing. Such a process functions through tension between reality and perception by transforming ignorance into awareness. Perception and reality thus appear as duals with conflicts and internal dynamics locked in dualistic continuum. Alternatively stated, there is the African perception space, and the space of African reality. The perception space presents experiential information through interpretations of acquaintance. There is also the African actual and African potential. The activity of knowledge acquisition by the African intelligentsia is to design a process where the elements in the African perception space are mapped onto the elements of the space of African reality and the elements in the space of the African actual are similarly mapped into the space of the African potential. At the level of knowing, the process is through systems of epistemic constructs that link elements of perception space to the elements in the space of reality through *conversions of logical categories* with the use of an appropriate paradigm of thought. At the level of transformation, the process is through systems of epistemic constructs of categorial moments that will link elements of *actual space* to the elements in the *space of potential* through *conversions of institutional* or *organizational categories* with the use of an appropriate paradigm of thought. These systems of epistemic constructs are embedded in the theory of categorial conversion. The core idea for reasoning is the establishment of actual and potential which together constitute the actual-potential polarity under the principles of the opposites with relational continuum and unity.

The conceptual power of the theory of categorial conversion is an operation where through thinking, the elements in the experiential information structure are gradually justified by thought to be equivalent to the elements of reality as the truth of existence (*what there is*). A claim of a knowledge item obtained by a methodology of constructionism is said to be established when an element in the experiential information structure from the perception space is shown by a theory to be collaborated with an element in the space of ontological reality under the same theoretical framework. These elements belong to a derived category by the process of methodological constructionism. Strong reasoning backed by a deceptive or non-deceptive information structure can convince one to accept, by belief, a conceptual element as the reality even though it is not. This is another way of looking at justified and unjustified beliefs where justified belief is information supported through reason by the use of methodological reductionism. In this respect, if the reasoning is weak, no amount of words can rescue reality from illusions in the perception space where illusory elements are equated to elements of reality by reasoning from the inferential information structure. Similarly, if the reasoning is strong, illusions in the perception space can be made into believable realities in the space of realities where phantom abstract ideas are rationalized as truth. The members of the awakened African intelligentsia have the task of solving this problem in the mirage-reality duality in order to rescue the African masses from living in an illusory world of the comfort zone of ignorance.

Examples of such illusions are: aid helps development, a positive relationship between democracy and development, there is a positive relationship between technical assistance and technological progress, African unity is an impossible task, no African philosophy and many other such illusory elements in the possible world space. The irony is that many Africans adopt these illusory ideas in the possible world space without a critical thinking. They internalize these ideas as truth and then become the active promoters in support of the imperial structure and the oppression machine. In the search for epistemic clarity, it must be kept in full focus that when ignorance is elevated to the throne of the kingship of human existence, knowledge is in trouble. Lies become elevated to the position of the deputy of ignorance, and the first casualty is truth, with relentless prosecution of knowledge whose only crime is nothing but the search for truth. Here, a paradox arises between ignorance and knowledge, thinking and knowledge, thinking and wisdom and wisdom and intelligence. It is this relational paradox among ignorance, knowledge, lies and truth that brought the centuries of African suffering, slavery and Africa's current demise. It is also through the understanding of this relational paradox in governance that complete emancipation of Africa can be achieved at the various stages of the African social set-up through the understanding of the theory of categorial conversion.

## 5.2 A Reflection on the General Logic in Categorial Conversion

Difficulties arise in all cognitive constructs that define the lines of communication within the perception-reality duality whose internal equilibrium, disequilibrium and processes of understanding and change are maintained by the organization of multiplicity of relational rules of thought. The difficulties tend to increase when the perception-reality duality is imposed on the actual-potential duality. Perception has the characteristic of subjectivity on the qualitative disposition of the perceiving individual, while reality has the characteristic of objectivity of existence (that is, "what there is") independent of perception and acquaintance. Reality is viewed in terms of ontology while perception is viewed in terms of epistemology. The conditions of perception are drawn from the elements in the experiential information structure. The perception space is thus a derivative of the experiential space that constitutes its primary category of knowing. Since perception is subjective and individual-specific, how do we know that a particular element from the perception space coincides with an element in the space of reality through theoretical constructs that are cognitive edifices of an individual or a group of individuals? How do we verify the validity

190

and consistency of our epistemic constructs that represent our *perceptual models* of what we perceive to be real and actual, and how do others come to accept them even given the acceptance of a common experiential information structure and a common logic of reasoning? All these lead us to consider the relational structure of time, actual, reality, potential, perception, quantity and quality in socio-natural processes.

### 5.2.1 The Time Trinity, Actual, Reality, Potential and Perception in Categorial Conversion

Actual has the characteristics of the present and is also related to ontological and epistemological spaces. In the logic of knowing, we have the ontological actual and the epistemic actual. The ontological actual relates to reality while the epistemic actual relates to perception and experiential information. The potential has the characteristics of the future and is also related to ontological and epistemological spaces. Similarly, in the logic of understanding, we have the ontological potential and epistemological potential. The ontological potential relates to the ontological actual. The epistemological potential relates to the epistemic actual. Alternatively, how does one distinguish conditions of colonialism, imperialism and neocolonialism from the conditions of independence and complete emancipation within the just-unjust duality or freedom-subjugation duality? The process of knowing the structure of the actual, reality, potential, perception, quantity and quality is dependent on the information structure and the paradigm for processing it into knowledge. The relational structure is presented in terms of cognitive geometry in Figure 5.2.1.1.

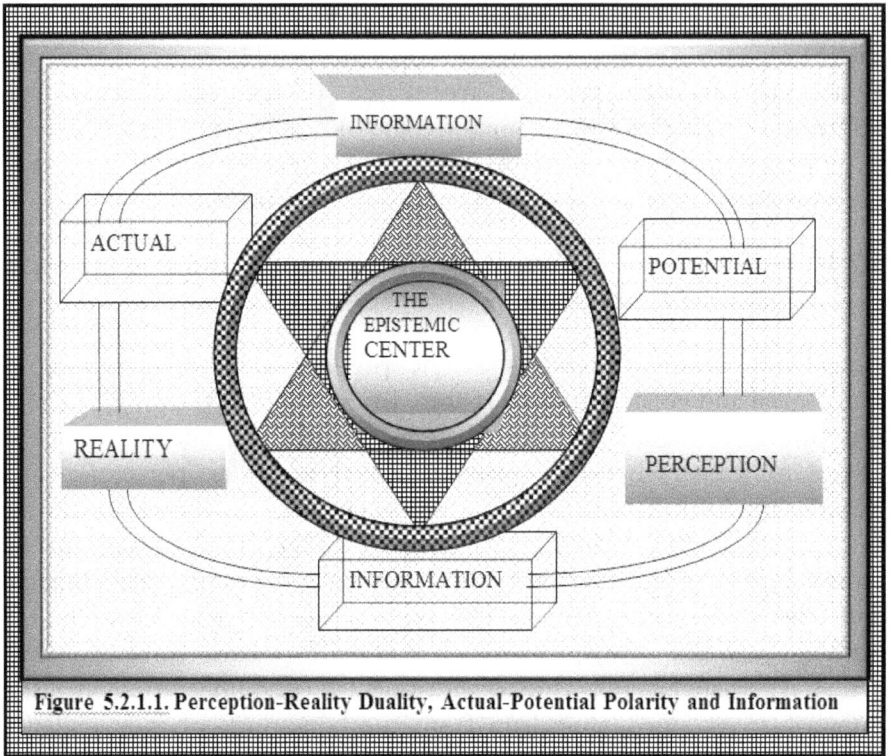

**Figure 5.2.1.1. Perception-Reality Duality, Actual-Potential Polarity and Information**

Similarly, a question may be raised as to how the elements in the perception space correspond to abstract ideas where such ideas may contain phantom elements as has been previously discussed. For example how statements do like *Africans have not create history, Africa was discovered* or other such statements as claimed by colonizers and oppressors and parroted by others correspond to true or phantom ideas? To answer these questions, the constructed theories must involve systems of arguments that convince one to accept the correspondence between elements of perception and elements of reality as shown by reason. The correspondence between perception and reality must then also be related to the actual and the potential by reasoning. Such a reason must be derived from a paradigm of thought that contains laws of thinking for the production of reasoning and mathematical systems composed of vocabulary and grammar to be used to process any given information structure. The epistemic center of the information processing that allows the construct of the relational structures between perception and reality, and between actual and potential is shown in Figure 5.2.1.1.

Perception is a subjective model of reality and the acceptance of any such model is through making a case from a theory under a given paradigm that allows the practice of decision-choice action under a defined rationality. The theory is a thinking system that must be governed by a collectively agreed system of inductive and deductive rules for reasoning and its construct. Potential resides in the possibility space and actual resides in the space of reality. The actualization of the potential and the *potentialization* of the actual demand the understanding of the theory of categorial conversion in both positive and negative directions under a system of rules of reason. The actualization of the potential is construction and the potentialization of the actual is destruction. In this way, the actual-potential duality finds expression in the creative destruction within the construction-destruction duality. In the process of categorial conversion, both the actual-potential duality and the construction-destruction duality may be subjected to the methodological complexity of construction-reduction duality with continuum under the principle of relational unity. This system of rules is called logic which is embedded in a paradigm of thought in construction-reduction duality to understand creation and destruction from within [R1.92] [R3.7] [R3.10].

Logic, therefore, is a science of thinking and reasoning that allows one to create the channels for relating elements in the perception space to the elements of the space of reality as well as allows one to interpret the truth and validity embedded in the crafted correspondence between abstract ideas and elements of ontological existence for social acceptance and practice whenever they may be needed. It is also a thinking and reasoning system for constructing communication channels between the potential and the actual in qualitative forward motion (which simply implies the fading of the actual into the potential), and between the actual and the potential in qualitative backward motion (which simply means the emergence of the potential from the actual) As seen, logic, as applied to socio-natural analysis and synthesis, is about creating conditions of acceptance of epistemic conclusions towards perceptions about reality from commonly agreed and/or received notions and rules rather than a search for social truth. Generally, however, logic in the final analysis is a special theory. It is a theory about channels for constructing theories that link perception to reality, and theories that link actual to the potential. It may, thus, be viewed as a meta-theory of the process of knowing and knowledge acquisition. Its subject matter is about rules of thinking and behavior over the epistemic process. As with any other theory, its development is affected by the content and character of the social environment in which it emerges. The social environment is a product of the culture of the social set-up. Thus, logic and the whole methodological process of the enterprise of cognition and activities of knowing must be viewed from social contentions of ignorance and knowledge within the process of knowledge creation. The thinking and knowledge creation affect the understanding of the theory of categorial conversion, which is the subject of study, analysis and synthesis about the dynamics within the perception-reality duality and the actual-potential polarity.

Every society has a corresponding epistemic system based on its social experiential information structure and social exigencies. Each of these epistemic systems may or may not collaborate with others since the subjective element of cognition is affected by social experience and conditions in interpreting the qualitative social existence as induced by the culture of knowledge creation. Different methods and rules of reasoning may arise from different social environments that contain different experiential information structures and cultural boundaries of reasoning. In general, therefore, an epistemic system, containing the rules of thought, which drives the process of knowing finds its most effective weapons from the social environment and cultural boundaries that give rise to it. It is here, that Nkrumah's advice to the members of an awakened African intelligentsia exhibits its liberating power of mental decolonization and cognitive subservience. It is, therefore, possible to find different logics of knowledge development from different social environments and conditions of existence. Each of the logics will have its own language composed of vocabulary and grammar that support the system of thought that is translated to action. This is also true of Africa's cultural environment and her people.

African cognition, cognitive process, epistemic development and the science of thinking must find their building blocks and weapons from Africa's social environment, experience and existence, and then fashion them in a manner that satisfies the effective search of solutions to the African problems of contemporary times within the African cultural confines. These building blocks as well as the needed weapons already exist in the African social set-up from antiquity. It has been stated that Nkrumah directly or indirectly used them in the development of his thoughts. The justification of this claim will be enhanced as one views the complete development of the epistemic structure of this monograph. Our only task is to explicitly restructure, sharpen and reveal them in a self-contained theory. This requires that they must be studied and use the derived knowledge for further development and refinement of the African thought system. The theory must relate to the terms of how they are used in the logical development of Nkrumaism and its application to socio-economic transformation with special reference to Africa. The manner in which these logical tools form analytical unity must then be made accessible to the general African populace and other interested persons for use. The complete structure is seen here as the Africentric reasoning system which is represented as polyrhythmicity supported by the logic of polyrhythmics [R1.92]. The integrated system of Africentric reasoning towards African emancipation, social life, preservation and progress was the driving force in the development of the conceptual system of Nkrumah. It is also the motivation for the development of Nkrumaism and the philosophical extensions that are pursued here. The objective is to uncover the underlying foundations of the Nkrumaist thinking system, its rules of reasoning and cognitive techniques, and lay down the methods of argument for the construction of Consciencism as an integrated philosophical and ideological system which guided Nkrumah's socio-political practices. Its conceptual framework must guide African social practices at the level of politics, national interest, social vision and economic construct for Africa's complete emancipation. The complete emancipation is composed of independence, freedom and nation building. In this way, the past will be linked to the present, and the present will be connected to the future for a full past-present-future continuity of the African conceptual and philosophical developments. Not only that, but the African perception will be linked to reality, while the actual will be creatively connected to the potential through an active application of the theory of categorial conversion. As stated, the axiomatic system and polyrhythmics of Nkrumaism are to reinforce the conceptual discussions presented in the previous chapters involving the primary and derived categories, categorial conversion and multiplicity of social forms that are linked by reasoning to discover and study the behavior of categories of *what there is* (the actual social conditions) and *what would be* (the potential social conditions) in terms of states and processes.

The primary intent is to discover how these states and processes become transformed into different categories of reality through an internal process. The secondary intent is to find out how to use this discovery of the logic of categorial conversion to emancipate Africa and African people from imperialist subjugation and the global power system of oppression and exploitation. In this respect, Nkrumaism may be viewed as an Africentric social philosophy developed through a general logical prism as a revolutionary philosophy required for the emancipation of Africa and her people. The need for its logic of reasoning and the understanding of the correctness of arguments, verifications, validity of propositions, perception-reality corroboration, actual-potential transformation and acceptance of conclusions about social states within the time trinity cannot be overemphasized. For one thing, Nkrumaism is presented as a modernized African social philosophy and a conceptual system on the basis of which the modern Africentric social formations must be reconstructed and molded, and human and non-human resources must be organized in support of the African and social management for nation building and Africa's progress, and security, and to hold off imperialist aspirations in Africa. Its concluding ideas and accumulated knowledge presents an analytical framework to define the African essence and personality that must continuously shape the actual which is pulled out of the potential, and where the potential emerges out of thought which is culturally derived. The implied logic of Nkrumaism must show how revolutionary thinking, reason and arguments proceed within its conceptual system leading to a representation of thought that captures human perception as a model of transformation of reality to the desired potential. In this framework, Africentric perception must be a cognitive reproduction of elements in the space of reality under African culture, environment, social experiences, and exigencies.

Perception as a model of reality cannot be divorced from social ideology and culture which collectively determine their mutual existence. Perception of reality varies over cultures and generations which are characterized by different collective and individual preferences. Just as perception is inseparable from ideology and culture, so also are philosophy and knowledge accumulation inseparable from ideology and culture. In fact, perception and interpretation of acquaintance are influenced by the available philosophy and accumulated social knowledge. The whole enterprise of science and social knowledge accumulation is culture-dependent and culture-determined, while the uses to which science and knowledge are put are ideologically determined. How do perception, ideology and culture influence and even determine theories of knowledge and acceptability of knowledge, and how does the growth of knowledge influence the development of ideology and culture? How does Nkrumaism shape and determine Africentric social knowledge for Africa's complete emancipation? How do the logical parameters of Nkrumaism and its extensions define the domain of revolutionary thinking? How do knowledge, culture and ideology shape the path of social transformation though the operations of the internal forces of society? In addition, how does the theory of categorial conversion help to explain socio-natural internal transformations that relate intra-categorial and inter-categorial changes? These questions bring us to the critical subject matter of the science of thinking within Nkrumaism where we must analyze, synthesize and present an integrated system of implied logic and the method of philosophical and ideological inquiry into how knowledge of nature and society is obtained within the construction-reduction duality with a continuum, and how the actual becomes potential and potential becomes the actual in the developmental process of nature and society.

## 5.3 Language, Vocabulary and Grammar of the Theory of Categorial Conversion

On the road to knowing, we are always confronted with the problem of the relationship between perception and reality, the perceptive information structure and the experiential information structure, phantom ideas and non-phantom ideas, reality and potential, possibility and probability, quality and quantity, and potential and progress. These are particularly relevant to the study of social formation and its institutional arrangements and qualitative transformations. They are also important in the study and understanding transformations in nature and in a general conceptual formation of unified theory of engineering sciences. Perception and reality on the one hand and reality and potential on the other hand constitute categories of polarity and duality, while potential and progress in turn constitute changes in the universal unity in terms of qualitative and quantitative dispositions that reveal the internal dynamics of categories. The whole enterprise of the theory of knowing and knowledge acquisition involves mappings of elements in the perception space into the elements in the space of reality as perceived in the ontological space. This is the information to knowledge mapping. The results of the process of knowing are then mapped out onto the space of engineering or human creation. This is the knowledge to

197

engineering mapping. The process of information-knowledge mapping is accomplished through a particular logical prism that allows claimed statements of truth to be made, verified and approved for social acceptance and action in the epistemological space. The process of knowledge-engineering mapping is also accomplished in a prescriptive logical prism which allows human internal creation in nature. All internal dynamics of the socio-natural process take place within and between categories under principles of opposite with relational continuum and unity. It is the principles of continuum and relational unity that give expression to overlapping categories. It is the opposites that establish actual and potential categories with contradictions and conflicts. The logic of Nkrumaism as a guide to the understanding of social transformation relies on assumed conditions, categories of reality, categories of potential, polarity, duality, multiplicity of epistemic rhythms, and relationality of opposites in continuum and unity to present the method of reasoning in qualitative dynamics for which the theory of categorial conversion projects.

### 5.3.1 The Logical Blocks of the Theory of Categorial Conversion

Let us now turn our attention to the essential building blocks of the logic of categories that will help to define the vocabulary and grammar in categorial conversion. The building blocks of the science of reasoning in the epistemics of categorial conversion are composed of the grand structures of categories of epistemological elements, primary category, derived categories, polarity, duality, unity, characteristic sets, rationality, and a multiplicity of rhythmic and counter-rhythmic behaviors in continuum that together present a unified epistemic entry into the knowing of states and processes in the ontological space through the cognitive activities in the epistemological space. These grand building blocks, enhanced by relevant sub-blocks, are conceptually linked together to constitute a unity of reasoning for the discovery of *what there is* (the actual), *what ought to be* (the potential), and the explanation of states and processes of the elements in the actual-potential duality and polarity in a continuum.

The initializing of the conceptual linkages proceeds with sets of postulates, principles and axioms that allow the universe to be initialized for either an explanation of *what there is* or a prescription for *what ought to be* in terms of states and processes from quantity-quality duality in a relational continuum and unity, which allows the appreciation of the analytical unity and epistemic power of Nkrumaism for understanding the construction-reduction *revolutionary thinking* in relation to the internal qualitative dynamics of nature as abstracted from the Africentric conceptual system and the applications that such understanding may help in social change, where the potential is purposely set against the actual in a never-ending social process. At the level of social organization, one may view the notion of the setting of the potential against the actual in terms of social engineering on the basis of the complex interplay of the negative-positive actions produced by the negative and positive characteristic subsets in contradiction and conflict. At the level of industry and human creation, the setting of the potential against the actual, in all areas of human activities, may be viewed as transforming the knowledge of the actual to artificial creation in all forms. This artificial creation of all forms may be seen as physical and social engineering while the practice of medicine may be seen as intervention in internal behavior of biological systems. In Chapter 4, a number of basic concepts were introduced, defined and specified as the bases of the development of the theory of categorial conversion and its connection to the use of fuzzy paradigm. Some of these concepts and ideas are now stated as postulates and axioms in terms of their entries into the theoretical construct.

### 5.3.1.1 Concerning the Universal Unity and Particular Unity in the Theory of Conversion

The universe is one continuum and contiguous phenomena that is composed of diverse forms in relational unity and in mutual existence. All the diverse forms are interdependent and are transformations from *irreducible primary existence* of an ontological reality and yet each form retains itself in a *particular unity* for identity and uniqueness. The universal unity is composed of categories of particular unity which find definitions from the structure of the universal existence. The universe is a collection of mutually non-exclusive and collectively interdependent categories of elements that project relational continuum and unity.

### 5.3.1.2 Concerning the Concepts of Ontology, Ontic, Epistemology and Epistemic in the Theory of Categorial Conversion

Distinctions are made between ontology and ontic, epistemology and epistemic, ontology and epistemology, and ontic and epistemic in the development of the foundations of the theory of categorial conversion,

Nkrumaism and the philosophical extensions presented in this monograph. These distinctions will allow us to critically examine the information structure that must be processed for the understanding of transformations as seen in the theory of categorial conversion and the applications that may be required of it for socio-economic change and politico-legal transformations. In the monographs [R3.7] [R3.9] [R3.10] [R3.13] a clear distinction was made between ontological and epistemological spaces which are respectively related to their information structures. This distinction is now being extended to relevant areas for more analytical clarity. Ontology concerns itself with existence. The ontological space contains ontological elements composed of states and processes. The states and processes of ontological elements are called ontic states and processes. The ontic states and processes are defined by *ontological characteristics sets* whose partitions define *ontological categories* under internal natural transformation on the basis of ontological information structure, while the ontic states and processes define the *ontological identities*. The epistemological space contains *epistemological elements* composed of states and processes created by cognitive agent for knowing. The states and processes of epistemological elements are called *epistemic states* and *processes*. The epistemic states and processes are defined by *epistemological characteristics sets* whose partitions define *epistemological categories* under internal cognitive transformation on the basis of *epistemological information structure,* while the epistemological information structure, the epistemic states and processes are cognitive derivatives of ontological identities. At any moment of time, the epistemological space is a sub-set of the ontological space in the process of knowing. Similarly, the space of the actual is a subset of the space of the potential in acquaintance and perception. The ontological elements and categories are natural creations that will exist independently of cognitive and non-cognitive ontological elements. The epistemic elements and categories are the formation of cognitive agents on the basis of information derived from acquaintances of the ontological identities of the elements. In this process of developing the theory of categorical conversion as a general dynamics on qualitative and quantitative disposition, the ontological space is seen as the identity and the epistemological space is seen as derived identity by cognitive action. The relational structure of essential vocabulary is presented in Table 5.3.1.

| ONTOLOGICAL SPACE: Primary Identity | EPISTEMOLOGICAL SPACE: Derived Identity |
|---|---|
| 1. Concerns Existence<br>2. Ontic States and Processes (Ontological Elements)<br>3. Ontological Characteristics<br>   A) Positive Characteristics in qualitative and quantitative attributes<br>   B) Negative Characteristics in qualitative and quantitative attributes<br>4. Ontological Categories<br>5. Ontological Information<br>6. Ontological Identity<br>7. Ontological Continuum and Unity<br>8. Natural Categorial Conversion<br>   A) A System of Actual-potential Polarities with a Primary Polarity and a system of Derived Polarities<br>   B) A System of Positive-Negative Dualities with primary Duality and a System of Derived dualities | 1. Concerns Knowing<br>2. Epistemic States and Processes (Epistemological Elements)<br>3. Epistemic Characteristics<br>   A) Positive Characteristics in qualitative and quantitative attributes<br>   B) Negative Characteristics in qualitative and quantitative attributes<br>4. Epistemological Categories<br>5. Epistemological Information<br>6 Epistemological Identity<br>7. Epistemic continuum and Unity<br>8. Epistemic Categorial Conversion<br>   A) A System of Actual-potential Polarities with a Primary Polarity and a system of Derived Polarities<br>   B) A System of Positive-Negative Dualities with primary Duality and a System of Derived dualities |

TABLE 5.3.1 The Essential Vocabulary in Understanding the Theory of Categorial conversion in a relational Structure of Ontology and Epistemology

## 5.3.1.3 Concerning Categories and Their Formations in the Theory of Categorial Conversion

The essential relational structure of the vocabulary in Table 5.3.1 needs some explanation. The ontological universal system is conceptually partitioned in the epistemological space into categories of being called epistemological categories that allow different forms of elements to be identified and distinguished from each other by their qualitative and quantitative dispositions in the epistemological space. The universe, of which the world is part, is thus analytically conceived to be made up of categories and sub-categories of being called ontic categories which appear as states and processes for identification knowing and naming. The states and processes are defined by matter, energy and information. The qualitative states are always in temporary equilibria, while the paths of the processes define permanent evolutions of change no matter how

201

long it takes. In this way, all categories are placed in the dynamics of equilibrium-disequilibrium duality. The categories constitute an infinite set or infinite family of sets of sets. Each category presents itself in a particular unity which is composed of elements with defined similar and dissimilar characteristics. There are two types of categories. They are the *ontological categories* defined by *ontological information structure*, and *epistemological categories* defined by *epistemological information structure*. Some ontological categories appear as epistemological categories through experiential information structures of quality and quantity.

### 5.3.1.4 Concerning Polarity, Duality, Oppositeness and Unity in the Theory of Categorial Conversion

The concept of *Polarity* characterizes every category of being where each polarity appears as right and left poles in *relational unity* under a continuum. Each pole has a residing duality where the components of the dual exist in opposites in a continuum and relational unity. The dual is simultaneously composed of qualitatively *positive and negative sets of characteristics* that exist in oppositeness and relational unity under a continuum principle. The positive characteristics set are qualitatively different from the negative characteristics set. The relative magnitude of the positive and negative characteristics sets in the duality defines whether a particular duality is negative or positive. Similarly, the character of the duality defines the nature of the pole in terms of negativity and positivity. The positive pole is made up of positive duality, where the magnitude of the set of positive characters far outweighs the set of negative characteristics in relational quantity. It is through the establishments of similarity and difference that qualitative and quantitative dispositions are confirmed as logically inter-supportive. Similarly, the negative pole is defined by its negative duality, where the negative characteristics set outweighs the positive characteristics set in relational quantity. Both the positive and negative poles exist in relational unity with each other, as well as they exist in relational unity with their internal duality. The right duality and the left duality are in relational unity and continuum. They are in turn established and identified by the negative and positive duals which are established and defined by the relative characteristics set of the negative and positive characteristics sets. The positive dual has the dominance of the positive characteristics set while the negative dual has the dominance of the negative characteristics set.

In the final analysis, the relative dominance of the negative and positive characteristics sets establishes the characters of the poles of a polarity. Both polarity and duality are linked together through categories of relationality such as complementarity, identity, supplementality, indifference, resemblance, and others of a give-and-take mode. It is these categories of relationality that establish conflicts and force in the development of socio-natural strategies for internal transformation of any category. It must be emphasized that the elements of the ontological space exist as ontological categories that satisfy the principle of identity in the development of the theory of categorial conversion and the paradigms of thought. These ontological categories are transformed onto the epistemological space by cognitive mapping to create epistemological categories. Some extra definitions will be useful in addition to those already given at this point. This is to reinforce the foundational frame that helps to relate the vocabulary, grammar and the language in the theory of categorial conversion for its understanding and usage through the laws of reasoning in the fuzzy paradigm as applied to cognitive activities in the quantity-quality space with neutrality of time where subjective and objective of information structures play important role in knowing.

## 5.4 Essential Definitions and Explications in the Language of the Theory of Categorial Conversion

The effectiveness of the development of a theory, its communication and understanding among the readers depends in an essential way on the key vocabulary. These key elements of the vocabulary used in theoretical construct need clear definitions and explications that will help to distinguish them from their usage in the common language of general communication. The general conceptual nature of definitions and explications, and the critical roles that they play in the theory of knowledge have been discussed in [R3.7] [R3.8] [R12.18] [R12.22]. Our attention is now turned to definitions and explications of some of the essential concepts in the development of the theory of categorial conversion

### 5.4.1 Definitions, Similarities and Differences of Dualism and Duality in the Development of the Laws of Reasoning in the Theory of Categorial Conversion

Dualism and duality as they relate to tension and change have been discussed intensely in all branches of the knowledge-production process. Both of them relate and derive their meanings from the principle of opposites [R1.8][R1.25][R1.35] [R1.48] [R1.56][R1.57][R1.243][R1.246] Duality has entered into mathematics, sciences, economics, operations research and the theory of optimization with some differences in interpretations in relation to computable systems as well as self-organizing systems. Dualism has been restricted mostly to philosophy and religion, and some aspects of social science such as political science. In our current discussion, we shall limit ourselves to logical conditions of duality and dualism as they relate to the construction and understanding of qualitative mathematics and transformation, and quantitative mathematics and change. Both dualism and duality have common roots and are based on a dual that projects the essence of binary and opposites, where the opposites may be viewed, either as both opposites are mutually exclusive and collectively exhaustive regarding a particular phenomenon and a category, or, the opposites may be viewed as mutually non-exclusive and collectively exhaustive in logical derivatives regarding the same phenomenon.

When the opposites are viewed as mutually exclusive and collectively exhaustive, then the object or the phenomenon is characterized by two discrete entities of distinction with lack of relational connectivity. It is here that the epistemic law of excluded middle in epistemological space denies validity of logical contradiction as well as denies the ontological continuum and universal unity. When the opposites are viewed as mutually non-exclusive and collectively exhaustive, then the object or the phenomenon is characterized by a continuum of infinitely relative opposites which may be mutually indistinguishable with relational connectivity at the same points. The idea here is that every phenomenon or entity in the universal object set is characterized by internal opposites in relational continuum and unity. These opposites may be viewed in terms of negative (cost) or positive (benefit) characteristics which provide a generic identity to each element for categorial distinction. The relational character of the opposites is used to place a distinction and a similarity between dualism and duality as they apply to category formation intra-categorial conversion and inter-categorial conversion. From the viewpoint of knowing in qualitative dynamics, every ontological element contains negative and positive characteristics sets in relative quantities that define and retain its identity in both ontological and epistemic spaces. The epistemological characteristics set may deviate from the ontological characteristics set as viewed in terms of the information content. An element is said to have been transformed if its relative quantitative negative-positive characteristics sets alter in favor of one opposite dual to bring in new qualitative identity that places the element in a different category. The transformation is said to be internal if the qualitative change is induced by the internal conflicts between the opposite powers that generate force of qualitative motion. The concepts of polarity, duality, relational unity and continuum are essential in the construct of the theory of categorial conversion and the applications that may be required of it. The description and the study of behavior of the elements are based on characteristics and characteristics sets. The needed paradigm and the corresponding laws of thought and mathematics for reasoning must accommodate the presence of duality, contradictions due to conflicts, continuum and relational unity. To explain ontological transformations the paradigm must be equipped to simultaneous accommodate qualitative and quantitative dispositions in the ontological space. It must also be able to deal with discrete and continuous process, as well as phenomena of exactness and inexactness.

## 5.4.2 Concerning General, Negative and Positive Characteristic Sets

From the *universal objective set*, there is the *universal characteristic* set that can be partitioned to specify and define universal entities and the categories of their belonging. The universal characteristic set embraces duality specified in terms of a *positive characteristic set* and a *negative characteristic set*. Individual objects in the universal object set carry this property of duality and hence every element in the universe is defined by a particular characteristics set which is composed of negative and positive characteristics sets in relational unity and continuum. The universe is made up of *universal duality* and *particular duality*. The universal duality finds expressions in universal polarity and exists in universal continuum and universal relational unity. The *particular duality* finds expressions in a particular polarity which also exist in a particular relational continuum and particular relational unity. The universal duality is a family of categories of particular duality. This discussion points to the digital-analog relational structure and discrete-continuous relational structure of the universe. The universe is under the continuum principle that allows a relational unity to be established at all-time points and periods. The analog is an enveloping of the digital while the digital is an approximation of the points in the analog states in the epistemological space. Similarly, the continuous is an enveloping of the discrete while the discrete project approximation points of the continuous in the epistemological space in relation to the ontological space.

### Definition 5.4.2.1: Characteristic Set

The set, $\mathbb{X}$ is said to be a universal characteristic set if there are subsets, $\mathbb{X}_{\mathbb{C}} \subset \mathbb{X}$ that may be assigned to any of the elements in the universal object set, $\Omega$, for identity and distinction of category $\mathbb{C}$ to which identical elements belong. The characteristic set, $\mathbb{X}_{\mathbb{C}}$ is said to be category-specific and defines the essential attributes of the classification of the category, $\mathbb{C}$.

### Note 5.4.2.1

As has been discussed, the characteristic set of any element in any category is made up of negative and positive characteristic sets. The relational structure of the positive and negative sets provides the character of the category. In this respect, we must have working definitions of both the positive and negative characteristic sets.

### Definition 5.4.2.2: Positive Characteristic Set and Positive Action

The positive characteristics set is composed of the elements whose activities seek to change the actual category in a direction which is healthy to them. The activities of the positive characteristics set produce positive actions for socio-natural categorial conversions composed of intra-categorial and inter-categorial conversions. The positive action in the categorial conversion is a fuzzy aggregation of positive forces of the individual elements of the characteristics set directed to destroy the actual and bring in a new actual from the potential. Positive action is always revolutionary seeking to alter the conditions of the actual while creating conditions for the potential.

## Definition 5.4.2.3: Negative Characteristic Set and Negative Action

The negative characteristics set is composed of the elements whose activities seek to maintain or reverse the actual category in a direction which is healthy to them. The activities of a negative characteristic set produce negative actions for socio-natural categorial conversions composed of intra-categorial and inter-categorial conversions. The negative action in the categorial conversion is a fuzzy aggregation of negative forces of the individual elements of the negative characteristics set directed to maintain or reverse the actual to the old order. Negative action is always reactionary seeking to create conditions that will maintain the actual or prevent the potential to emerge.

## 5.4.3 The Relativity of the Negative and Positive Characteristic Sets and Corresponding Actions in Categorial Conversion

It must be noted that both the positive and negative actions have differential expressions in social systems and natural systems. These differential expressions find meaning in self-excitement, self-organizing self-correction, and the manner and process in which these expressions are manifested in the socio-natural systems. Their understanding in relation to social systems is easier for cognitive agents. They are, however, not an easy task to understand in natural systems of animate and inanimate objects. It is, however, through the understanding of the nature of categorial conversions in natural systems that an improvement can be brought to self-exciting, self-organizing and self-correcting social systems by mimicking nature. The natural categorial conversion is a process to create a new element that belongs to a different category of nature by creating conditions that destroy the actual, as well as creating conditions that will allow the potential to become an actual. Similarly, the social categorial conversion is a process where natural categorial conversion is mimicked to change the existing order of socio-economic relations by creating the conditions of its destruction, as well as creating conditions that will be favorable for an element in the potential to become the new actual. The theory of categorial conversion claims that the processes of categorial conversion are the same for nature and society. In fact, it forms the foundations for the possible construction of a general theory of unified engineering sciences. It also offers us an important analytical and logical toolbox in understanding social transformation and the construction of general theory of socio-economic

207

development. It presents a framework in posing and solving the transformation problem. It is the foundation of understanding Nkrumah's conceptual system and socio-economic practices as applied to the African conditions. It is also claimed at this point that the framework of the theory of categorial conversion is the proper methodological approach in the construct of the theory of economic development where quality, quantity and time are the basic elements for transformation which must involve qualitative and quantitative equations of motion.

### 5.4.3.1 The Concepts of the Actual and Potential in the Theory of Categorial Conversion

The theory of categorial conversion starts its development from the notion that every element and every category reside in an actual-potential polarity. Every pole of a polarity is defined by a positive-negative duality. Every actual has corresponding characteristic sets through its duality and every potential also has a corresponding characteristics set through its duality. For the purpose of verifying positive action and its effects on actual-potential polarities and positive-negative dualities, the characteristics sets may be identified as actual, $\mathfrak{A}$ and potential, $\mathfrak{P}$. The positive and negative elements have already been identified. The actual is always a singleton set which will serve as the primary category in the transformation process. Its identity is revealed by the relational structure of the negative and positive characteristic subsets. The potential is a set that may be a finite or infinite set whose elements must be compared to the actual for initializing a transformation. The identity of each element in the potential is revealed by the internal relational structure of the negative and positive characteristic sets. An actual category is $\mathbb{C}_{\mathfrak{A}}$, an actual characteristic set is $\mathbb{X}_{\mathfrak{P}}$, a potential category is $\mathbb{C}_{\mathfrak{P}}$ and the potential characteristic set is $\mathbb{X}_{\mathfrak{P}}$. In the space of the actual, the characteristics set of each category may then be identified as $\mathbb{X}_{\mathbb{C}_{\mathfrak{A}}}$ while in the space of the potential, the characteristic set of each category may be identified as $\mathbb{X}_{\mathbb{C}_{\mathfrak{P}}}$.

Let us keep in mind that the actual element and the potential element have different characteristic sub-sets of the category to whom they belong. In this way, a completely qualitative transformation requires the establishment of differences between the identity of the actual and identity of the potential elements such that, $\mathbb{C}_p \neq \mathbb{C}_a$, and hence the potential category is not the same as the actual category. The relational structure of the actual-potential polarity and positive-negative duality with corresponding characteristic sets is shown as cognitive geometry in Figure 5.2.2.2.1 in terms of zonal analysis.

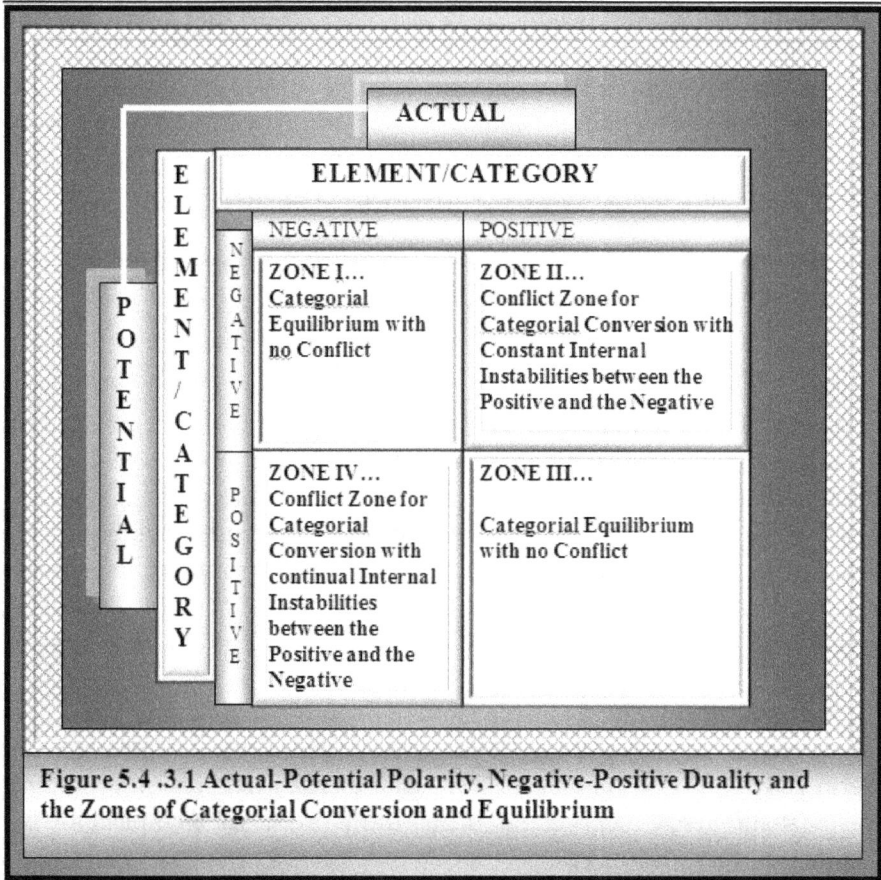

| | | ACTUAL | |
| --- | --- | --- | --- |
| | | ELEMENT/CATEGORY | |
| | | NEGATIVE | POSITIVE |
| | NEGATIVE | ZONE I... Categorial Equilibrium with no Conflict | ZONE II... Conflict Zone for Categorial Conversion with Constant Internal Instabilities between the Positive and the Negative |
| | POSITIVE | ZONE IV... Conflict Zone for Categorial Conversion with continual Internal Instabilities between the Positive and the Negative | ZONE III... Categorial Equilibrium with no Conflict |

*(Left margin labels: POTENTIAL, ELEMENT/CATEGORY)*

**Figure 5.4 .3.1 Actual-Potential Polarity, Negative-Positive Duality and the Zones of Categorial Conversion and Equilibrium**

The potential category emerges from the actual though the internal dynamics of the actual. In this way a potential becomes the actual and the actual becomes a potential through the actual-potential internal dynamics. The actual is under its own dynamics to become a potential that is to be derived from the actual. Every actual element serves as a primary category for an unknown potential. Similarly, every actual is a derivative from a previous primary category which has become a derived potential. The working mechanism in the potential-actual structure of the socio-natural process reveals itself as the disappearance of the old and the emergence of the new through the system's internal dynamics. It is under these conditions that the studies of self-organizing stochastic and self-organizing fuzzy-stochastic systems find important utility in understanding the theory of categorial conversion. This also holds for self-correction and self-exited systems. The results of these processes are the complex works of synergetics and energetics of matter, energy and information. At the level of cognition, the science and art of knowing is information that works through knowledge, wisdom, and intelligence in the activities of innovation, *datamatics*, analytics and decision. Linguistic representations and symbolism of the information of the characteristic sets project differential activities of these characteristic sets over the action space.

The development of the theory of positive action is to create a framework, techniques and methods to explain how through the internal dynamics of an element a qualitative development of an internal force is generated to move an element in $\mathbb{C}_p$ into $\mathbb{C}_a$, and an element in the actual into the potential. The whole process is simply the destruction of the actual and the creation of the potential. It is also *potentialization* of the actual and the *actualization* of the potential. In other words, from the space of the actual, the main concern involves the understanding of the dynamics of the behavior of the relative negative-positive characteristics sets to the destruction of the actual and the movement of it into the potential by altering its identity. The main concern in the space of the potential is the understanding of the dynamics of the behavior of how the relative negative-positive characteristic sub-sets become altered in favor of the potential and bring it into the space of the actual with a new identity. The process may be viewed as an identity transfer of a potential to actual and actual to potential. When these processes take place by the system's internal dynamics, we speak of *categorial conversion*. The theory that explains these internal dynamics is what is being developed here as *the theory of categorial conversion*. At the level of natural categorial conversion, examination of matter-energy production is to reflect the information content of natural actual-potential variables in order to understand how the natural potential is set against the natural actual. At the level of social categorial conversion, the knowledge-ideological production process is to examine the *information content* of the actual-potential variables in order to understand how to set the *social potential* against the *social actual*. From the space of ideology to the space of the actions, the main social concern

210

shifts to the examination of the knowledge that is relevant for the transforming variables which reveal the goals and constraints imbedded in national interests and social vision. In the ideological plane, the test of transformation is on the credibility of the information content in terms of quality and quantity, while in the action space, the test of conversion is on the knowledge content for the acceptance of social vision, national interest and the social goal-objective set in terms of categorial distance in the actual-potential polarity.

## Proposition 5.4.3.1: Existence of Categories

Every element belongs to a category and is defined by its opposites. Its identity is specified by a duality where every dual is specified by a non-empty negative characteristic set, $\mathbb{X}_C^N \neq \varnothing$, and a non-empty positive characteristic set, $\mathbb{X}_C^P \neq \varnothing$, such that its complete characteristics set is specified as, $\mathbb{X}_C = \left( \mathbb{X}_C^N \cup \mathbb{X}_C^P \right)$ with the condition that $\left( \# \mathbb{X}_C^N \lesseqgtr \# \mathbb{X}_C^P \right)$ in order to establish the identity of the element and its category in the ontological space. Its relative characteristic set is defined as $\rho_C = \left( {}^{\# \mathbb{X}_C^P} \big/ {}_{\# \mathbb{X}_C^N} \right) \gtreqless 1$. The potential and the actual present themselves as a polarity with every pole defined by its residing duality and every dual defined by its residing relative negative-positive characteristics set.

## Note: 5.4.3.2

As it has been previously pointed out, the negative and positive characteristics sets divide a unit into two opposites. They also unite them into relational unity. The negative and positive characteristics sets may be taken as discrete entities of attributes with nothing in common in order to define the qualitative essence of the elements and the category to which they belong. They may also be taken as residing in a continuum that expresses a smooth transition between the extremes to establish the identity of the elements through linkages. The linkages of the negative and positive characteristic sets are through relations that must be established, depending on the categorial qualitative disposition. The relative characteristics set defines the categorial quantitative disposition of each element and points to the home category. The relations are complementary, supplementary, give-and-take (reciprocity) and others such as asymmetry of force and governance. All types of relations fall

211

under the analytical concept of relationality which establish the continuum and unity of the opposites. The relations may change as time proceeds and transformations of characteristics take place to alter both the categorial qualitative and quantitative dispositions. The essential elements of thinking in the theory of categorial conversion are that both negative and positive characteristics interact in fulfilling and supplying something that both the negative and the positive characteristics lack, to ensure their mutual existence and the survival of the element in the category of their residence. For example, the existence of neo-colonialism is such that the neo-colonialist supplies something to maintain its status as an imperialist while the neo-colonial state supplies something to maintain its status as an occupied and a slave state.

It is this relationality that ensures the quality, identity and integrity of the elements for distinction from others in the relational space through informational acquaintance. The relationality of the positive and negative sub-sets of characteristics is essential in defining the concepts of dualism and duality as they relate to the paradigms of categorial conversion with the development of their respective internal logic and corresponding mathematics needed in the construct of revolutionary philosophy and ideology for social change as well as for the understanding of the dynamics of the emergence of the new and the disappearance of the old. Let us keep in focus that every element in the universal object set presents to cognitive agents both *real cost* and *real benefit* or negative and positive characteristic sets as seen in terms of harm and usefulness. At this point and from Figure 5.2.2.2.1, we may specify the actual and potential negative characteristic sets as $\mathbb{X}^N_{C_a}$ and $\mathbb{X}^N_{C_p}$ respectively, and the actual and potential positive characteristic sets as $\mathbb{X}^P_{C_a}$ and $\mathbb{X}^P_{C_p}$. Since the space of the actual is given, we shall concentrate on the space of the potential. The proposition of the existence of categories combines the essential elements of qualitative and quantitative disposition of the universal elements. Using the concepts of negative and positive characteristics sets, we have specified the analytical distinction and similarity between the concepts of dualism and duality. Dualism relates to separation of opposites through non-relationality while duality relates to unity of the opposites through relationality. This idea has been extended to distinguish the classical paradigm with an excluded middle from the fuzzy paradigm with a continuum. As has been argued the principle of excluded middle denies contradiction and the possibility of continual categorial conversion. In fact it denies transforming activities in the ignorance-knowledge polarity [R3.13].

## 5.5. Information, Knowledge, Wisdom and Intelligence in Systemicity

Given the basic elements of matter and energy, the theory of categorial conversion is developed from information flows and the internal order of the organization of the elements in the system's relational unity on the basis of which ontological elements acquire their identity for cognition. The information is the third basic element for universal relational unity. Information is the linkage between matter and energy where either matter is taken to be a primary category from which energy emerges as a derived category, or energy is taken as the primary category from which matter emerges as a derived category for the universal categorial conversion.

### 5.5.1 Concerning Information, Knowledge, Wisdom and Intelligence in the Theory of Categorial Conversion

As has been discussed, duality contains dualism as extreme cases in the grammar and language of categorial reasoning. Every dualism is, thus, a sub-set of duality which is an infinite set in the actual-potential space. The conceptual relationship between dualism with relational separation without relational unity and duality with relational continuum and unity leads to the statement that for every conceived socio-natural dualism there is a conceived covering in the socio-natural duality. However, not all transformations in duality have correspondence in dualism. Alternatively stated, the principle of continuum in the theory of categorial conversion suggests that for every activity in the actual-potential dualism there is a categorial covering in duality with the principle of relational continuum and unity. Viewed in terms of symbolic reasoning, if we construct any set $\mathbb{A}$ of the actual-potential transformation phenomenon on the principle of dualism, and then construct a set $\mathbb{B}$ of the same actual-potential transformation phenomenon on the principle of duality, then $\mathbb{A}$ is always contained in $\mathbb{B}$ $(\mathbb{A} \subset \mathbb{B})$, and in fact is just a point in $\mathbb{B}$. We shall speak of *categorial-conversion dualism* and *categorial-conversion duality*. As stated, the logic of reasoning in categorial duality is such that, every true actual has an

213

inseparable potential as its support and every potential has a true actual support. Translated in the logic of sets, every negative actual or potential characteristic set has a positive actual or potential characteristic set that mutually defines its individual identity in relational continuum and unity. These discussions relate to the continuum phenomenon in nature and society and internal dynamics for continual *transformation-substitution* phenomena and their relationship to socio-natural cost-benefit rationality as an integral part of the theory of categorial conversion. Here, every transformation has a cost-benefit configuration and every substitution has a cost-benefit configuration on the basis of which socio-natural transformations are formed.

In the universal system of transformations, the socio-natural categorial conversions are the works of information, knowledge, wisdom and intelligence. *Information* is abstracted by the thinking component of the system and stored for processing into a useful input called *knowledge* to be used by the categorial conversion unit. The effective use of this knowledge input to transform elements in the system depends on the system's *wisdom* in selecting a potential among the set of potentials in order to create cost-benefit comparison with the actual. The wisdom part of the system prepares the various parts of the system for the manufacturing of the categorial moment needed for categorial conversion. The *datamatics* is action process on quality of knowledge and wisdom while analytics is a cognitive process on quantity and quality of information and knowledge in the system. The system's intelligence acts to create innovations with datamatics and analytics to produce input refinements into socio-natural decision-choice systems in a relationally integrated organic super system. The results of the work of the decision-choice systems come out again as information which then rejuvenates the system of transformations in continual modes. The quality, direction and speed of the categorial conversion depend on the order of the system's internal arrangements. The integrated organic super structure is presented as a cognitive geometry in Figure 5.5.1.1. The relational structure of information, knowledge and representation has been discussed in [R1.92] [R3.7] R3.10] [R3.13].

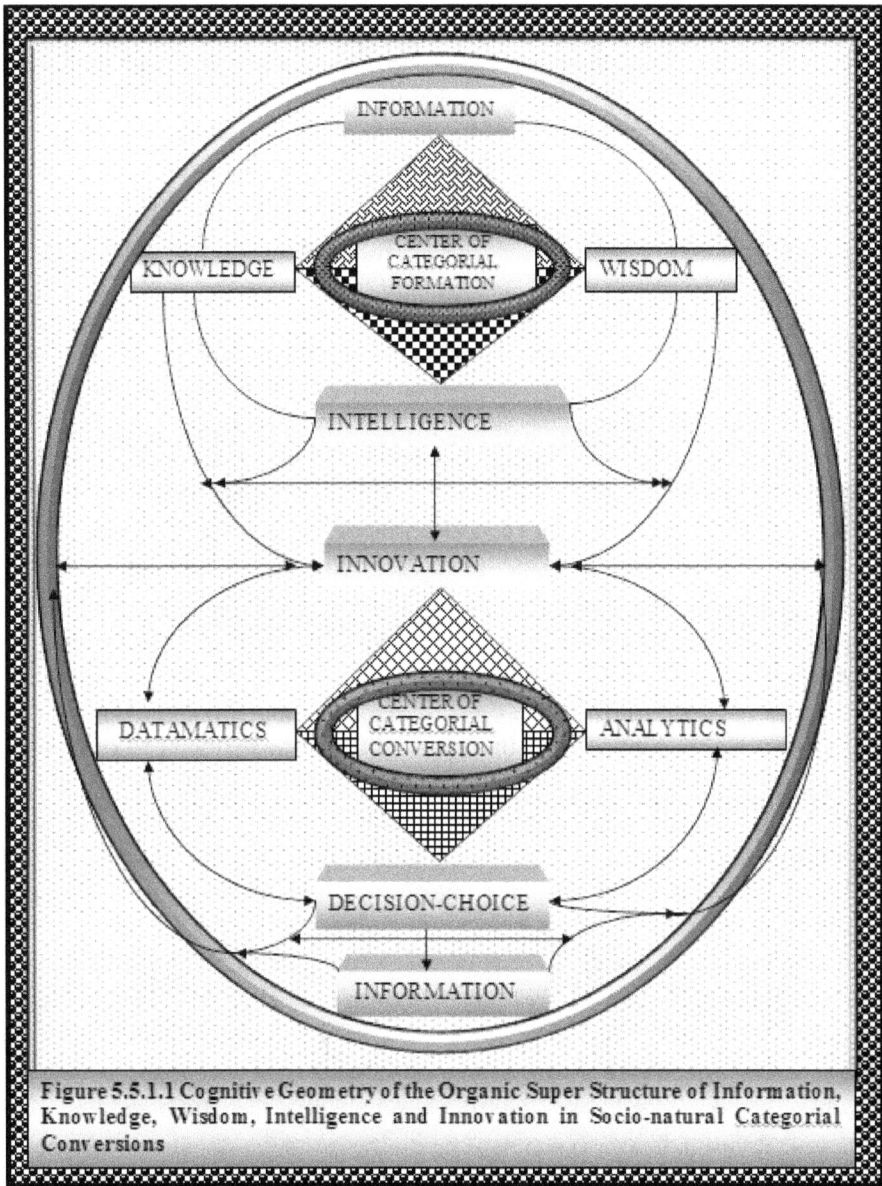

Figure 5.5.1.1 Cognitive Geometry of the Organic Super Structure of Information, Knowledge, Wisdom, Intelligence and Innovation in Socio-natural Categorial Conversions

## 5.5.1.1 Concerning the Continuum and Quantum in the Theory of Categorial Conversion

In the development of the theory of categorial conversion, all the opposites in the universal system are characterized as existing in continua. This is the principle of the universal continuum. The concept

215

of continuum needs some clarification and explication for the understanding of the universal behavior of categories and the paradigm associated with the theory. The concept of continuum is such that the categories have no excluded middle and hence they exist as overlapping categories. Each category is a polarity where the poles are defined by residing dualities which exist in a continuous series that blend into each other in such a way that the overlapping categories define an *oneness* of the universe where it is impossible to qualitatively separate the ends from the beginnings except by some subjective decision-choice action. The poles of the polarities which are continuous are linked by interlocking dualities which themselves are linked by interlocking negative and positive characteristics sets in contradictions under tension.

By the categorial continuum, every polarity goes through a gradual transformation from one qualitative state to another qualitative state without any changes. In other words categories qualitatively fade slowly into one another. The logical structure for viewing the universe in terms of the collection of overlapping categories is such that every negative characteristics set in a defined polarity has a supporting positive characteristics set and vice versa without which their respective identities are unrecognizable. It is just like the fuzzy-paradigm idea that every true proposition has a supporting false proposition, and hence, true or false is obtained by decision-choice action with fuzzy conditionality [R3.9] [R3.13]. The principle of continuum allows the justification of the quantum phenomenon that allows the qualitative linkages of categories to be specified in the interactive process of quantitative and qualitative dispositions where the qualitative linkages may be quantized. The quantization allows the claim that behind every qualitative disposition is a quantitative disposition in the quality-quantity duality. The quantization process allows the development of categorial relative negative-positive characteristic set which can be used to specify the *categorial transversality condition* between a category of actual and categories of the potential. The minimal relative negative-positive characteristic set for categorial transversality will be called *quantum relative characteristic set* in categorial conversion The quantum relative characteristic set defines a point of observable transition in the categorial conversion. It will vary from polarity to polarity.

216

## 5.5.1.2 Concerning Analog and Digital in the Theory of Categorial Conversion

From the concepts of continuum, quantum, and overlapping categories, it is held that the ontological universe is analog. The ontological space is infinitely infinite. It is continuous and infinitely self-contained in preservation of matter and energy but under a continual categorial conversion. It is self-correcting and self-transforming through infinite categorial conversions in terms of a quality-time phenomenon, where every qualitative disposition has a corresponding quantitative disposition and vice versa. The ontological universe is a creation outside of the human direct experience, but it incorporates humans as ontological elements. It is the attribute of the analog of the universe, as structured in overlapping categories that ensures continual categorial conversions including the disappearance of the old and the emergence of the new. It is this property that ensures the permanency of the universe and temporality of categories in the qualitative equilibrium-disequilibrium processes in the actual-potential polarities. Permanence is the ontological universe and temporalities are the ontological elements which exist in continuum of the actual-potential polarity in differential time structures, where the flow of time is permanent and acts as a neutral element.

The epistemological space is a creation of the cognitive process within the ontological space in an attempt by cognitive agents to know and understand the ontological elements that surround them as well as the environment in which they function. Time, energy, liquid, wind, air and information flows are ontologically continuous. These flows can be represented and studied with digital processes as epistemic approximations between and among intervals of qualitative-quantitative dispositions with continuously connected points in the epistemological space. Importantly, a minute, an hour, a day, a month and a year are human calibrations to understand complexities of time flows. Similarly, a unit volume, a unit length, unit energy and many others must be viewed in the same epistemic structure, and must be related to the consumption-production duality and the phenomenon of exchange. The discreteness and digital are purely epistemic methods which allow the construction of epistemic reality but not ontological reality. The epistemological space that is either discretely or continuously represented is finitely infinite and under a continual expansion and revision depending on the epistemic successes of cognitive activities and behavior of cognitive agents with the epistemic elements. The differences and similarities of digital and analog are due to methodologies of knowing since the epistemological process may be viewed as residing in the discrete-continuum or digital-analog duality. In continuum, we find discreteness and vice versa and in analog we find digital and vice versa in the epistemological space. From the view point of the classical paradigm that has taken hold of our knowledge-production process, the only way to represent, describe, model and theorize about the ontological continuum is through the methods of discreteness and digital. It is useful to keep in mind that the ontological space is the identity, and the epistemological space is a derived identity whose structure is axiomatically established, such that the epistemological space is simply an epistemic model of the ontological space whose nature is affected by the methodology of knowing. The nature of such an epistemic model is qualitatively and quantitatively dynamic

## 5.5.2 Relationality, Polarity, Duality, Information, Matter and Energy

It is useful now to reflect on the relational structure and the continuum of matter, energy and information as they relate to polarity, duality and opposites. In the theory of categorial conversion, matter is taken as the primary category and energy is taken as a derived category with relational unity under the principle of relationality. Matter and energy are linked by information which assures the type of relational unity that may exist. As conceived energy resides in matter where the internal activities of matter generate energy, and the forces that give rise to qualitative motions.

## 5.5.2.1 Concerning Relationality in Polarity and Duality in the Theory of Categorial Conversion

The *relationality* establishes the channels, interpretation, meaning and response of either intra-communications or inter-communications within polarity and duality, among polarities and dualities and between the set of categories of polarity on one hand and the set of categories of duality on the other hand. Relationality is the essential element that links categorial forms together into a particular and universal unity in both ontological and epistemological spaces through information activities. Basically, categorial transformation is internal decision-choice action for any given information environment. Without relationality and channels of communication categorial conversion is impossible. Categories of events, states and processes are in relational unity with each other in continuum. Similarly, the positive and negative poles of polarity, the positive and negative duals of duality, as well as the positive and negative characteristic subsets of duality are in relational unity. This is consistent with the claim that the universe is analog but not digital. The nature and type of relationality and the particular and universal unity that they engender are defined and established by the *multiplicity of rhythmic behaviors* of stability, equilibrium and change in the quality-quantity duality and polarity with neutrality of time.

### 5.5.2.2 Concerning Dualization of Universal Elements and Poles and Polarity

Every universal element is dualized by its negative and positive characteristic subsets producing *elemental duality* with intra-relational conflicts in continuum and unity. The negative characteristic sub-set is the *negative dual* and the positive characteristic subset is the *positive dual*. Every duality is also dualized into *negative duality* and *positive duality* with inter-relational conflicts in continuum and unity. The negative duality is one in which the negative dual dominates the positive dual while the positive dual is one in which the positive dual dominates the negative dual in relationality and unity. Every *polarity* is dualized by negative and positive dualities into *negative pole* and *positive pole*. The negative pole is defined by the negative duality while the positive pole is defined by positive duality under rhythmic conflicts in relational continuum and unity.

### 5.5.2.3 Concerning the Multiplicity of Energy in the Theory of Categorial Conversion

Energy appears in a multiplicity of forms that define the internal characteristics of states and processes of categories of polarity and duality, given the primary category of matter for intra-categorial and inter-categorial conversions in disequilibria and equilibria in both ontological and epistemic spaces. The multiplicity of rhythms also establishes the characteristics of a complexity of communications through the categories of relationality in terms of signal production, reception, interpretation and response of categories of particular and universal forms. We must keep in mind that rhythms are generated by energy; in fact, rhythms are surrogates of energy within the primary category. The internal organization and arrangements of a multiplicity of rhythms establish and maintain the transformational equilibrium and disequilibrium of each categorial state in duality and polarity. Changes in the internal organization and arrangements of the multiplicity of rhythms define the intra-dynamics and inter-dynamics of the system, behavior of categories of duality and the direction of qualitative changes of the derived and primary categories of polarity in terms of *pyramidal logic* and the *universal trinity*.

## 5.5.2.4 Concerning the Conceptual Pyramid and the Trinity in the Theory of Categorial Conversion

The pyramidal points constitute a logical trinity and define the foundational points of epistemic constructs for analysis and synthesis in categorial conversion. The intersections of the dividing lines that connect the pyramidal points to the middle of each pyramidal side constitute the epicenter of cognition in the epistemological space, and the creative process in the ontological space based on polarity, duality, continuum, categories, relationality and internal qualitative transformations viewed in terms of changes of one category to another. The internal qualitative transformation takes place first as intra-categorial conversion and secondly as inter-categorial conversion. The pyramidal points also represent the matter, energy and information in the triangular structure, where the interactions in the unity at the epicenter define the forces of conversion on the basis of actions in intra-inter polarity and duality leading to the cognition of *what there is*, the ontic state that reveals the actual and its transformation, to *what ought to be*, the epistemic state that reveals the potential. The explanatory structure and prescriptive structure are locked in axioms, postulates, principles and laws in terms of laws of nature and laws of reasoning. The interactions of matter, energy and information define states, processes and their corresponding conditions and behavior in the quality-quantity duality under the principles of continuum and relational unity. The epistemic geometry of the path of reason is provided in Figure 5.3.1. Here, the categorial conversion is about negation of the actual and the negation of the potential. In other words, the theory of categorial conversion is about the construct of explanatory and prescriptive theories of continual negation of negation in the actual-potential polarity and duality. The logic of the actual reveals itself in the methodological constructionism while the logic of the potential reveals itself in the methodological reductionism.

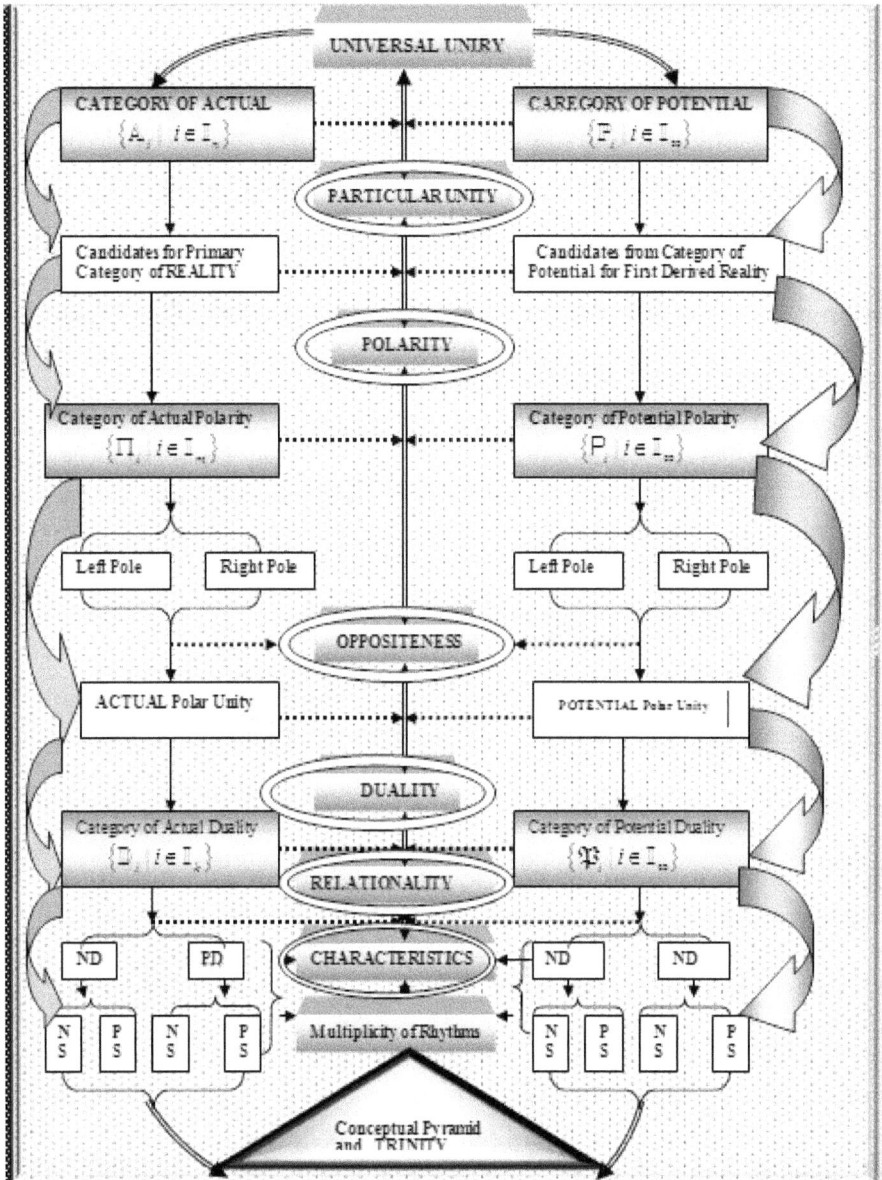

Figure 5.5.1 The Logical Geometry of Categorial Conversion
ND = Negative Duality, NP = Positive Duality, NS = Negative Set, PS =Positive Set. The sets are in the form $X = \{x_i \quad i \in I_n\}$, I = index set

The epistemic structure represented in Figure 5.3.1 for the construct and understanding of the theory of categorial conversion may be used in logical reasoning as an information processor given the information representation in the ontological space. It may also be used in logical reasoning as an information processor given the information representation in the epistemological space. When it is used in the ontological space, the required information structure with its representation is an ontological information structure. Similarly, when it is used in the epistemological space, the required information structure with its representation is an epistemological information structure as has also been presented in [R3.8] [R3.10]. As discussed, the ontological information structure constitutes the identity as well as the primary category of information. The epistemological information structure constitutes a derivative on the basis of the ontological identity and the primary category of knowing. Similarly, the relationship of the actual to methodological constructionism and the potential to methodological reductionism accepts the proposition that in transformations the potential is the primary category and the actual is the derived category.

## 5.6 Principles and Laws of Reasoning in the Theory of Categorial conversion in Systemicity

The logical construct of the reasoning leading to the discovery and explanation of *what there is* and its transformation into the potential, and *what ought to be* and its transformation into the actual, is such that the grand structures of discovery, explanation, prediction and prescription are supported by held principles, postulates and laws of evolution and change. The grand structures present foundations of all forms of human engineering as learning from the model of the natural order of creation. The theory of categorial conversion may also be viewed as a theory of socio-natural creation. These principles and postulates are grouped under categories and conceptual structures in order to establish the important logical system in the theory of categorial conversion required in the understanding of the Nkrumaist conceptual system in theory and practice concerning both social and natural transformations. The supporting laws of universal transformations for explaining states, changes and events are stated for reasoning and understanding of relationships between philosophy and ideology, between the actual (*what there is* ) and the potential (*what there is not*), between the possibility space, and possible-world space, and between positive-negative actions and the possible-world space. These laws of reasoning allow further understanding of global and specific enveloping of universal socio-natural events in the past-present-future dynamics and the possibility of different forms of actual and potential existence in other areas of the universe besides ours. The past-present-future is the *time trinity* which is presented as *Sankofa-Anoma* in an African philosophical thought system [R1.92].

## 5.6.1 General Africentric Postulates in the Theory of Categorial Conversion

There is a set of minimal postulates required to initialize the theory of categorial conversion for the understanding of qualitative changes, where a movement from one category to the other may be understood in terms of the relationships between the primary and derived categories within the potential-actual polarity and duality. Let us keep in mind that a movement of any element from one category to another is induced by changes in the qualitative disposition as well as changes in the relative positive-negative characteristics set of the category.

### 5.6.1.1 *The Postulate of the Existence of Infinity and the Transformation from Nothingness*

There exists a *nothing-something polarity* as the primary category of reality where matter was formed from nothingness by self-evolution and that all categories of reality are evolutions through the processes of categorial conversion. Both the nothingness and somethingness poles contain a somethingness-nothingness duality which in turn contains the negative and positive sets (Nothingness is the primary reality while somethingness is a derived category of reality of continuous transformation of categories of particular unity within the universal unity and through the internal dynamics of the nothingness-somethingness polarity working through the residing dualities and the positive-negative characteristic sets under the relational unity and continuum. Matter is the first derived category that becomes the primary category for other derived categories. It houses energy and information.

### 5.6.1.2. *The postulate of conversion through internal dynamics and self-evolution*

The universal conversion is a continuous and never ending process where the beginning and end appear as opposites in nothingness-somethingness (actual-potential) polarity linked together by energy and information. In this epistemic construct, nothingness is the actual and

223

somethingness is the potential. They are reversible under the principle of opposites. The conversion is through internal conflicts of the residing positive and negative characteristics sets that generate energy which then produces positive and negative actions for forces of self-conversion activities in the primary category of reality which then gives birth to sequences of derived categories of reality through energy and information sharing under relational unity and continuum. The system is closed under categorial conversions in the sense that nothingness resides in infinity which also contains the seed of somethingness, such that the discovery of its offspring presents a problem for human cognition and other cognitive agents if any. It is within the nothingness-somethingness polarity involving categorial conversions that the concepts of ontology, epistemology, ontic, epistemic, matter, energy, information, discrete and continuum become useful vocabularies in describing exact and inexact events within the actual-potential duality where the categorial conversion process is about qualitative changes brought about by *qualitative equations of motion* that affect the qualitative dispositions of elements of ontic states and processes.

### 5.6.1.3 *The Postulate of Cognitive Independence*

The existence of any reality is independent of human cognition and this independence covers all categories of reality including primary and derived categories that bring forth actual-potential transformations and the discovery of *what there is* and its categorial conversion into *what ought to be.*

### 5.6.1.4 *The Postulate of Actual-Potential Transformations and Substitutions*

The categorial conversion processes involve actual and potential elements. The actual and potential elements appear as opposites in inseparable positive-negative characteristics sets, duality and polarity in continuum and relational unity where there is a continuous transformation from the potential to the actual and the actual to the potential through their internal dynamics that are generated by the behavior of a multiplicity of internal and external rhythmic energies, as well as inducing partitions in the universal space through the interplay of information sharing in give-and-take processes where the actual is substituted for the potential and the potential for the actual in transformations.

## 5.6.2 General Africentic Axioms in the Theory of Categorial Conversion

These Africentric postulates are supported by a minimal set of Africentric axioms which together constitute part of the vocabulary and grammar of the theory categorial conversion that formed the foundations of Nkrumah's conceptual system.

### 5.6.2.1. *The Axiom of the existence of Nothingness-Somethingness Polarity as the Primary Category of Universal Reality*

At the beginning of creation (the beginning of beginnings) there existed the category of *nothingness-somethingness polarity* where at the beginning of time the *primary pole* residing in the nothingness-somethingness polarity was *nothingness.* Each pole contains the nothingness-somethingness duality. Each dual in duality contains the nothingness-somethingnes characteristics sets in differential relative proportions that give the respective identities of the duality and the polarity. Nothingness is the reality and somethingness is the potential from which matter, as something, came to be. In each pole there is a duality while in each dual there is the presence of negative-positive characteristics sub-sets. In this axiom of existence, negative is associated with nothingness and positive is associated with somethingness.

### 5.6.2.2 The *Axiom of Categorial Conversion and Existence of Matter*

Matter emerges from nothingness through the internal rhythmic *dynamical energy* of the nothingness-somethingness duality in the nothingness pole due to the conflicts in the nothingness-somethingness duality where the characteristics sub-set of somethingness quantitatively outweighs the characteristic subset of nothingness in the nothingness pole and transforms it into somethingness pole where the characteristic set of somethingness outweighs the characteristics set of nothingness through the residing duality with the corresponding negative and positive characteristic sub-set. This is the essence of categorial conversion of matter from nothingness to acquire the primary properties of somethingness.

## Note 5.4.2.1

The continual categorical conversion that projects a never-ending self-transformation within the universal unity and the constituent categories of particular unity of oneness of matter, energy and information form the analytical foundation of Africentric account of the universal creation in relational continuum and unity on the basis of which theology, philosophy and sciences developed . In other words, it forms the foundation of human knowledge and its accumulation. An important example of this Africentric conceptual foundation of continual categorical conversion of ontological actual-potential polarity is provided from the text of the Papyrus of Any [R1.56, p. xcix] [R1.67, pp. 35-123]

> I am he who evolved himself under the form of the god Khepera, I, the evolver of the evolutions evolved myself, the evolver of all evolutions, after many evolutions and developments which came forth from my mouth. (I developed myself from the primeval matter, which I had made).
>
> No heaven existed, and no earth, and no terrestrial animals or reptiles had come into being. I formed them out of the inert mass of watery matter, I found no place whereon to stand…. I was alone, and the gods Shu and Tefnut had not gone forth from me; there existed none other who worked with me. I laid the foundation of all things by my will; "all things evolved themselves therefrom." I united myself to my shadow, and I sent forth Shu and Tefnut out from myself; thus from being one god I became three, and Shu and Tefnut gave birth to Nut and Seb, and Nut gave birth to Osiris, Horus-Khent-an-maa, Sut, Isis, and Nephthys, at one birth, one after the other, and their children multiply upon this earth. [R1.56, p.xcix - c].

There are a number of key words and concepts in the above quotation that that have been underlined for emphasis and clarity as they relate to the phenomenon of categorical conversion in Africentric thought system. The logical strength of the nothingness-somethingness polarity in Africentric universal analytics of categorical conversion is such that nothing is external to our universe which is infinitely closed under transformational dynamics of categories. Here the Creative Force is internal but not external to the universal system. From nothingness-somethingness actual-potential polarity, the Creator *evolved* himself or herself and hence self-created. This process of self-creation is categorial conversion that is the result of the relational process of interplay of the Creative Force, Spirit Force and Light Force of many years of evolutions through the relational tension and contradiction of negative and positive characteristic sub-sets of the residing duality in the nothingness pole of the actual-potential polarity. In this account, one may associate the Creative Force with *matter*, the Spirit force with *energy* and the Light Force with *information*.

Keep in mind that each pole has a duality where each duality is composed of negative and positive characteristic sets as defined from the respective duals. The quantitative relativity of the negative and positive characteristics sets defines the qualitative disposition of each pole and the nature of the polarity as well as each dual and the nature of the duality. The duality of the nothingness pole contains the characteristic set of nothingness and the characteristic set of somethingness where the quantitative relativity favors the qualitative disposition of nothingness that gives rise to the nothingness pole. Similarly, the duality of the somethingness pole contains the characteristic set of nothingness and the characteristic set of somethingness, where the quantitative relativity favors the qualitative disposition of somethingness that gives rise to the somethingness pole. It is here that we find the creative interaction of the qualitative and quantitative dispositions where there exists a *quantum set of characteristics* that defines the identity of reality, and enveloping points of categorial conversions. The quantum relative characteristics set will be used to specify the *categorial transversality* conditions in order to ensure the existence of conditions of inter-categorial convertibility.

### 5.6.2.3. *The Axiom of Self-motion, Creation, Evolution and Transformations*

The Creative energy within nothingness which is self-created through internal self-dynamics that brings about qualitative self-evolution transforming nothingness into somethingness which gives rise to matter. All other categories of reality are derived therefrom and inherit this property of qualitative self-motion, evolution and dynamics through the behavior of a multiplicity of rhythmic energies, information structures and categorial moments not observed from the beginning to the end. In this way, matter inherits the property of qualitative self-motion and self-transformations that allow categorial conversions.

227

### 5.6.2.4. *The Axiom of Beginning-End Polarity, Actual-Potential Duality and Categorial Conversion*

The beginning resides in the end and the end resides in the beginning. They exist in an inseparable unity and mutually determine their identities and uniqueness. The primary category resides in the derived category and the derived category resides in the primary category as seen from the actual-potential duality under the principle of continuum, continual internal conflicts and relational unity.

## 5.7 Essential Postulates, Principles and the Theory of Categorial Conversion

The postulates which follow are associated with the development of the conceptual system of Nkrumah in his work to explain qualitative changes in a society and nature through the interplay of their internal forces. Some of these postulates are the author's abstraction to develop the logical continuum on the principles of polarity, duality, conflicts and internal qualitative motion inherent in elements that allow self-transformation from one qualitative state to another, or to change the category to which an element originally belongs. The development has also benefited from the works in [R1.92] [R3.7] [R3.10] [R3.13].

### 5.7.1 Essential Postulates in the Theory of Categorial Conversion

These are essential postulates and principles that support the general Africentric postulates and axioms needed to enhance the vocabulary and grammar of the theory of categorial conversion and the applications that may be required of it in prescriptive sciences.

#### 5.7.1.1 *The Postulate of Universal Partition*

**5.7.1.1(a)** The universe is completely partitioned into a family of sets of *fuzzy categories* of reality and potential with fading but interconnected boundaries with respect to states, processes and particular unity, where the primary category must be identified to initialize the self-transformation process. Each element is described by its degree of belonging to a particular category (This can rigorously be defined in a fuzzy mathematical form in terms of sets with grade functions of belonging. See [R4.17] [R4.37] [R4.38] [R4.48] [R4.41]).

**5.7.1.1(b)** The universal partition is made up of two components of the ontological partition that generates the ontological information structure about ontic states and processes. There is also the epistemological partition that generates the epistemological information structure about epistemic states and processes.

**5.7.1.1(c)** Every element exists in a temporary qualitative equilibrium state that is going through an internal qualitative process. Every element belongs to either a primary category or a derived category or both in accordance with the information structure.

**5.7.1.1(d)** Every information structure of an element is established by its qualitative characteristics which exist in dualistic sets. Every element in either the ontological or epistemological space is identified by its quantitative relativity of characteristics set that provides its qualitative disposition. The characteristics initialize the channels of possible knowing.

### 5.7.2 *The Postulate of the Existence of a Primary Category*

The set of categories of reality has a *primary category of reality* from which all other categories of reality evolve themselves through internal transformations into sequences of *derived categories of reality* composed of matter and non-matter. There is the ontological primary category for ontological transformations and there is the epistemological category that initializes the knowing process (see the exposition in [R1.92] [R7] [R3.13]).

### Note 5.7.2

We may view the process in terms of the primary category of reality that is to be converted into derived potential, and the primary category of the potential that is to be converted into derived actual under appropriate conditions of convertibility and *categorial transfer function* that must overcome the negative impulse function. There are relationships among the ontology, actual, epistemology and potential. These are linked together by matter, energy and information in the categorial conversion processes.

229

### 5.7.3. *The Postulate of Non-sole Primary Reality of Matter*

Matter is not the sole primary category of reality as it exists in a category of inseparable duality for categorial conversion. Energy is a candidate for the primary category where information reveals the distinction energy and matter. Furthermore, in the categorial conversion process, every derived category also serves as a primary category for its successor. Every derived category is both a parent and a successor.

### 5.7.4 *The Postulate of General Categorial Conversion*

Every category of reality, whether primary or derived, is under internal forces for self-transformation from one category to another through the internal actions of the negative and positive characteristics sets. The organic path of transformations is the path of conversion and development as well as an enveloping of the derived categories of reality, with the primary category constituting the initial point of self-transformations. This is the process of *categorial conversion* through construction-destruction processes or transformation-substitution processes of polarity, duality, continuum and relational unity in the actual-potential space.

### Definition 5.7.4.1: Categorial Conversion

Categorial conversion is an internal process where an element in one category of either ontological or epistemological nature emerges from another category by internal qualitative transformation, where the process is governed by qualitative motion and a transfer function which is induced by energy generated by the internal conflicts of the opposing characteristics which define the identity of the element at all transitional points.

### Note 5.7.4.1

As defined, every elemental actual emerges from a category of a potential and every elemental potential may be actualized as a current reality by an internal process. The history of any qualitative particle is the enveloping of categorial conversion. To explain the process, we must locate the energy and the forces that generate it through the information that is provided by the characteristics set of the particular element and the category. This is easy to understand if we view the universe as infinitely closed and every qualitative transformation is internally induced by internally generated energies which are produced by internal conflicts. The internal conflicts are produced by internally opposing rhythms from the activities of the residing negative and positive characteristics sets. In general, we have the ontological categorial conversion and the epistemological categorial conversion which are linked by information and paradigm of thought. The categorial conversion acknowledges the existence of both qualitative disposition and quantitative disposition of elements and categories where identities of elements are defined by qualitative dispositions. The relative structure of quantitative disposition provides information on transient points of change, indifference, thresholds, reversibility and equilibrium conditions. An example of categorial conversion is the process where mind emerges from matter through the internal conflicts and forces of matter as provided by Nkrumah [R1.203], other examples are given in Akan-Dogon traditions and Nile Valley civilization where death emerges from life and life emerges from death through internal process [1.8][R1.35] [R1.36] [R1.60b] [R1.218a] [R1.218b] [R1.243][R1.248]. In this framework, every element has internal contradiction and conflicts which establish it as under plenum of forces where there is a continual competitive struggle between the duals for dominance.

**5.7.4(a):** *The Postulate of Categorial Conversion at the Level of Ontology*
The ontological conversion comes to us as an ontological categorial conversion that presents the ontological transformation in the ontological space where the ontological conversions relate to the ontological primary category, ontological derived categories and the conversion paths.

**5.7.4(b):** *The Postulate of Categorial Conversion at the Level of Epistemology*
At the level of epistemology, the categorial conversion comes to us as an epistemic categorial conversion of the epistemological information in the epistemological space where the epistemological conversions relate to the epistemological primary category, derived categories and the paths of the epistemological conversions which are established by the path of knowing the epistemic realities. Every epistemic reality is obtained by resolving the conflicts and contradictions in ignorance-knowledge

231

polarity where ignorance is the actual pole and knowledge is the potential pole in the actual-potential polarity, where the residing dualities are defined by paradigms of knowing.

### 5.7.4(c) *The Postulate of Categorial-Conversion Identity*

The ontological categorial conversion is the identity in the sense of *what there is,* the ontic state. The epistemological categorial conversion is a derived category of knowing of the elements of identity in the sense of an epistemic state.

### 5.7.4(d) *Concerning the Postulate of Categorial Transversality*

There is a critical point in the inter-categories beyond which an element is qualitatively transformed and enters a new category. There is a critical point of qualitative traversing called categorial transversality point, and the conditions of this qualitative traversing are the *categorial transversality conditions which are specified by the quantum relative characteristics set.*

### 5.7.5 *Africentric Principles of the Opposites of the Universal system*

In the Africentric conceptual system, every element in the universe is defined by its opposites which provide its qualitative identity and category of residence. These opposites are reflected by opposite characteristics sets as are revealed in the Dogon conceptual ideas of relational unity and conflicts in continuum, the Akan conceptual ideas of conflicts, relational unity and qualitative change in relational continuum within the *Adinkralogy* and *Adinkramatics*, as well as the Pharaonic and Dogon conceptual ideas of opposites, conflicts, tension and forces in continuum.

### 5.7.5.1. *The Principle of Duality and Unity:*

The existence of every elemental reality and potential always presents itself in oppositeness, polarities and dualities operating in a relational unity and continuum to define the nature and character of the poles in polarity.

### 5.7.5.2. *The Principle of Polarity and Unity of the Poles*

Duality in an inseparable unit resides in every pole of polarity where the dual and the poles are seen in a relational oppositeness and united by a multiplicity of conflicts in rhythms, energy and information in a relational continuum. The characters of the residing dualities in poles are defined by relative negative-positive characteristic subsets.

### 5.7.5.3 *The Principle of Relationality and Multiplicity of Rhythms*

Relational unity is held together by the organizational arrangements of the multiplicity of rhythmic forces produced by the activities of the residing negative and positive characteristic sets in the constituent opposites to generate tension and energy flows for internal qualitative motion for intra-categorial and inter-categorial conversions.

### 5.7.6 The Law of Motion, Conversion and Transformations

These principles and postulates in the theory of categorial conversion are supported by a law of motion that governs qualitative transformations, substitution, categorial conversions and evolution of kinds due to the behavior of the multiplicity of the internal and external rhythms induced by energy to generate conflicts. The law of motion is composed of categories of:

### 5.7.6(a) The Law of Interdependence of Dual and Poles in Duality and Polarity.

The poles of polarity, the duals of the dualities and the negative and positive characteristics sub-sets exist in mutual interdependence in an essential way and yet they retain their individual identities for recognition and activities.

### 5.7.6(b) The Law of Interdependence of Mutual Negation of Poles and Duals

Each pole seeks to preserve itself by negating the characteristics of the other pole through the changes in the behavior of its residing duality for dominance.

### 5.7.6(c) The Law of Quantitative and Qualitative motion:

Each element or category is identified by its characteristically qualitative disposition in positive and negative relational structure. The qualitative characteristics of each duality in each pole are under continual qualitative and quantitative motions that bring about categorial differences of the poles and polarities as well as the duals and the duality. The qualitative motion is seen in transformation in the residing dualities while the

233

quantitative motion is seen in terms of shifts of the measures of the relative characteristic sets.

### 5.7.6(d)   The Law of Rhythms

Rhythms appear in multiplicity of forms and their behaviors are the driving force of the relationality which appears as categories of interdependence, conflicts, cooperation, energy, give-and-take and the final motion to establish the path of categorial conversions in the primary-derived categories of dualities.

The stated principles, postulates, laws and axioms are combined together with the grand structures to design the science of thinking and the logic of reasoning in the theory of categorial conversion as an important framework to understand the revolutionary philosophy underlying Nkrumaism, its politico-legal practice and the possible extension that would allow collectively structured discussions to be undertaken on the road to discovery of revolutionary thinking and social knowledge (truth) in society and nature. Let us expand on some of these ideas and show how they operate in the logic of the theory of categorial conversion as a revolutionary Afrocentric philosophy and ideology as an aspect of the African thought system in understanding change as applied to society and human creative works. The axioms, postulates, principles and the laws of motion are presented in a relational geometry in Figure 5.5.6.1.

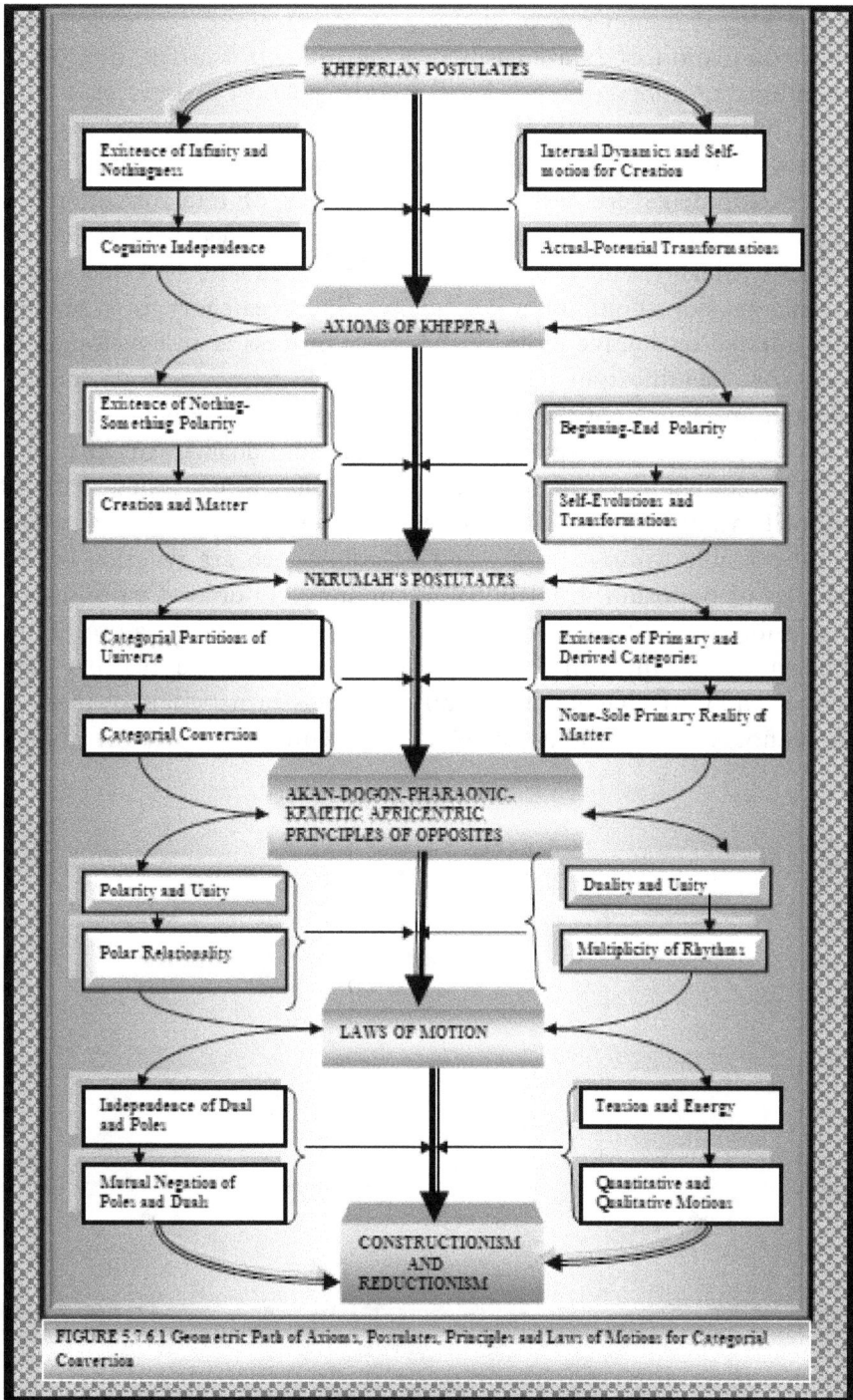

FIGURE 5.7.6.1 Geometric Path of Axioms, Postulates, Principles and Laws of Motion for Categorial Conversion

The logical framework of the theory of categorial conversion as a critical foundation of Nkrumaism may be used to examine the logical foundations of moral theory, legal theory, political theory, economic theory, theory of social formation, management, and others that are consistent with the Africentric theory of knowledge and ideology where their developments are constructed by the way of transformations of categories. Logically, every state is a category. Every category is defined by polarity, duality, and negative and positive characteristics sets under relational tension, continuum and unity. Every category is a process under internal qualitative motion, and every process is an enveloping of states for identification under quantitative motions and internal qualitative motion. Every state and every category are under a relational plenum of forces in the equilibrium-disequilibrium duality. Any change is not possible without energy. Energy cannot be generated without tension. Tension is not possible without conflict and contradiction which are not possible without duality and polarity which are revealed by the principles of opposites in relational continuum and unity. Opposites are not possible without categories which reveal the internal nature of the opposing forces through the internal behavior of the negative and positive family of characteristics sub-set. Transformation is qualitative, and cannot occur without actual and potential existence of things.

# CHAPTER SIX

## The Theory of Categorial Conversion:
## The Analytical Building Blocks

Axiomatic foundations initialize the development of the theory of categorial conversion. This is to help to establish the logic of revolutionary thought and action algorithms for revolutionary transformations. The axiomatic foundations provide us with some relevant information for epistemic processing with a defined paradigm of thought. It is an axiomatic information structure and not an empirical information structure. Both information structures are characterized by defective information structure which is made up of information vagueness and information limitation. The essential elements of these information structures have been discussed in [R1.92] [R3.7]. Given the axiomatic foundations of the theory of categorial conversion, viewed in terms of boundaries of reasoning, there is a need to specify the analytical building blocks of the theory. These building blocks are given further expansion, analytical clarity and the explicit role they play in the development of the theory of categorial conversion under the applications of fuzzy paradigm with its laws of thought and mathematics. By specifying the information input as defective containing vagueness, the information input is placed in an inexact space, thus the classical paradigm with exact information structure is inappropriate. The fuzzy paradigm is one that offers the needed analytical processing tools of logic and mathematics.

## 6.1 The General Conceptual Structure

Categorial conversion for the understanding of internal qualitative changes in the conceptual system of Nkrumaism are category formation and categorial moment. The development begins with the conception that the universal system of states and processes is classified into fuzzy categories (overlapping categories) of actual and potential existences that satisfy the postulate of category formation in both the ontological and epistemological spaces. These categories constitute the diversity in the global unity satisfying the postulate of universal unity. Each category is fuzzily unique satisfying the postulate identity and uniqueness. The universe is one and is classified as analog rather than digital [R3.13]. Every process is seen as an enveloping transformation of sequential forms of a category as well as a path of categorial states. The sequential transformation forms are made up of interconnected digital qualitative states where each digital state may be viewed as a primary or derived category. The number of categories is technically infinite since transformations, as induced by the creative force of energy and information, are continuous without end. Each process defines a set of categories that are fuzzy partitions with connectedness through a fading process for a defined continuum for the emergence of the new and the disappearance of the old. Within the set of categories, there is a *primary category* of reality that must be identified to distinguish it from the rest which are derived categories of reality from the primary category. The primary category of reality is taken as the actual and the identity. All other categories are considered as potentials that can be actualized to become derived categories of reality through internal processes of categorial conversion in the actual-potential polarity and duality under the principles of continuum and relational unity.

Each category is distinguished by its qualitative and quantitative characteristics that are established by grades of belonging (for example, shades, grades of color, or the degrees of wellness) where categories may fade into each other by an internal process. Viewed in this logical descriptive structure, every dual of duality is identified by its relative negative-positive characteristic structure that establishes its grade of belonging and the character of the duality and by a logical extension, the character of the polarity. Thus, given a process, there is a primary category of reality from which a first, second, third and other categories of reality are internally derived. The set of categories of reality may be arranged in a hierarchical order of transformations by means of logical constructionism as illustrated in Figure 6.1.1. In the figure we initialize the process of explanation by the primary category of reality $\{\bullet = \Omega\}$ which may be specified to contain nothingness in which the energy-mater duality resides. Here, the energy may be equated with nothingness and matter may be equated with somethingness in the nothingness-somethingnes polarity. There is one thing which is a fundamental law of the categorial conversions. Nothing is lost in all categorial conversions except a change of categories where matter and energy are retained in different categories of being. An alternative diagrammatic representation

238

of the categorial conversion process in the knowledge production within the ignorance-knowledge polarity is presented in [R3.13].

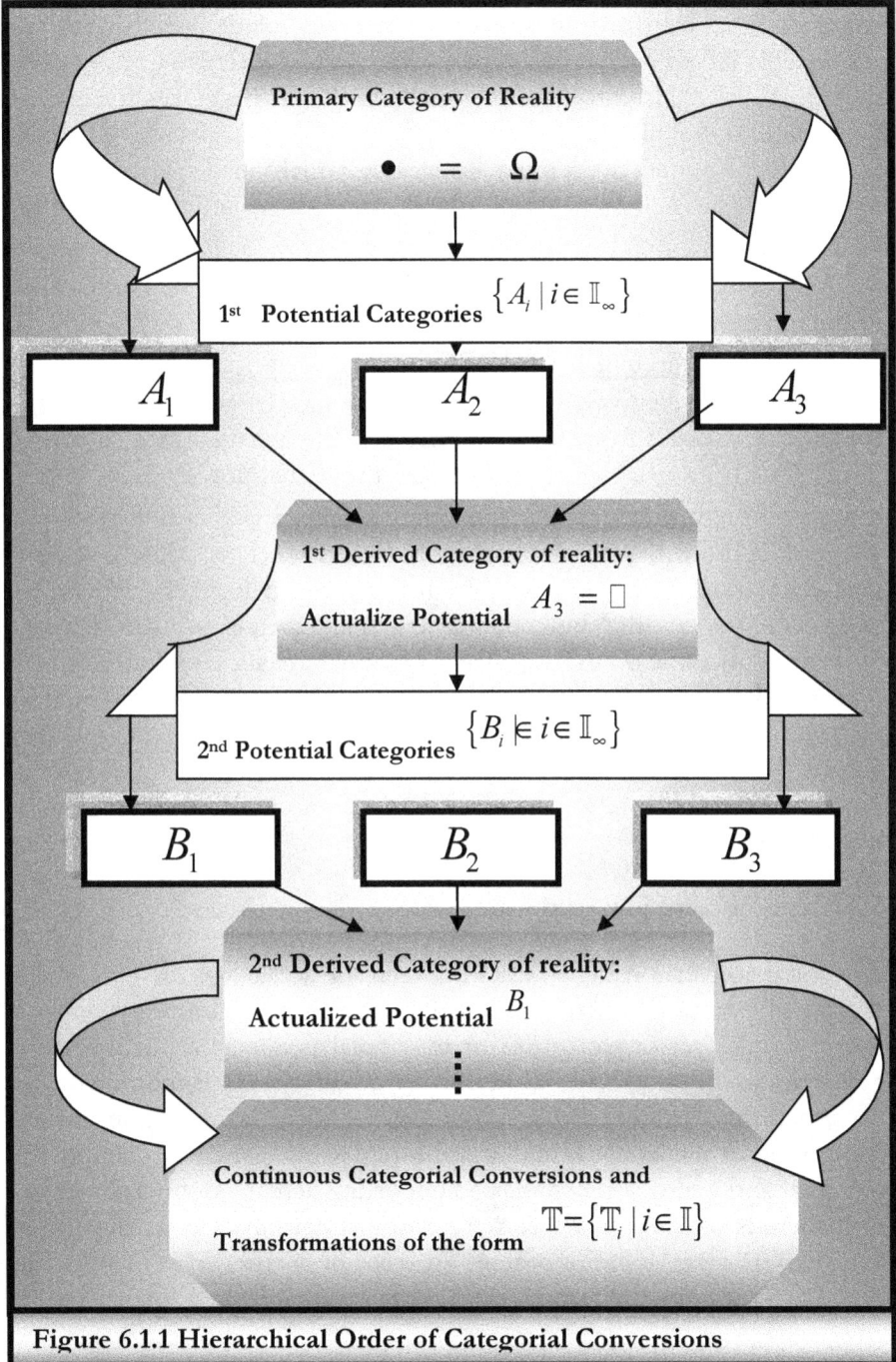

**Figure 6.1.1 Hierarchical Order of Categorial Conversions**

In the Nkrumaist conceptual system that we are trying to reconstruct with philosophical and mathematical extensions, we have the *energy force* and *material force* in unity to produce energy-matter polarity and duality in continua. The elements in the energy-matter polarity and duality are relationally linked by an *information force*. The interactive processes of these three forces lead to the categorial conversion processes for internal qualitative self-evolution. This is followed by the first level set of potential categories of derived categories which may be symbolically represented as $\mathbb{A} = \{A_i \mid i \in \mathbb{I}_\infty\}$ with an infinite index set $\mathbb{I}_\infty$ from which one $\{A_3\} = \mathbb{T}_1$ is actualized by a categorial-conversion process to be a first derived category of reality. This is followed by a second level set of potential categories of the form $\mathbb{B} = \{B_i \mid i \in \mathbb{I}_\infty\}$, also with an infinite index set, $\mathbb{I}_\infty$ from which $\{B_1\} = \mathbb{T}_2$ is actualized by self-evolution as a second derived category of reality by a process. The set of categories of reality as actualized from the potential may be specified as a set of the form $\mathbb{T} = \{\mathbb{T}_i \mid i \in \mathbb{I}_i\}$ which constitutes the actual-potential enveloping process through time. The $\mathbb{T}_i$'s are the actualized elements that may or may not be known to humans since their existence as *what there is* and *what would be* are independent of human cognition from the Africentric axioms. How are the potential and actual categories identified and distinguished? How do the actual and potential dualities and polarities get seen and described? In other words, what are the information structures that are available for a logical processing in terms of explanatory and prescriptive sciences? The reader may go back to chapter 4 and refer to Figures 4.2.2.1 and 4.2.2.2 in an illustrative mode of constructionism-reductionism duality.

## 6.2 Characteristic Sets and Categories

Categorial conversion is a conflict between the actual and the potential in the actual potential polarity. From the actual-potential interactions and processes, the set of categories of the actual is contained in the set of categories of the potentials while the set of categories of the potential is contained in the set of categories of reality in the time infinity composing of the past, present and future. Every actual has a supporting potential and every potential has a supporting actual. At any moment of time there is one actual whose perception formation of cognitive agents may differ from one another. To see how the logic works, let $\mathbb{A} = \{\mathbb{A}_i \mid i \in \mathbb{I}\}$ be the family of *actualized categories* (that is, the categories of reality) with a finite index set $\mathbb{I}$, and $\mathbb{B} = \{\mathbb{B}_i \mid i \in \mathbb{I}_\infty\}$ as the family of all potential categories with an infinite index set $\mathbb{I}_\infty$, given the primary category of reality. It is easily observed that at any moment of time, the set of the family of actual categories of reality is contained in the set of the family of potential categories to be actualized, that is $\left(\mathbb{A} \subset \mathbb{B}\right)$ and $\left(\mathbb{I} \subset \mathbb{I}_\infty\right)$ as viewed either in the ontological space or epistemological space. It has been explained in the previous chapters that reality expresses itself in the two spaces as the ontological reality and the epistemic reality. The ontological reality is infinite while the epistemic reality is finite. At infinity, however, it is the case that $\left(\mathbb{A} \subseteq \mathbb{B} \text{ and } \mathbb{B} \subseteq \mathbb{A}\right)$ and $\left(\mathbb{I} \subseteq \mathbb{I}_\infty \subseteq \mathbb{I}\right)$. It must also be noticed that the actual category of reality $\mathbb{A}_i$ exists independently of the knowledge of human mind and cognition. The actual category of reality can only be known through epistemic activities of cognitive agents using a toolbox of a paradigm of thought in the epistemological space. In each case, the ontological reality is the identity and primary category while the epistemic category is a derivative by cognitive action.

Categories are general classifications imposed on the universal system of things, states and processes for identity, similarity, distinction, analysis and synthesis for knowledge acquisition through the understanding of universal qualitative and quantitative motions with intra and inter-categorial dynamics for explanatory and prescriptive processes towards knowing. The process of knowing and understanding categorial conversion involves the application of the joint methodologies of nominalism, constructionism and reductionism as applied to categories in the universal ontological space under the organic principles of continuum and relational unity. But, the principle of universal continuum is vague with *fuzzy boundaries* that allow one category of potential to fade slowly into a category of actual reality and vice versa through a process. Each category is endowed with a family of sets of characteristics $\mathbb{Z}$, which establishes its existence, identity and behavior in a fuzzy logical plain with shades of belonging and dynamics of internal qualitative transformation. In other words, the universe is composed of a family of families of overlapping categories. Thus corresponding to each category $i$, we have $\mathbb{Z}_i = \{z_{ij} \mid i \in \mathbb{I} \text{ and } j \in \mathbb{J}\}$ where $\mathbb{I}$ is an index set of categories and $\mathbb{J}$ is an index set of the characteristics which present themselves as a multiplicity of rhythmic forces of quantity and quality in a local unity,

241

such that any of the qualitative characteristics has quantitative disposition and vice versa. The index sets may be finite or infinite depending on whether we are considering the categories of actual or potential. The family of actualized categories, the family of potential categories and the corresponding families of families of sets are essential building blocks in the construction of the theory of categorial conversion.

The category of ontological reality is *what there is,* the ontological actual, and the category in the process of making is *what would be,* the ontological potential. In the organic process of transformation, substitution and categorial dynamics, *what would be* fades into *what there is* and *what there is* fades into *what would be* through a process of internal self-transformation. In other words, a potential is transformed into the actual and the actual becomes a potential through the dynamics of categorial conversion defined in the ontological quantity-time space [R3.13]. Let us keep in mind the definition of categorial conversion as a qualitative transformation induced by the internal activities of elements. A question arises as to how do cognitive agents come to know *what there is* and the process of *what would be* through the logic of epistemic action in the theory of categorial conversion as a foundation of Nkrumaism and its application to socio-political dynamics and economic transformation. This is the *knowledge   acquisition process* or the process of the discovery of truth or the development of epistemic reality within the ignorance-knowledge polarity in the epistemological space [R3.10] [R3.13].

### 6.2.1 Ontology, Epistemology and Categorial Conversion

The knowledge acquisition process in understanding categorial conversion and to develop an explanatory or prescriptive theory proceeds by identifying two conceptual sets: A) *categories of reality* $\mathbb{T} = \{\mathbb{T}_{ij} \mid i \in \mathbb{I} \text{ and } j \in \mathbb{J}\}$ defined in the ontological space and B) a family of *categories of perception of reality* defined in the epistemological space. For each element in the set of primary and derived categories of reality there corresponds a family of categories of perception of reality in the form $\mathbb{P} = \{\mathbb{P}_{ij\ell} \mid i \in \mathbb{I}, j \in \mathbb{J} \text{ and } \ell \in \mathbb{L}\}$ where $\mathbb{L}$ is an index set of perceiving agents and $\mathbb{I}$ and $\mathbb{J}$ are as defined above. The categories of perception are models produced by agencies of the senses of living and organic bodies, at least, from the view point of human cognition and learning. These categories of perception come to cognitive agents as experiential information structures that define the space of epistemic action. Categories of primary and derived reality send information signals through their characteristics that reveal the degree of their existence in accordance with the accumulated information set to cognitive agents. The information set is then processed into categories of individual or collective perception as models of reality. In other words, there is a general set of information signals from which a knowledge base is humanly constructed by a logical process in accordance with an individual or collective model of perception. The knowing involves the perception of states as categories and the perception of behaviors of the categories in terms of qualitative and quantitative dispositions through which conversions take place under the principle of neutrality of time.

In this logical system of reasoning, it is important to note that the ontological reality is independent of human cognition while perception as a model of reality is dependent on human social characteristics, cognition and channels of knowledge acquisition in the epistemological space. There is also a space of realities independent of humans .This is the ontological space and the elements are ontological elements and categories with ontological categorial conversions. The descriptive elements in the ontological space present the *ontic reality*. There is also the perception space which is dependent on human cognition. This is the epistemological space and the elements are epistemological elements and categories with epistemological categorial conversions. The descriptive elements in the epistemological space present the *epistemic reality* which may be different from the ontic reality. The knowing process is such that the elements in the perception space $\mathbb{P}$ is then mapped onto the elements of the reality space, $\mathbb{T}$ through the connecting processes of decision-choice rationality as applied to the epistemological information. Some form of knowledge or *degree of truth* about reality is said to be acquired if there is an establishment of either degrees of identity, similarity or resemblance relation between the elements of $\mathbb{P}_{ij}$ and $\mathbb{T}_{ij}, i \in \mathbb{I}$ and $j \in \mathbb{J}$, by a perceiving agent $\ell \in \mathbb{L}$. The implication here is that the identity, similarity and resemblance relations are established by the *cognitive process* within the methodological constructionism-reductionism duality with relational continuum and unity. It is here that the development of theories and the test of theories acquire content and meaning over the epistemological space. This establishment is between the *characteristic set of reality* and the *characteristic set of perception*. Here, every transformation that brings into being a particular category of reality needs a logical explanation from within the structure of a particular set of laws of thought that provides a consistency in reasoning with epistemic transformation-substitution duality. It is here that the relational continuum and unity find analytical expressions in methodological constructionism and methodological reductionism over the epistemological space in forward-backward process of reasoning in knowledge construction.

## 6.2.2 Polarity, Duality, Continuum, Unity and Categorial Conversion

By the development of the logic of categorial conversion, each element belongs to a category and each category is composed of polarity and duality that exist in universal and specific unity with the presence of actual and potential poles. Each pole has a residing duality and each dual of duality is described by both negative and positive characteristic sets. The actual pole is distinguished by the fact that the magnitude of the positive characteristics set of the dual is greater than the negative characteristics set of the duality that resides in it. The potential pole on the other hand is distinguished likewise, where the magnitude of the negative characteristics set in the dual is greater than that of the positive characteristics set of the pole's duality. After identifying the primary and derived categories of reality, we then proceed to identify the poles and the dual in each category of reality for explanatory and prescriptive analyses and syntheses to learn and understand their relative qualitative and quantitative elements of the characteristic sets of the duality. Each polarity has a primary pole which is identified with the actual while the other pole is identified with the potential. Categorial conversion or transformation is impossible without the existence of actual-potential polarity under the principle of a system of dualities defined by negative and positive characteristic sub-sets under contradiction and conflict in relational continuum and unity

The poles are in relational tension induced by two different sets of plenum of internal forces which are generated by differential dualities. There is also a tension in each duality. The tension is induced by a multiplicity of rhythmic forces between the negative and positive characteristics sets. Both the polarity and duality exist in relational unity, in contradiction and conflicts, which allows them to interact and share information through giving-taking processes in a number of different ways. It is the nature of these contradictions and tensions that generate energy which then creates transfer function, impulse function and resistance function to produce a categorial moment for categorial conversions. The actual pole and the potential pole are identified by the nature, structure and relative strengths of the qualitative and quantitative magnitudes of positive and negative characteristics sets in the pole-specific duality. Each dual in a duality contains relative magnitudes of positive and negative characteristic sets that tend to distinguish the positive dual from the negative dual. Every dual contains sets of positive and negative characteristics in differential magnitudes that exist in a relational unity and mutual negation under the forces of a multiplicity of rhythmic energy which maintains *equilibrium* for *identity* and *disequilibrium* for conversions of categories and changes of qualitative states. The logic, which is the process of reasoning, locked in the theory of categorial conversion, requires us to analyze and synthesize the relations between the actual and potential poles and their connections to the negative and positive duals as defined by the relativity of the positive to the negative sub-sets of characteristics, in order to understand the categorial conversion that forms the basis of Nkrumah's conceptual system for philosophy and ideology for decolonization and development. Nkrumah provided the concepts of category, conversion, qualitative and quantitative dispositions. He placed the epistemic development of the connecting theoretical work on us. The development of this theoretical connection and framework is the work that is being undertaken in this monograph.

In every pole reside multiplicities of internal rhythmic forces through the duality in accordance with the nature of the relational contradictions and tensions of the behaviors of the negative and positive characteristics sub-sets. The nature of the organization of these internal rhythmic forces provides the pole with its properties which then allow the actual and potential poles to be distinguished in uniqueness and classification of belonging through the behavior of the duality. Here, uniqueness and classification of belonging are in grades or shades of representation and understanding due to overlapping categories. It has been pointed out that the actual is a singleton set at any time but the potential is a set of potential elements whose actualization and characteristics are many. There are intra-communications and inter-communications of polarity and duality which establish rhythmic exchange in terms of giving-taking processes in an interdependent relational mode. These communications take place through information and internal processing of information. Intra-communications of each pole or dual are the result of unique arrangements and rearrangements of a multiplicity of internal rhythmic forces that are generated by the respective negative and positive characteristics sub-sets.

The behavior of the organizational structure of the polyrhythmic forces provides temporary equilibria and disequilibria for internal stability, instabilities and relational unity of the duality through which the opposite rhythmic forces operate to generate energy in support of internal qualitative motion and transformation of the polarity into another polarity. The intra-communications and conditions of the actual-potential relationality are the result of rhythmic behavior in each pole. The temporary equilibrium and categorial stability require the actual pole to develop either a *containment strategy* or *reverse transformative strategy* favorable to its dominance and existence. The reasoning process of the theory of categorial conversion is first to identify the category of polarity. This must be followed by an identification of the residing duality in each of the poles of the polarity, the tension and conflicts that they induce, and the internal energy that they produce to bring about a *categorial moment* for internal transformation and a change of state and category. The concepts of negative and positive characteristics sets and their analytical role must always be interpreted in relation to the category of interest. For example, in medical sciences, negative is good and positive is bad; in decolonization positive action is good and negative action is bad; and many other examples may be given. Care must therefore be taken in the specification of the relevant building blocks. The concepts of negative and positive are interchangeable where such interchangeability depends on the nature of actual-potential polarity and the energy-material medium of analytical and action concerns.

The process of cognition of categorial conversion requires us to establish the principle of identity of existence as well as necessary and sufficient conditions of presence of polarity and duality in the categories of reality. In this framework of reasoning, every position of equilibrium is a state under a category under categorial conversion, and hence every state is in temporary categorial equilibrium undergoing internal transformation of the actual and a substitution of a potential. There is no permanent categorial equilibrium and there is no permanent categorial state under the logic of categorial conversion as applied to simultaneity of qualitative and quantitative motions. Every position of categorial disequilibrium, on the other hand, is a process in transformation and every process in transformation is a categorial disequilibrium. The

246

thinking process, here, is to instruct us how to observe, analyze and synthesize the dynamic behaviors of categories of polarity and duality through the spectacle of the multiplicity of rhythmic forces generated by the internal contradictions and tensions of the negative and positive characteristics sets. The universe is thus characterized by a set of temporary qualitative equilibria of reality and permanent qualitative disequilibria of reality. At the temporary categorial equilibrium, the nature of organization of the multiplicity of rhythms allows us to distinguish between properties and categories of relationality and their impacts on transformational dynamics of categories of polarities as viewed from the theory of categorial conversion.

The positive characteristic set provides the actual existence of the pole while the negative characteristic set provides the potential existence of the pole through its relative negative-positive characters in the duality. Polarity appears as a unity between the actual and potential poles while the duality appears as a relational unity between the positive and negative sets of the defining characteristics; all of which must be identified, analyzed and synthesized. The *actual* pole is identified with *what there is*, while the *potential* pole is identified with *what would be*. The nature of relationality induced by the internal organization of polyrhythmic forces creates internal conflicts in the duality and in each pole. The conflicts are expressed in *degrees of intensity* which generate and produce degrees of energy forces as a necessary requirement to create a *categorial moment* for actual-potential negation and negation of negation. The categorial moment defines the conditions and foundation of intra-qualitative motion that seeks to establish degrees of conversion of categories or transformations and substitutions, by altering the nature and identity of the poles through the qualitative internal self-motion in the duality by reversals of relationships and properties.

The qualitative motion presents itself in two categories of change in terms of relations and changes in properties. Such changes come about through processes of grades or shades of property and relational reversals depending on the degrees of intensity of struggle of the polyrhythmic forces as exhibited by the negative and positive characteristics sets. The process of gradual alteration allows the logical establishment of internal motion of categories of polarities and dualities. A simple example of qualitative motion is decay, aging, chemical change, geomorphologic evolution or socioeconomic transformation of societies for which we shall deal with in conditions of application of the logic of

247

categorial conversion as applied to colonialism-decolonization polarity and neocolonialism-independence polarity. To define and establish the trajectory of categorial conversion, the primary categories of polarity and duality must be established and then abstracted from within the principle of transformation elements that allow a logical explanation of how a category of reality becomes derived from the primary category of reality through a substitution. The substitution principle is a logical extension of the principle of the construction-destruction process where the actual is destructed and forced into the potential space, and the potential is constructed into a new actual to be substituted in the place of the previous actual. In this respect, the primary category of polarity and duality must logically be shown to have the capacity and ability of internal self-motion in terms of changes of relation and properties. The thinking structure in the theory of categorial conversion demands that the relations and characteristics in duality and polarities be defined and established through the multiplicity of the system's internal rhythmic forces in relation to the external rhythmic forces, in order to bring about the conflicts associated with categorial stabilities (equilibria) and categorial instabilities (disequilibria) in relation to the relative negation-negation activities of the negative and positive characteristics sets.

Here, there is the quality-quantity duality where quality is always viewed as a quantitative disposition whose measure exists but unknown for the category of reality. The reasoning position of the logic of the theory of categorial conversion is that every identifiable category of reality is either a primary category of reality or a derived category of reality which is completely reducible to the primary category of reality by methodological reductionism. The cognitive deduction is such that first there is the primary category of reality, and then, there is the time-ordered sequence of derived categories of reality from the initial category of reality through categorial conversion. In the logic of the theory of categorial conversion, the transformations, substitutions and conversion processes are such that the primary category of reality of a polarity must be specified in terms of an actual-potential  polarity where the primary pole of reality is the primary category in which the actual-potential duality resides in accordance with the nature of the organization of the multiplicity of rhythmic forces that define the properties and relationality of the actual-potential elements for identity, stability, equilibrium, disequilibrium, conditions of conversion and self-motion in relation to qualitative disposition.

In this paradigm of reasoning, ontological reality is the unquestionable actual and the ontological derivative is the potential, so that reality-derivative duality also exists as actual-potential duality while reality-derivative polarity also exists as actual-potential polarity with overlapping categories. The process of categorial conversion or transformation-substitution activities from the reality pole to potential pole are induced by the categories of relationality between the set of the properties of the ontological reality and the set of properties of the ontological potential, where reality seeks to convert the potential into reality and the potential simultaneously seeks to convert reality into potential by gradual alteration of the characteristics of the duality in the state of temporary equilibrium. The process of change is a categorial conversion where categories of potential are transformed into categories of actual (reality) and in their places categories of actual are transformed into categories of potential through the internal organization and reorganization of the multiplicity of internal and external rhythmic forces that generate tension, energy, conditions of categorial moment and force of motion to affect the quantitative and qualitative characteristics of elements under categorial conversion.

The categorial conversion is the process of universal transformation and evolution in nature and society where the organic trajectory of categorial conversion is established as a set of universal realities of *what there is* for cognitive discovery over the epistemological space. The theory of categorial conversion views the universe as self-exiting, self-correcting and self-organizing in order to execute the program of internal self-motion and continual creation. The knowledge-accumulation process requires us to pattern our cognition and reason along this process. In fact, as it would later be shown, this logic forms the theory of unified engineering sciences. Social transformation and change also require us to follow these lines of thought for the understanding of developmental processes whether intentionally or unintentionally conceived. Social intentionality implies that the social categorial moment is not spontaneous but by the will of the people to create appropriate conditions which must be necessary for the categorial-conversion process. Non-intentionality implies that the categorial moment is undirected and may not be related to the will of the people. It may be pointed out that categorial conversion demands transformation and substitution without which change is impossible. The categorial conversion acts on the transformation-substitution possibility frontier of the creational process in order to change the potential by converting it into the actual and then the actual into the potential. Any external intervention may or may not work. The conditions for external intervention to work lie in the internal acceptability of the intervention by a pole of the polarity. Intervention works if it is internalized by a pole of the polarity. This seems to be the case for the practice of medical science where the treatment must be internalized by the patient's immune system or the practice of political domination where the policy of domination must be internalized by the dominated.

The system of reasoning induced by the logic of categorial conversion with the primary category of matter-energy polarity and duality is revealed within the conceptual system of Nkrumah for universal socio-natural transformations, and is derived from the Africentric philosophy of universal creation where nothing-something polarity and duality are the beginnings (for further historical and conceptual discussions on Africentric universal beginnings (see [R1.20] [R1.24] [R1.24] [R1.26] [R1.35] [R1.37] [R1.56] [R1.56][R1.67] [R1.92] [R1.83b] [R1.86][R1.79][R1.180[[R1.181]). The matter-energy polarity, derived from the nothingness-somthingness polarity becomes the primeval polarity while either matter or energy is taken as the primeval pole. The energy-matter duality is the primeval duality with defining characteristics that establish the positive sub-set and negative sub-sets of the duals. The concept of transformation must be broadly interpreted within Nkrumah's conceptual system. It is part of the Africentric theory of knowledge, philosophy of life, and human behavior. It is an approach to the understanding of the universal system of "*what there is*" and "*what would be*" for socio-natural transformations. First, one may take energy as the primary category of reality. The concept and reality of energy are abstractions that may be traced to the Nile-valley conceptual system of beginning and creation where "infinity", "nothingness", "primeval" and "somethingness" initialized the explanation and search for "what there is" and how "what there is" can be brought to human cognition [R1.20] [R1.26] [R1.180] [R1.53] [R1.56] [R1.67]. Again the Coffin Texts IV suggest that the "Universal Lord was simultaneously infinity, nothingness, inertness and invisible" [R1.57] [R1.67], where the universal Lord may be equated to energy. The concepts of Infinity, Nothingness, Nowhere-ness, Everywhere-ness and Darkness are repeated in the Coffin Texts IV which became adopted into Judeo-Christian theology and philosophy. It may be noted that without zero and infinity our modern mathematics in terms of Arabian numbers with base ten has no closure. These are some of the foundations of the Africentric claim of the existence of the primary category in Nkrumah's conceptual system.

The conceptual beginning defining the explanatory and prescriptive analytical and synthetic processes of transformation through the method of categorial conversion is that, in the category of energy resides the characteristics of infinity, inertness, invisibility and matter in unity. This unity is put together by an internal organization of a multiplicity of rhythmic forces which establish the character and internal arrangements of the characteristics of nothingness that give the unity its quality and identity where matter resides in energy and energy resides in matter. The property of invisibility or darkness projects matter that exists independently of human existence and cognition. The properties of infinity and inertness project nothingness that also exists independently of human existence and cognition. Both energy and matter are the defining characteristics of the primeval polarity and duality under internal organization of a multiplicity of rhythmic forces for categorial conversion. As it is presented, the existence of derived categories of reality constitutes the *explanandum* (that which is to be explained in the categorial conversion) while the behavior of the duality as induced by deferential organization due to the dualistic relationality constitutes the cognitive path of *explanans* (that which explains the categorial conversion) on the basis of conceptual understanding of arrangements and rearrangements of the multiplicity of internal rhythmic forces which establish equilibrium and disequilibrium processes in the equilibrium-disequilibrium duality for categorial conversions. Here, energy, matter and information become the inputs into the categorial conversion process. Information is taken as a linkage between matter and energy as shown in Figure 6.2.2.1. If matter is taken as the primary category of ontological reality and energy is considered as a derived category then it must be shown how energy is a derivative by methodological constructionism and how energy is reducible to matter by methodological reductionism.

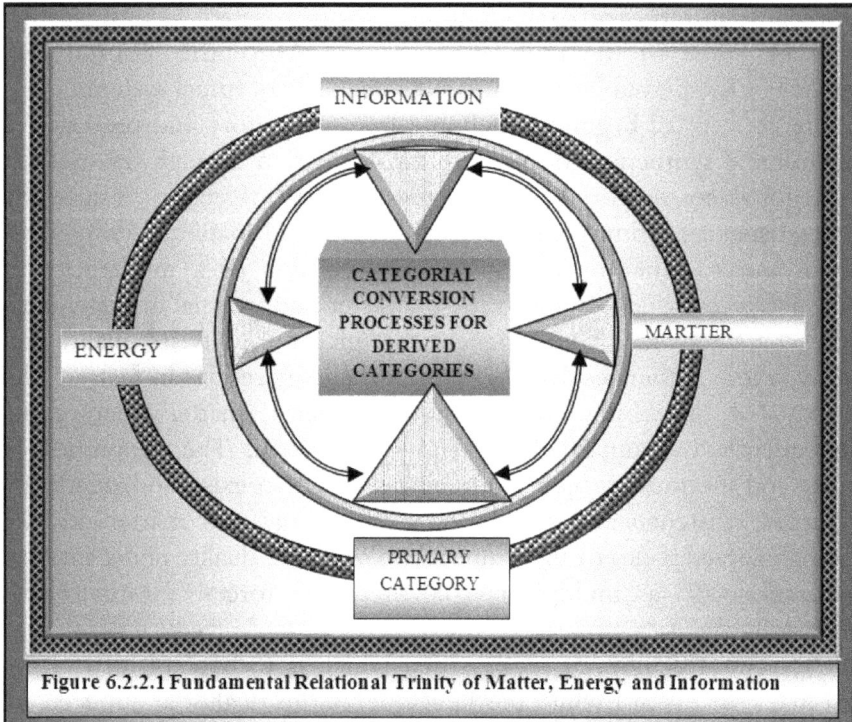

Figure 6.2.2.1 Fundamental Relational Trinity of Matter, Energy and Information

## 6.3 The Problem of Categorial Conversion Revisited

The problem that must be resolved in the theory of categorial conversion as well as in Nkrumah's conceptual system toward the categorial conversion is to identify the primary category of reality and then show how the derived categories of reality emerge out of the primary category through the forces of the categorial moment. This requires the identification of the primary polarity and the use of the methodology of constructionism. For example, in social formation, one may take freedom as belonging to the primary category of human rights and then show how by a process, either an oppression, slavery or colonialism emerges out of freedom to enter into the derived category of human subjugation. The first task is to identify the primary category of reality and to establish the primary category of reality as part of a derivative of the primary category of reality. If the identified primary category of reality is either part or derived from the identified primary category, then we conclude that it is self-caused and hence has capability of self-motion through the categorial-conversion process. Every other derived category of reality from the primary category will be endowed with this property of self-motion and self-conversion. It must be noted that an ontological category of reality is said to be primary if there exists no known process to reduce it to something else. The next task is to show that any derived category can be traced to the primary category by methodological reductionism. In the epistemological space, the choice of the primary category may vary among cognitive agents. This will affect the choice of methodology within the methodological constructionism-reductionism duality.

## 6.3.1 The Conceptual Solution to the Categorial Conversion Problem

The initialized conditions of knowing for the application of the categorial-conversion process of *what there is*, the actual, to *what would be*, the potential, are the conditions of energy, and since energy is part of energy it is self-caused. By the logic of categorial conversion, all other categories of reality must be shown to be derivable from or reducible to the primary category of reality through the internal changes of the organization of the multiplicity of the internal rhythmic forces. Matter is energy and energy is matter. The existence and changes of the external rhythmic forces merely define the conditions for change, while the existence and rearrangements of internal rhythmic dynamics are the true basis for categorial conversion. As stated, the impression of an external force cannot cause a categorial conversion except if it is internally accepted by at least one of the poles. Thus, categorial conversion projects internal destructive-constructive dynamics for the evolution that involves transformations and substitutions in the actual-potential polarity. Here, the mode of reasoning affirms the independent existence of the multiplicity of rhythmic forces from human existence and cognition. The behavior and organizational structure of the multiplicity of rhythmic forces generated by internal activities of the negative and positive characteristics sets allow cognitive agents an access to a reasoning path through which cognitive agents are able to state the *necessary* and *sufficient* conditions for categorial conversion for a forward-progressive logic of constructionism from the primary category of reality to the derived categories and backward-regressive logic of reductionism from the derived to the primary category of reality.

In the framework of the theory of categorial conversion and the Nkrumaist conceptual system, logical analysis and interpretational synthesis, derived categories are surrogates of the category of energy which is also matter, to which everything is reducible to without a residue, and where energy and matter reside in an inseparable unit. The primary categories of energy-matter polarity and duality exist with internal rhythmic forces under relational tension that provides the necessary and sufficient conditions for mutually continuous and gradual conversions of attributes of the internal characteristics of the positive and negative sets. The tension resulting from conflicts in the multiplicity of rhythmic forces provides continuous interactive flows of energy which gradually change the qualities and quantities of dispositions that establish equilibrium between the positive and negative characteristics sets in the respective duality of each pole in order to shift the balance that helps to define the identity of the category of reality and in favor of the potential in mutual transformations, substitutions and negation of negations in the actual-potential polarity. These changes are mutual, gradual and in degrees of positivity and negativity as specified in overlapping socio-natural categories. They proceed by altering the internal arrangements of the relative relationality of the negative and positive sets as established by the organizational arrangements of the multiplicity of rhythmic forces, until a breaking point is reached at which the actual category of reality fades into the potential, and then the existing potential acquires the dominating characteristics to fade into the actual by shedding off its obscurity to reveal its new category of reality. The breaking point is the *point of indifference* and the point of *categorial transversality conditions* as specified by the relative characteristics set is the point of categorial conversion. . The structural dynamics of the enveloping of categorial conversion is such that through the interplay of the multiplicity of rhythmic forces of categories of polarity and duality energy fades into matter through a positive categorial conversion and matter fades into energy through negative categorial conversion. The concepts of negative and positive transformations are used in the sense of the logic of duality but not in the sense of judgment. When an actual-potential polarity is transformed, a new actual-potential polarity emerges and a new struggle is ensued.

It must be emphasized that the necessity for change of the existing equilibrium of the duality or polarity is established by the nature of the changes of the organization of the multiplicity of the external rhythmic forces. The sufficient conditions and the basis for change are defined by the nature of alterations of the organization of the multiplicity of internal rhythmic forces which provide the categorial moment of incipient motion or transformation where qualities and mutual relations are altered. The nature of the gradual conversion and the degree of perception of reality and truth are such that the mathematical reasoning and the equations of qualitative motion governing the dynamics in the categorial relations and characteristics may be analyzed and synthesized by the use of the fuzzy paradigm composed of methods and techniques of mathematics and logic for fuzzy phenomena and approximate reasoning that are currently available to us [R2][R3] [R4] [R5][R6].

The current developments of fuzzy mathematics, technology and approximate reasoning have made it possible to establish the needed qualitative motion that allows us to trace the trajectory of the categorial conversion and its enveloping. They also allow us to define the problem of the forces of conflict between the negative and positive characteristic sets and to solve it to find the points of indifference [R3.13]. It may be noted that the methodological approach within the theory of categorial conversion and the Nkrumaist conceptual system, by accepting from the outset the category of energy-matter polarity and duality as primary elements of reality with the unity of internal and external organizations of the multiplicity of rhythmic forces which simultaneously maintain temporary equilibrium for identity and permanent disequilibrium for categorial conversion, makes it duly possible to develop the *convertibility conditions* to the *categorial moment* of the conversion of the initialized primary category of reality to derived categories of reality, and the reduction of derived categories of reality to the primary category of reality in the actual-potential space with neutrality of time. The *explanandum* of the need for transformation emerges from the outcomes of categorial conversions while the *explanans* is developed from categories of polarity, duality, negative-positive characteristics sets, rhythmic forces, relationality and unity of existence.

## 6.3.2 More Reflections on the Africentic Roots of the Logic in Categorial Conversion

The concept of energy-matter polarity, as a surrogate of nothingness-somethingness polarity, as well as actual-potential polarity and the concept of energy as the primary category of reality from which all things are derived, may seem to be paradoxical and a categorial absurdity but they are not. Similarly, when one examines the text on Africentric creation in the Papyrus of Ani (quoted below) one encounters something that identically seems to be paradoxical and logically absurd. The logical absurdity evaporates from the statement:

> I am he, who evolved himself under the form of the god Khepera, I, the evolver of the evolutions evolved myself, the evolver of all evolutions, after many evolutions and developments which came forth from my mouth. (I developed myself from primeval matter, which I had made).
>
> No heaven existed, and no earth, and no terrestrial animals or reptiles had come into being. I formed them out of the inert mass of watery matter, I found no place whereon to stand…I was alone, and the gods Shu and Tefnut had not gone forth from me; there existed none other who worked with me. I laid the foundation of all things by my will; All things evolved themselves therefrom. I united myself to my shadow, and I sent forth Shu and Tefnut out from myself; thus being one god I became three, and Shu and Tefnut gave birth to Nut and Seb, and Nut gave birth to Osiris, Horus-Khent-an-maa. Sut, Isis, and Nephthys, at one birth, one after the other, and their children multiplied upon this earth [R1.56, pp xcix-c] ([Note Tefnut=water, Shu= air, Geb= earth]).

The paradox of transformations and evolutions is resolved in Africentric philosophical traditions through the interplay of opposing patterns of the multiplicity of rhythmic forces and their organizational mobilizations, arrangements and behaviors. Nothing is lost in categorial conversions and nothing is gained in categorial conversions but, only transformations and substitutions of states, categories and processes in particular, and universal unity as projected in actual-potential polarity and duality with the supporting negative-positive characteristics sets. By the Africenteic principle of opposites, the universe is seen in terms of opposites composed of a families of polarities and dualities with negative-positive characteristics sets, all of which define overlapping categories in categorial equilibrium-disequilibrium processes in quantity-quality polarity.

From the above quotation we find that in unity are Khepera and primeval matter (the choice of the primary category) existing in a temporary categorial equilibrium. The temporary equilibrium of the unity state of Khepera and primeval matter is maintained by a unique organization of the multiplicity of rhythmic forces in accordance with the logic of the theory of categorial conversion. In terms of categorial conversions and creational evolution of things, energy is identified with the primary category, and matter is identified with the derived category within the actual-potential polarity. From the quoted text, Khepera is identified with energy which is self-created, energy is the primeval matter which was the basis of self-creation, and hence, by the principle of identity the primeval matter was self-created with self-motion through its internal dynamics. Matter is reducible to energy without a residue and energy is reducible to matter without a residue. Within the theory of categorial conversion, therefore, energy and primary matter exist in an inseparable unity. Just as energy and primeval matter are in an inseparable unity, so also are nothingness and somethingness as seen within the conceptual system from which the theory of categorial conversion is abstracted as a foundation of Nkrumaism. In this Africentric philosophical tradition, initial categories of polarity and duality are established for the abstracting methodological constructionism and reductionism to relate the primary category to the derived categories given nominalism. They are energy-matter polarity and duality which are identified and established through their respective relative negative and positive characteristics sets. The energy-matter polarity becomes the primary category of polarity and so also the energy-matter duality as the primary category of duality. Energy becomes the primary category of reality from which all other categories of reality are internally self-derived or evolved by processes, as indicated by *All things evolved themselves therefrom* in the quotation. The processes are through the internal dynamics of organization of rhythmic forces that define the elementary properties and the relational structure of the multiplicity of rhythmic forces and their behavior from the activities of the negative and positive characteristics sets. Nkrumah's conceptual system is thus derived from the Africentric conceptual system as he points out from the beginning that our philosophy must find its weapon from the African tradition if it is to be useful as a revolutionary philosophy to guide transformations. The development of the foundations and extensions of Nkrumaism rejects the unqualified statement of the association of Nkrumah's conceptual system with Marxism. There are similarities that one can point out, but the roots are fundamentally Africentric based on the principle of opposites with relational continuum and unity. It is this relational continuum and unity that demand the use of a fuzzy paradigm that accepts contradictions and simultaneous existence of true and false as a legitimate logical value different from the classical paradigm of thought.

257

As it stands, it does not make any analytical difference whether energy is identified with spirit, soul, mind, or primeval matter in accordance with our current linguistic systems, because the reasoning holds at all higher logical planes. Energy and matter are inseparable in the unit in which they exist. The primary and derived categories of reality with self-motion are fully indicated in the Africentric philosophical traditions as well as fully reflected in Nkrumah's conceptual system. The statement: *All things evolved themselves therefrom* is captured by internal self-motion and logically represented by the theory of categorial conversion. The inseparability of energy and matter as nothingness and somethingness is indicated by the phrase *I developed myself from the primeval matter which I had made* and *I united myself with my shadow...* The system of reasoning indicates the principles of inseparability and relationality of opposites. This principle of inseparability, the principle of opposites and the principle of self-motion are important conceptual foundations of the logic of categorial conversion and Nkrumaist transformation dynamics, where polarity, duality, unity, relationality, rhythmic interplay of forces from the negative and positive characteristics sets and categorial conversions are tools of reasoning to discover and explain *what there is* (state and processes) as well as *what would be* (state and processes) through the transformation-substitution processes over the epistemological space.

## 6.4 Relationality, Opposites, Conflicts, Energy and Self-Motion

The principle of inseparability of dual in duality and poles in polarity is maintained by the principle of the multiplicity of rhythmic forces in both equilibrium and disequilibrium states over the trajectories of the transformational dynamics. The structure of reasoning is such that, given the poles and duals in the polarity and duality respectively, the logic of categorial conversion demands that we identify the relationships within each polarity and duality that tend to establish conditions of relational unity, conflicts, mutual existence and mutual negation of the poles and duality through information sharing and the responses from information channels under the principle of continuum.

### 6.4.1 Relationality and Information

The categories of the relationality are part of the attributes of the polarity and duality that exist in universal and particular unity. The relations, as properties of categories of polarity and duality, are established by the nature and organizational structure of the multiplicity of rhythms and their behavior as shown by the characteristics sets for any given information structure. The relationships that these multiplicities of rhythmic forces engender tend to produce temporary categorial equilibria, continual categorial disequilibria and internal dynamics in the quality-quantity duality that bring about mutual negation depending on the relative rhythmic strength of the positive and negative characteristics sets in each dual as well as the relative strengths of the duality in the actual and potential poles. Categorial equilibrium consists or tends to remain the same in qualitative sense while categorial disequilibrium has the character of changing to became something qualitatively different in accordance with the relative rhythmic strength that generates the mutual give-and-take processes for similarity, difference, conflict and unity in both categories of duality and polarity for any given information set.

There are communications between the actual and potential poles, the negative and positive dualities in the polarity, and the positive and negative characteristics sub-sets in the duality. The information communications establish the mutual give-and-take processes which allow the mutual existence and transformations of the polarity and duality. The nature of the changes in the organizational structure of the multiplicity of rhythms induces the reversals of give-and-take processes which then bring about changes in the relative quality and quantity of the negative and positive characteristics sets leading to a negative, neutral or positive categorial conversion, first of the duality and then of the polarity over the particular and universal enveloping of the path of categorial conversions. The categories of relationality induced by the information-transformation processes are the essential reasoning tools in the process of discovery of cognitive truth via the use of the logic of the theory of categorial conversion.

## 6.4.2 Energy, Communication and Self-Motion

Changes in the categories of relationality brought about by changes in the relative structure of the negative and positive characteristics sub-sets alter the information structure, create disequilibria, bring about tensions which generate categories of degrees of energy, initiate, manufacture and maintain increasing categorial moments for categorial conversions in duality and polarity. In the analytic and synthetic structures leading to the universal cognitive unity, we must, therefore, identify and establish the *relational processes* that are established by the complex structural organizations of the multiplicity of rhythmic forces, in order to ascertain knowledge about the internal dynamics and the conflicts of the opposites. The internal dynamics and the conflicts of the opposites allow the law of opposites to be maintained in both duality and polarity and in the unity itself, in order to establish the qualitative law of motion which the theory of categorial conversion projects. The organizational structure of the multiplicity of rhythmic forces establishes other forms of characteristics that define the mutual existence of the polarity and duality in addition to establishing various categories of relationality such as complementarity, identity, supplementality, interdependence, reciprocity, information asymmetry and others in the give-and-take processes. These categories of relationality are such that the opposites are in relational unity rather than mutually exclusive. It is, therefore, useful to speak of degrees of relationality such as independence, interdependence, similarity, supplementation, and complementation and information asymmetry. Thus the characteristic set of energy contains some elements of the characteristics set of matter just as the characteristic set of matter contains some elements of the characteristic set of energy to define the energy-matter duality and polarity in particular unity and universal unity. The logic of the theory of categorial conversion points to the condition that it is useful to keep in mind that there is the presence of conflicts, unity and tensions of qualitative and quantitative dispositions, as either qualitative or quantitative motion is analytically and synthetically implied. The point of categorial indifference requires us to develop and analyze the *categorial transfer function* and *categorial transversality* conditions that we have previously discussed. The theory of categorial conversion, by advancing a hypothesis that qualitative transformations are internal to categories and that external forces, provide necessary conditions for change but cannot affect any sustainable change. The sufficient conditions for qualitative transformation come from the conditions of internal self-motion. It is only when the internal characteristics come to accept and integrate the conditions of the external force that internal motion is externally possible. The theory, therefore, provides analytical conditions for social transformation from within societies in line with transformations in nature. It also provides a general theory of unified engineering science.

# CHAPTER SEVEN

## The Theory of Categorial Conversion:
## Mathematical Foundations, Symbolic Representation and
## Conditions of Convertibility in Socio-Natural Systemicity

This chapter is an exploration of the mathematical and symbolic structure of the theory of categorial conversion and its usefulness in understanding Nkrumah's conceptual system as it relates to transformational dynamics of both nature and society. It has two sequential parts of the mathematics of category formation and mathematics of categorial conversion. The mathematical structure that we seek may have some similarities and differences from category theory. The symbolic representation, mathematics of the problem and logical analyses of the theoretical system begin with the elemental trinity of matter, energy and information as they have been discussed in the previous chapters. From the viewpoint of knowing, the elemental trinity comes to cognitive agents from the interactive process of the spaces of ontology and epistemology. There are two elemental trinities. There is the *ontological elemental trinity* that functions as the real identity which is the source of knowing through cognitive transformation functions. Then there is the *epistemological elemental trinity* which is a derived identity that constitutes the epistemic target. Each of these trinities exists in a relational unity where energy and information are inseparable properties of matter. The energy and information properties of matter define the quality and the qualitative characteristics in addition to the internal arrangements of matter. The elemental trinities in both spaces present the internal relational structure of a matter-energy-information trinity, and superimposed on it is the quality-quantity-time trinity whose relational interactions allow forces to be internally created to bring about a resultant direction in the quality-quantity space. Each of these trinities exists in relational continuum and unity. The conceptualization of their unity is through the understanding of the complex dualities with continua that are linked together in relationality which is established by an information structure as projected by the negative and positive characteristics sets.

## 7.1 Discussions on some Conceptual Difficulties and Solutions

Some philosophical difficulties, epistemic questions and mathematical problems tend to arise in the conceptual systems of the theory of categorial conversion as seen in advance by Nkrumah in terms of quantity-quality duality in relation to natural and social transformations. These difficulties and problems are projected onto the space of science for resolutions and answers. As it has been pointed out in the previous chapters, Nkrumah introduced the concepts of category but offers no theory of category formation. He also introduced an analytical tool of categorial conversion but not a complete theory of categorial conversion. His job was to demonstrate that internal qualitative motion is a property of matter and this property is inherent in all elements of the universal object set including socio-political systems whose transformation from within is his main concern. Categorial conversion, at the level of ontology for any given time, presupposes the existence of ontological categories and a process by means of which they are formed. At the level of epistemology for any given time, there are the epistemological categories by means of which the ontological categories are known. As has been discussed, the ontological categories are overlapping qualitative structures that are locked in systems of structural particularities and within the system of structural universalities that defines the ontological space as a space of a family of overlapping structures satisfying the conditions of universal continuum for any time point. The formation of the epistemological categories must satisfy this property of overlapping structures in terms of systems of particularities and universalities in discovering *what there is* as well as designing explanations to their behavior in the qualitative-quantitative static states. At the level of ontology, category formation is a natural mapping under the principle of structure-preserving for elemental identity in the quality-quantity space for any given time point.

At the level of epistemology, category formation is done through cognitive mappings that are also structure-preserving in the quality-quantity space for the discovery of *what there is* (the actual) and *what would be* (the potential) as well as the understanding of the behavior of *what there is* relative to *what would be* at any given time point. The *ontological category formation* is naturally structure-preserving mapping of any ontological element into an ontological category in the ontological space. The epistemological categorial formation is cognitively structure-preserving mapping of epistemological elements into epistemological categories in

262

the epistemological space and into an ontological category for a test of knowledge. Category formations are static phenomena under the *principle of sorting*. The idea that the epistemological space is a sub-set of the ontological space has been discussed [R3.7] [R3.10] [R3.13]. It is important to conceptualize the static condition in category formation as it is being discussed here. Static conditions merely apply to the qualitative disposition of elements to allow effective sorting where structure-preserving mappings are assured in both the ontological and epistemological spaces. Categorial formation function is a static concept in the quality space where quantitative dynamics may or may not take place.

Categorial conversions are structure-transforming through internal intra-qualitative and internal inter-categorial mappings. The mapping is not structure-preserving but structure-altering of elements by the behavior of the internal arrangement of their characteristics. The categorial conversion functions in dynamic states in the quality-time space. They work with quantitative statics and dynamics where quality cannot be held constant and where every category is in temporally qualitative structural equilibrium. There are *ontological categorial conversion functions* and *epistemological categorial conversion functions*. The principles of knowing bring them in relational unity through epistemic activities in the epistemological space.

At the level of ontology, the ontological categorial conversion is to take an element from one category and by the means of intra-categorial changes of the relative composition of the negative and positive characteristics sets alters the residing duality through an intra-categorial conversion function and hence alters the actual-potential polarity to create an inter-categorial conversion function that is structure transforming from within the element to have inter-categorial traveling. The concept of structure-transforming implies changes in the quality of a categorial element or some categorial elements. It also implies quality transforming. The nature of the ontological categorial conversion function is a definition of the global structure of the infinitely closed universe under continual internal transformations as its permanent feature. At the level of ontology, the ontological categorial conversion is natural transformational mappings due to the internal dynamics of an element that allow it to migrate from a category which is the source called the *primary category* , and to enter into a new different category which is the target, called the *derived category*. The categorial conversion mapping is of two compositions of intra-categorial mapping and inter-categorial mapping in the quality-quantity space with neutrality of time. The intra-categorial mapping takes the element into the threshold of qualitative change by creating the conditions of convertibility. The inter-categorial conversion function takes the element from the primary category into a new category which is the derived category. The ontological categorial conversions are natural transformation processes that are independent of cognitive agents. They are induced by forces which are universally produced by internal contradictions and conflicts. It is on this basis that the usefulness of the development of theories of contradictions and conflicts enter into the knowledge production. For example, death emerges from life or mind emerges from the brain and others as they are considered in dualistically mutual negation and existence. The objective of ontological categorial conversion is a continual creation as the permanent character of the universe.

Things are different at the level of epistemology with the behavior and requirements of the needed categorial conversion function. At the level of epistemology, the epistemological categorial conversion is an epistemic construct. The objective is to present an epistemic process to discover the process through which the ontological categorial conversions are manifested in the ontological space. This implies the epistemic discovery of the process where an ontological element from one ontological category and by intra-categorial changes of the relative composition of the negative and positive characteristics sets exits from parent and enters into a new category which is now a new parent. The epistemic discovery involves a critical examination of how the residing duality, through an intra-categorial conversion function, develops mutual negation to alter the actual-potential polarity. From the alteration of the actual-potential polarity, a further examination is undertaken to explain how inter-categorial conversion functions are created to become structure transforming from within the element to have inter-categorial traveling. Inter-categorial traveling means the movement of an ontological element from the primary category into a derived category. The objective of the epistemological categorial conversion is to explain the structure-transforming nature of ontological activities where qualitative dispositions of elements to allow inter-categorial traveling. It also implies an explanation to the internal transformative activities of the quality of ontological elements.

The nature of the epistemological categorial conversion function is also a definition of the cognitive structural dynamics as an attempt to understand how the universe is infinitely closed under continual internal transformations and how every element is under temporary qualitative equilibrium with qualitative disequilibrium as its permanent feature. At the level of epistemology, the epistemological categorial conversions are cognitive transformational mappings due to the epistemic processes in the ignorance-knowledge polarity that must be related to the ontological actual-potential polarity under relational continuum and unity. In a sense the theory of categorial conversion must explain the internal dynamics of an ontological element that allow it to migrate from a category which is the source called the *primary category* , and to enter into a new different category which is the target, called the *derived category*. The epistemological categorial conversion mapping as implied in the theory must mimic the structure of the ontological categorial conversion function which is of two compositions of intra-categorial mapping and inter-categorial mapping in the quality-quantity space with neutrality of time. The logic and the implied mathematics must resolve the ontological categorial conversion problem which is a relational structure between the primary category (the source) and the derived category (the target) The theory of categorial conversion must show how the intra-categorial mapping takes the element into the threshold of qualitative change by creating the conditions of convertibility. Similarly, it must show how the inter-categorial conversion function takes the same element from the primary category into a new category which is the derived category in terms of from a source to a target which is different from but a child of the source in a transformational process. Cognitively, it must show how structures are formed, old structures are destroyed and new structures are created under the principle of creative destruction. In this respect, the source may be viewed as the real cost and the target as the real benefit with a possible reversibility. It is also here that the theory of categorial conversion offers a pathway to develop a *general theory of engineering sciences*.

The epistemological categorial conversions are epistemic transformation processes that dependent on the paradigm of thought of cognitive agents. They are induced by forces of thinking to create a universally epistemic structure of thought in relation to internal contradictions and conflicts in categories. It is on this basis and in a given paradigm of thought that the development of theories of contradictions and conflicts enters into the knowledge production to understand the behavior of energy, force and directions of qualitative motions. For example, the theory of categorial conversion must show how from internal behavior death emerges from life, mind emerges from the brain, and tumors grow from healthy human and non-human cells and others as they are considered in dualistically mutual negation and existence. The objective of the theory of categorial conversion is; thus, to offer an explanation to a continual creation as the permanent character of the universe in equilibrium-disequilibrium processes. The objective is the discovery of internally induced qualitative motion and the direction of the motion including the speed and acceleration within the universal construction-destruction processes. In this respect, the relationships among the theory of categorial conversion, primary category and derived category are such that categorial conversion is forward looking from the primary category relative to the derived category and backward looking from the derived category relative to the primary category. The epistemological categorial conversion function is forward-looking and obtained by methodological constructionism where the derived category is shown to emerge from the primary category. The inverse of the epistemological categorial function is a backward-looking process and is obtained by methodological reductionism where the derived category is traced to the primary category by a logical process within a given paradigm. Any category is said to be derived if the inverse of the categorial conversion function exists in a way that allows the primary category to be found. The problems of the theory of categorial conversion are 1) the construction of categorial conversion function with inverse, 2) deriving conditions of convertibility and 3) constructing the conditions of relationships among primary category, derived category and epistemological categorial conversion functions with the direction of conversion. The outline of this theoretical structure forms the foundational thought of the Nkrumah's conceptual system and its socio-political practice. The analytical structure of internal process of self-transformation requires one to use the principle of duality formation where the characteristic set of every ontological element is dichotomized into negative and positive sub-sets to establish the corresponding negative and positive duals of the duality with a relational continuum and unity. It, further, requires every ontological element not only belongs to a category but dichotomized to exist as *actual-potential polarity* where each pole has a residing duality and identified as negative or positive depending on whether the residing duality is negative or positive. The epistemic progress of this path for understanding transformations and general mechanism of change requires the development of *theories of opposites*. The development of theory of opposites must contain sub-theories of duality and polarity with clearly established relational structure between the two as well as connected to the characteristic sets in the quality-quantity space.

## 7.2 A Reflection on the Theory of Categorial Conversion and Nkrumah's Conceptual System

Within the conceptual system as seen by Nkrumah in terms of quality-quantity duality in relation to natural and social transformations some philosophical difficulties tend to arise where these difficulties are projected onto the spaces of sciences and mathematics for formulations and resolutions. These difficulties present themselves in terms of questions about the primary category and the constituent properties. A few fundamental questions may be asked. Is energy a property of matter where information establishes their mutual relationality? Similarly, is matter a property of energy where information establishes their mutual relationality? Do both matter and energy exist independently of one another or are they mutually determined and measurable? In this framework, what is information and what role does information play in understanding the concepts of relationality and information? Can information exist without matter or energy? These questions are troubling. The answers to them may help pave a smooth path to understanding the concepts of the primary and derived categories and the theory of categorial conversion over the epistemological space. If energy (matter) is seen as a property of matter (energy) then matter (energy) may be selected as the primary category where all other categories are derived through a series of compositions of categorial conversion functions that show the alterations of internal contents and arrangements of the constituent energy (matter) properties where each category is distinguished and identified by the emitted information.

In Nkrumah's conceptual system, matter is axiomatized to have absolute and independent existence as well as to have capacity of spontaneous self-motion. This matter is postulated to emerge from nothingness-somethingness polarity in [R1.92]. In this respect, energy is seen as a property of matter and different arrangement of the energy characteristics establish different qualitative dispositions of matter leading to different categories of matter. A movement from one category

267

of matter to another is induced by categorial conversion, as for example, the relational structure of the geomorphological evolution of earthly elements. The theory of this categorial conversion allows one to explain, for example, the emergence of mind from matter and put to rest the mind-matter problem in terms of divisibility, independence and interdependence on the principle of duality and polarity with relational continuum and unity to establish interactions between the duals. The Aristotelian dualism with excluded middle basically gets rid of interactions between duals of any duality as well as prevents give-and-take relations, mutual negations and qualitative changes due to internal contradictions and conflicts. The universe is postulate to compose of mutually exclusive and collectively exhaustive categories. The principle of excluded middle and non-contradiction in the Aristotelian framework in the epistemological space is dropped when dealing with the theory of categorial conversion. This is replaced by the principle of relational continuum and unity where the universe is seen to be composed of overlapping categories that are mutually non-exclusive and collectively exhaustive thereby allowing give-and-take relational sharing, conflict of the opposites, contradictions and the creation of energy from the relational unity. In this respect, there is the quality-quantity duality with a continuum seen in terms of relationality. The conceptual framework is stated by Nkrumah as:

> According to philosophical Consciencism, qualities are generated by matter. Behind any qualitative appearance, there stands a quantitative disposition of matter, such that the qualitative appearance is a surrogate of the quantitative disposition. [R1.203, P.87]

The reasoning mode of this statement can be found in the use of the *fuzzy paradigm* of thought. The concept of continuum implies that each transformational point is a duality where the duals are relationally inter-supportive to give meaning and identity to any category and any element. Each negative characteristics set has a supporting positive characteristics set and vice versa at any point in the continuum; just like every truth has a false support and vice versa in a continuum of knowledge production. The inter-supportiveness is expressed in terms of conflicts, contradictions and negations at each point of the relational continuum and unity. For the theory of categorial conversion polarity is distinguished from duality. The presence of dualities defines the relational continuum and unity of the poles of the polarity. Each point of conversion is viewed as actual-potential polarity under relational continuum and unity. The presence of negative and positive characteristics sets allows the specifications of the relational continuum and unity of the dualities. It is these phenomena that make it possible to assert the internal self-motion of matter as the primary category from which other categories are derived. The qualitative characteristics defined by atoms, electrons and other quantum particles produce constant vibrations which generate sound waves in polyrhythms. The polyrhythms underlie the internal structure of categorial elements where there are negative polyrhythms generated by the negative characteristic sub-set, and positive polyrhythms generated by positive characteristic sub-set. The negative and positive polyrhythms create contradictions and conflicts creating asymmetric power relation between the duals in the game of categorial dominance and transformation. There is the existence of matter-mind duality and other dualities. The general polarity in all forms is the actual-potential polarity where the actual must be defined in knowing and the preferred potential element is specified. The epistemic relational geometry involving the path of categorial conversion is shown in Figure 7.2.1.

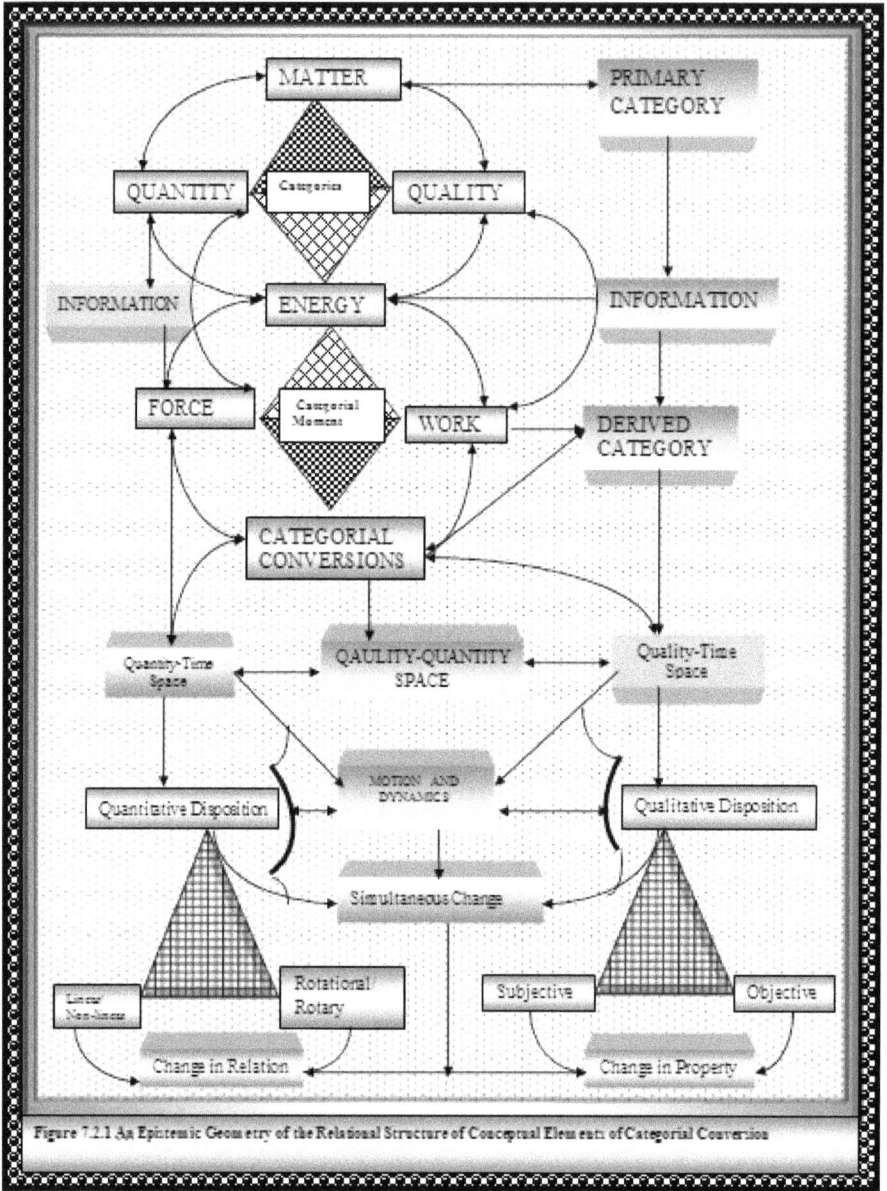

Figure 7.2.1 An Epistemic Geometry of the Relational Structure of Conceptual Elements of Categorial Conversion

The concept of relational continuum and unity of the duality is concerned with interdependence between divided self, which is the negative and positive duals in the space of transformational. It is integrative in the process of transformation. The concept of relational continuum and unity of the polarity is concerned with interdependence

270

among the actual and potential elements, which is the actual and potential poles translated into negative and positive poles of the polarity in the space of transformational. It projects a form of battle field of separation and integration over the space of categorial conversion. The relational continuum and unity between duality and polarity concerns the interdependence of decision-choice actions by the poles and duals in the game of transformation in creating differentiations of the same category by the method of part-variation. The process is a separation and unity formation to maintain the continual creation where duals of duality and poles of actual-potential polarity of categorial elements are maintained in differences and unity in a relational continuum. The categorial-conversion process points to simultaneous existence of permanent and temporary equilibria. The unity defining the Africentric oneness with elemental differentiations in continua presents a permanent equilibrium while every categorial element exist in a temporary qualitative equilibrium.

From the viewpoint of African principles of opposites, the universe is conceived in terms of dualities and polarities with families of continua that allow categorial conversion when conditions of convertibility are met by categories to generate the needed categorial moments. A number of essential definitions have been given in the previous chapters. The definition of categorial conversion is being given here to relate it to that of Nkrumah.

## Definition 7.2.1: Categorial Conversion

A categorial conversion is a process where, by internal forces, one category emerges from another category through changes in the relative qualitative characteristics, induced by internal struggles and contradictions to establish either negative identity and dominance or positive identity and dominance.

Nkrumah presents his own definition where he states:

> By categorial conversion, I mean such a thing as the emergence of self-conscious from that which is not self-conscious; such a thing as the emergence of mind from matter, of quality from quantity [R1.203, p.20].

From the definition of categorial conversion, the epistemic demand is to develop a theory that will establish the process through which one category emerges from another category. This is another way of looking at the problem of the theory of categorial conversion. The demand of the theory is the development of required conditions of convertibility. This is the problem of convertibility in the theory of categorial conversion. The convertibility conditions include how forces are internally generated and conversion moments are internally created. These are intra-categorial conversion problems. The final requirement of the theory of categorial conversion is to develop the conditions of how categorial transfer functions are internally manufactured to transfer elements from one category to another. The theory of categorial conversion is about the development of the structure of creative destruction in the constructionism-reductionism duality.

## 7.3 Dynamics, Motion and Progress in the Theory of Categorial Conversion

To appreciate Nkrumah's analytical preoccupation with the concept of motion and types of motion in understanding change as applied to matter, it is useful to revisit the phenomenon of the quality-quantity-time process as has been discussed in the previous chapters and also in [R3.10] [R3.13]. The conditions of this preoccupation have been extended and intensified in this monograph. The extension and intensification involve the development not only of the structural foundations applicable to Nkrumah's analytics but to present analytical foundations for socio-natural construction-destruction phenomenon that applies to social and non-social engineering under cost-benefit rationality where new resources are created from the destruction of other resources. Here, the concepts of costs, benefits and resources must be broadly interpreted in terms of sciences of organicity, systemicity, synergetics and global complexities. The understanding of similarity and differences of types of motion allows one an analytical access to pull out the type of motion that is applicable to the social transformation required to explain how freedom emerges from oppression and vice versa, independence from colonialism and vice versa, emancipation from neo-colonialism and vice versa, and justice from injustice and vice versa under the organic principle of mutual negation and negation of negation. To make sure that national independence (sovereignty) is not identified with neo-colonialism, and that one form of oppression is not substituted for another form with a different face, the concept of identity of meaning is introduced. This *principle of identity of meaning* is equivalent to the *principle of categorial identity* which introduces its opposite of the *principle of categorial difference.* The categorial identity and difference are not applicable to all types of motion and dynamic systems.

Different types of motion present themselves under different conditions of matter, broadly defined, in terms of processes which may be identified with structural dynamics in the quality-quantity space as seen from vibrations, sound waves and polyrhythmic colors. Let us keep in mind, matter, energy and information where information structures define the corresponding environments as well as become inputs for *positive* or *negative action.* Nkrumah identifies, and rightly so, two categories of motion that correspond to qualitative motion and quantitative motion with *semi-negation* for any given matter and energy. The quantitative motion is identified with two sub-types of linear and non-linear motion where the non-linear motion is made up of rotary and rotational motion as presented in Figure 7.2.1. The quantitative motion may be related to ridged and non-ridged bodies of matter. These types of motion have epistemic relation with matter, energy and information. The quantitative motion is related to changes in relation while the qualitative motion is related to change in property. The most complex motion is that of simultaneity of qualitative and quantitative motions. To understand these particular and general types of motion, let us relate them to matter as Nkrumah does. The understanding of linear or curvilinear motion and the corresponding stabilities requires that one holds on to the epistemic assumption of frozen qualitative characteristics of matter whether it is rigid or non-rigid [R3.10] [R3.13]. Similarly, pure qualitative motion must assume fixed quantity of matter behind its description. The complex simultaneous motions of matter have no assumption on the state of quality or quantity. The matter may be rigid or non-rigid under the conditions of simultaneous motions.

The analytical concern of these types of equation of motion is to be able to describe the rate at which the concerned matter is travelling or transforming in terms of either quantitative or qualitative distance for any given time. Collectively, these are the attributes of the behaviors of the organic dynamic system of any given matter and its relation to energy and time. The dynamic attributes involve speed, acceleration and distance that allow the identification of time needed for a destination to be reached if it is identified and relevant with analytical need.

In the analysis and understanding of quantitative motion, one speaks of quantitative velocity, acceleration and distance with fixed and given qualitative characteristics under the conditions of either linearity or curvilinear structure. Under conditions of rotational or rotary structure, one speaks of, for example angular velocity, acceleration and number of revolutions in terms of distance.

## 7.3.1 Concerning the Structure of Quantitative Motion

In terms of quantitative motion, one must examine the behavior of rigid and non-rigid bodies of matter relative to energy use, forces, distance and time. In the study of quantitative motion of either rigid or non-rigid bodies, the sets of qualitative characteristics are held as given and constant within the relevant period. The constancy of the quality of matter in motion simply means that the content of the matter will not go through any transformation during the period of the study of the motion, and that every qualitative characteristic defining the identity of the matter is under the influence of the same force in the same direction with equality of energy content and distribution in the same direction. In this respect, the equation of motion of the rigid body satisfies the conditions of the classical logic and mathematics of exact rigid determination where the rigid body holds its shape. The equations of non-rigid bodies such as fluids may also be examined with classical logic and mathematics to the extent to which the essential qualitative characteristics are held constant and invariable with time even though the non-rigid body may assume different shape.

## 7.3.2 The Structure of Qualitative Motion

Things are different and complex when we turn our attention to the phenomenon of qualitative motion. Here, one speaks of velocity and acceleration of quality under transformational force. Corresponding to qualitative motion, one has categorial-conversion (transformational) velocity, acceleration and distance with fixed quantity of matter under conditions of assumed quantitative disposition. In the more complex system of motions dealing with a system of simultaneous changes of quality and quantity in relational unity under two different energies and forces, one speaks of changes in relation and changes in characteristics (properties) under velocity, acceleration and distance induced by forces of different energies where such forces may be external or internal or both. In all elements involved in qualitative motion, there is a clear relational continuum and unity in the quality-quantity duality where quantity may be viewed as surrogate of qualitative disposition and vice versa.

It is essential to have a clear scientific and philosophical understanding of the relational structure of the phenomena of rigid bodies, non-rigid bodies, internal organizational complexity, quality, quantity, energy, force, change and transformation if one is to understand and appreciate the theory of categorial conversion. The complex relational structure projecting continuum and unity of the theory of category formation and the theory of categorial conversion with the essential conceptual blocks is shown in Figure 7.3.2.1.

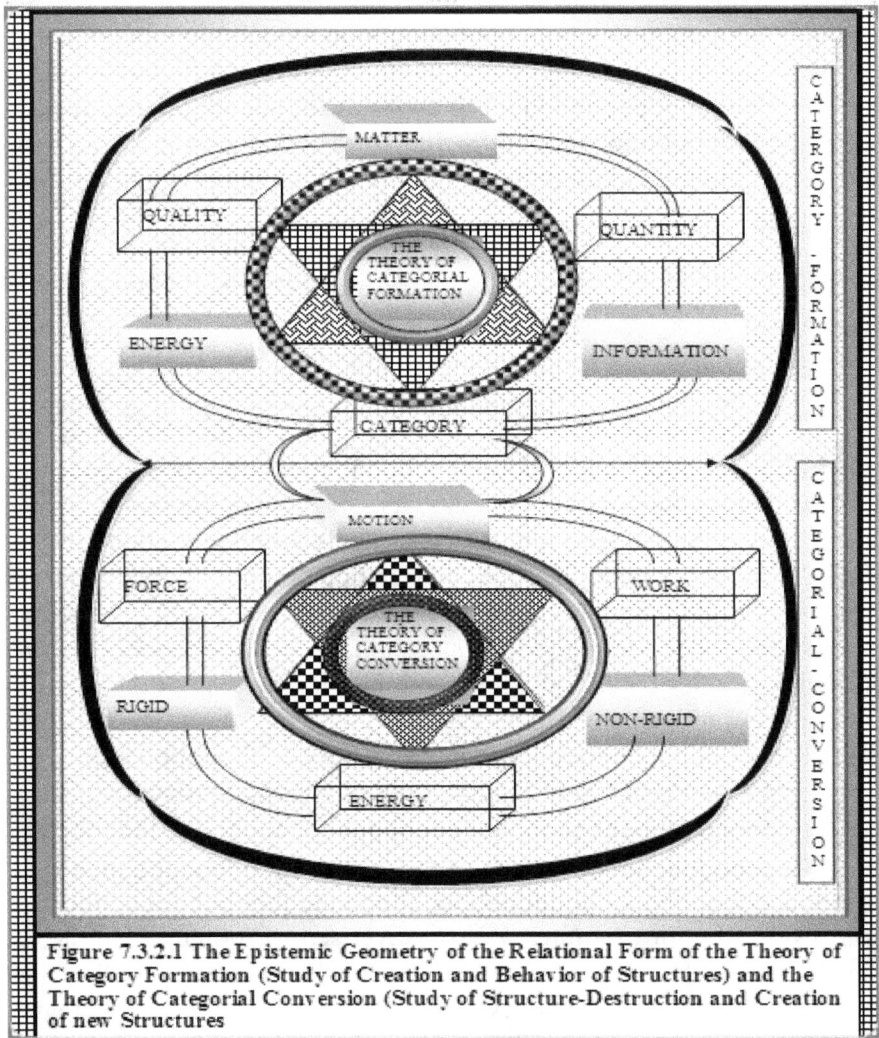

Figure 7.3.2.1 The Epistemic Geometry of the Relational Form of the Theory of Category Formation (Study of Creation and Behavior of Structures) and the Theory of Categorial Conversion (Study of Structure-Destruction and Creation of new Structures

275

The epistemic geometry presented in Figure 7.3.2.1 presents two inter-supportive relational sets of logically pyramidal reasoning structures. One set of the relational structure is matter, energy and information, and superimposed on it is the structural pyramid of category, quality and quantity where quality and quantity provide information for the study of formation and behavior in qualitative static states. The other relational set of the logically pyramidal structure is energy, force and work, and superimposed on it is motion, rigid bodies and non-rigid bodies. The relational set of the pyramidal structures provides a thinking environment to study and analyze structures in qualitative statics, as well as transformation dynamics of quality and quantity irrespective of whether natural or social phenomena are under study. Changes and transformations of quality and quantity are viewed in terms of motions generated by complex interactions among matter, energy and information. A work in [R3.13] offers discussions about the differences and similarities of exact and inexact science where discussions on information representation are offered and related to natural and social sciences, and the logical unity that exists between them.

The study and analyses of logical structures and the mathematics of qualitative motion are very important in the understanding of social change and welfare of the people. They offer us another way of understanding all forms of engineering. In the study of qualitative motion, one concentrates on transformations of categories where the categories are characterized and distinguished by their sets of characteristics that define their identities. Comparatively, while quantitative motions are seen as space-time phenomena connected to change in space-time relation, qualitative motions are seen as characteristics-time phenomena that allow the study of internal transformations of categories in destruction-construction mode in terms of destruction of existing categories and the formation of new categories as connected to transformation-time relation. Thus, the characteristics-time phenomenon is related to and deals with transformational activities in the actual potential duality with relational continuum where the actual may be dislodged (destruction) into the potential space and the potential may be brougt into the actual space (creation). In other words, a potential category may be transformed into an actual category through changes in the proportion of a relative set of potential to actual characteristics which are revealed by the content of information that they carry and their internal dynamics. To deal with qualitative motion that is explicit in the theory of categorial conversion, it is useful to present information definition, content and representation of the characteristics that allow categories to be distinguishably formed and connected to qualitative motion. It may be kept in mind that the problem of dealing with quality and quantity presents important challenges in the progress within all areas of knowledge development. It may also be pointed out that it is the problem of qualitative and quantitative dispositions in information representation in terms of exactness and inexactness that forms the central disagreements among the intuitionist paradigm with its logic and mathematics, the classical paradigm with its logic and mathematics in the Brouwer-Russell debate and the current development of fuzzy paradigm with its logic and mathematics [R3.10] [R3.13] [R12.5] [R12.6] [R12.4][R12.16] [R12.8] [R16] [R16.5] [R16.9] [R16.37] [R16.58] [R16.63].

## 7.4 Information Definition and Representation in the Theories of Category Formation and Categorial Conversion

The concept of information from a decision-choice process for space-time and characteristics-time phenomena must be viewed from two interdependent components of *properties of objects* and *relationships among objects*. On one hand, information constitutes the general set of the overall properties of objects, states and processes on the basis of which identities are abstracted and categories are formed giving rise to language of thought. On the other hand, information is a set of relationships among objects as they pass through states and processes by means of sender-recipient modulus. The former is objective in the sense that they exist independently of the awareness of other entities. The latter is subjective in that the relationships and their types require the awareness of other objects. This is essentially the case in socio-natural dualities and polarities that relate to the establishment of foundations in the actual-potential polarity. Information may thus be defined in terms of *objective-subjective* duality as relationally viewed in term of *properties* and *relations* that characterize a universal system of objects, states, processes and events where these properties may involve among other quantum characteristics, sounds and colors in supper-relational polyrhythms to produce stability as well as instability in categorial elements. Some important questions arise from the basis of the dual character of information. Under what set of conditions is information a *set of the overall properties* of objects, states and processes? Similarly, under what set of conditions does information constitute *a set of relations* among objects given their states, processes and events? Let us look at the required conditions while we keep in mind the evolving arguments of category formation and categorial conversion. Category formation and categorial conversion are shaped by the internal dynamics of categorial elements operating with objective-subjective duality of information. The information structure creates the environment of socio-natural decision processes under various types of rationality and irrationality. The assumed information structure over the epistemological space dictates the needed paradigm for analysis. The epistemic geometry of the interconnectedness for the development of the theory of category formation is presented in Figure 7.4.0.1.

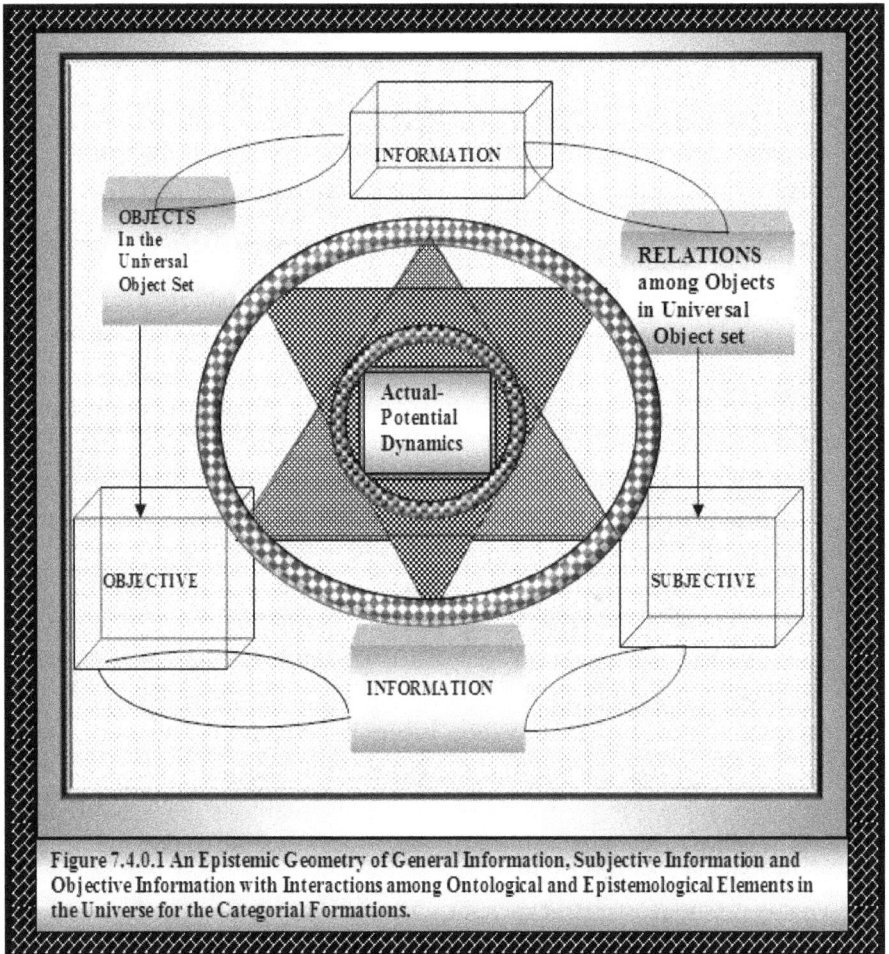

Figure 7.4.0.1 An Epistemic Geometry of General Information, Subjective Information and Objective Information with Interactions among Ontological and Epistemological Elements in the Universe for the Categorial Formations.

## 7.4.1 The Concept and Nature of Information in the Theories of Category Formation and Categorial Conversion

Information structure and the knowledge structure derived from it are essential in all socio-natural decision-choice actions. All socio-natural states, processes and transformations are action-determined on the basis of ontological information structure and its processing into knowledge for any action. At the level of epistemology, the definition and representation of information are relevant to the understanding of the socio-natural decision-choice processes and rationality or irrationality with the postulate that objects, states, processes and events exist as realities in the ontological space. The character and identity of any object, state, process and event in the universal system are completely defined by a set of properties which present themselves as characteristics. Each entity is viewed uniquely as composed of a bundle of characteristics (quantitative and qualitative). The core idea of the postulate is that each entity in the universal system is identified by a set of characteristics under the principle of nominalism to develop the set of vocabulary and language of knowledge. Variations and non-variations in the set of characteristics establish differences and similarity in entities for logical comparison. The universe presents itself as a collection of entities that appear in *variety*. The properties of objects are naturally presented as *something* that helps to identify the objects and classify them into similarities, dissimilarities and resemblance, and hence place them into categories. This something is a set of attributes, which presents itself as *objective information*. The attributes become established through cognitive processes as *something* that generates awareness called *information relations* among objects. The information relations present themselves as *subjective information*. The *somethingness* that gives evidence of identity and existence of objects, states, processes and events for a given awareness is called information which is a unity of objective and subjective information. The objective information determines the nature and direction of states, processes and events. The subjective information determines the nature of knowing of the directions, processes and events through acquaintances and knowledge creation.

## 7.4.2 Essential Definitions and Explication of Information and Category Formation

To develop the theory of category formation and categorial conversion, it is useful to have a formal definition of information. Over the epistemological space the information presents us with inputs that may be empirical or axiomatic on the basis of which a paradigm may be applied to derive epistemic reality. The definition is couched in both verbal and set-theoretic representations which provide structures for clarity. Let us keep in mind that category formation is about structural classification of the universe at any given time while categorial conversion is about structural destruction and creation of new forms with information structures. There is also an important conceptual idea that must be kept in mind. There are two types of categories in the quality-quantity space in relation to the characteristic set for epistemic analytics. The quality establishes the family of categorial superstructures while quantity establishes categories within any qualitative superstructure to preserve the interactions, relational continuum and unity of qualitative and quantitative dispositions of categorial elements. Let us begin with the concept of the universe and then its representation where the elements of the universe may be seen as social, natural or both.

### Definition 7.4.2.1 -: Universal Object Set

The universe is composed of the collection of all objects, states, processes and events that are exhaustive, complete and infinite which is referred to as the *universal object set*. If $\omega$ is the generic elemental representation of objects, states, processes and events in the universal system then the collection of all $\omega$ constitutes the global unity and is simply the *universal object set* $\Omega$ written as: $\Omega = \left\{ \left( \omega_1, \omega_2 \cdots \omega_\ell \cdots \right) \mid \ell \in \mathbb{L}^\infty \right\}$ where $\mathbb{L}^\infty$ is an infinite index set of $\omega$.

The universal object set is also the *universal object space*. It is the ontological space. Each object, $\omega$, in the object space, $\Omega$, $[\omega \in \Omega]$ is or has a potential to be well-defined and identifiable by attributes that allow actual or possible naming, concept-formation, relational ideas and thought to be formed about the elements in the object set by cognitive objects operating in the epistemological space. The elements in the universal object set are divided into cognitive objects that have awareness and non-cognitive objects that have no awareness. Both cognitive and non-cognitive objects reside in the universal unity. The entities in the universal object set composed of objects, states, processes and events are infinite in number. Their existence is objective and defined by *objective information*. Their awareness is subjective and defined by *subjective information*. The concept of objective and subjective information will be explicated in terms of their philosophical and scientific unity as it relates to category formation and categorial conversion. The distinguishing factor for real existence and identification of the elements of the universal object set requires a definition of information that follows. The objective information is that which exists as an ontological identity. The subjective information is that which is perceived by acquaintance and which is to be processed into an epistemic identity which is to be compared to the ontological identity. The difference between the ontological and epistemological identities is the gulf of cognitive ignorance.

## Definition 7.4.2.2: Information

Information is a set of *characteristics* that provide evidence regarding existence and identity of the elements in the universe in an *objective sense*. It is also a set of *relations* that create awareness possibilities among objects in a *subjective sense*. If $x$ is the generic elemental representation of *attributes* on the basis of which similarities, differences, sameness and resemblances of objects, states, processes and events in the universal object set are naturally defined, identified and separated then the collection of all $x$ constitutes the *total space* called the *universal characteristic set* $\mathbb{X}$ that may be written as: $\mathbb{X} = \left\{ \left( x_1, x_2 \cdots x_j \cdots \right) \mid \Omega, \text{ and } j \in \mathbb{J}^\infty \right\}$

where $\mathbb{J}^\infty$ is the infinite index set of all attributes, $x$.

The set $\Omega$ defines all elements in the universe while the set $\mathbb{X}$ defines *objective information structure* in terms of attributes associated with elements in $\Omega$ independently of the awareness of any of the objects in the universal object space (the ontological space) for category formation for the study of categorial-conversion dynamics. All cognitive and non-cognitive objects, processes, states and events belong to the universal object set, which is infinitely closed under object collections. The set $\mathbb{X}$ is also infinitely closed under attribute collections. Substantial elements in the universal object set and substantial characteristics in the in the objective information structure may not be known by cognitive agents operating in the epistemological space. Relationally, we have objects defined by characteristics that present objective information in the ontological space. The relational mapping defining the objective information structure is illustrated by an epistemic geometry in a rectangular structure as in Figure 7.4.2.1. From definitions 7.4.2.1 and 7.4.2.2, conditions are available to define a partition of the characteristic set $\mathbb{X}$ that imposes categories of reality in the universal object set $\Omega$.

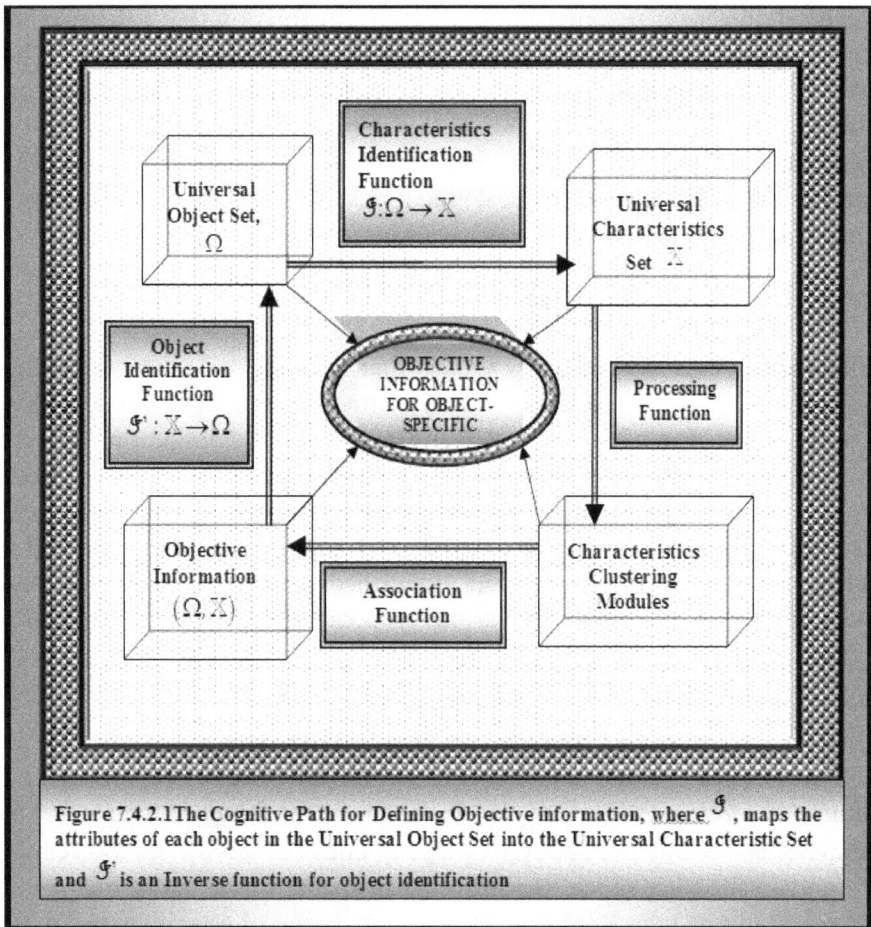

Figure 7.4.2.1 The Cognitive Path for Defining Objective information, where $\mathcal{G}$, maps the attributes of each object in the Universal Object Set into the Universal Characteristic Set and $\mathcal{G}'$ is an Inverse function for object identification

## Definition 7.4.2.3: Partitioned Characteristics Set

The partition of the universal characteristics set is the collection of non-empty groups of attributes that give sameness and difference, which impose groupings or categories on the elements in the universal object

282

space $\Omega$ . A partitioned characteristic set $\mathbb{X}_\ell$ is a collection of attributes, $x_{\ell j}$ about any fixed element $\omega_\ell \in \Omega$ in the ontological space, such that

$$\mathbb{X}_\ell = \left\{ (x_1, x_2 \cdots x_j \cdots) \mid j \in \mathbb{J}_\ell \subset \mathbb{J}^\infty \text{ and } \ell \text{ is fixed in } \mathbb{L}^\infty \right\} \text{ where } \mathbb{J}_\ell \text{ is a}$$

finite index set of attributes that define the identity $\omega_\ell \in \Omega$, $\mathbb{X} = \bigcup_{\ell \in \mathbb{J}^\infty} \mathbb{X}_\ell$,

$\mathbb{J}^\infty = \bigcup_{\ell \in \mathbb{L}^\infty} \mathbb{J}_\ell$, $\mathbb{J}^\infty$ and $\mathbb{L}^\infty$ are infinite index sets.

By combining definitions, (7.4.2.1 – 7.4.2.3.), we may define the *objective universe*, $\mathbb{U}$ as a schedule in terms of *universal object set* $\Omega$ and the *universal characteristics set* $\mathbb{X}$ that meets the conditions of partitioning in term of categories.

## Definition 7.4.2.4: Category of Reality

The categories of reality $\mathbb{C}_\ell$'s are collections of all identical elements $\omega_\ell \in \Omega$ where each of the $\ell_{th}$ categories is identified by a partitioned characteristic set $\mathbb{X}_\ell$ in the form:

$$\mathbb{C}_\ell = \left\{ (\omega_\ell, x_{\ell j}) \mid j \in \mathbb{J}_\ell \subset \mathbb{J}^\infty, \omega \in \Omega, x_{\ell j} \in \mathbb{X}_\ell \text{ and } \ell \text{ is fixed in } \mathbb{L}^\infty \right\}$$

where $\mathbb{X}_\ell$ is the full attribute condition of a particular reality $\omega_\ell \in \Omega$. Similarly, it may be written as $\mathbb{C}_\ell = \left\{ (\omega_\ell, \mathbb{X}_\ell) \mid \ell \in \mathbb{L}^\infty \right\}$ where

$$\mathbb{X}_\ell = \left\{ (x_1, x_2 \cdots x_j \cdots) \mid j \in \mathbb{J}_\ell \subset \mathbb{J}^\infty, \ell \in \mathbb{L}^\infty \right\}.$$

## Postulate 7.4.2.1: Objective Universe

The objective universe, $\mathbb{U}$, is an exhaustive, mutually exclusive and infinite collection of categories, $\mathbb{C}$ , whose elements appear as schedules in the form $\mathbb{U} = \left\{ \mathbb{C}_\ell \mid \ell \in \mathbb{L}^\infty \right\} = \left\{ (\omega_\ell, \mathbb{X}_\ell) \mid \ell \in \mathbb{L}^\infty \right\}$ where exclusivity is defined in terms of duality with relational continua that present universal unity.

## Postulate 7.4.2.2: Identity of Belonging to a Category

Given the universal object set $\Omega$ and characteristic set $\mathbb{X}$, let $(\approx)$ be an *identicality* relation defined over $\mathbb{X}$, then the group $\mathbb{C}$, is said to be a category if and only if there exist $(\omega_1, \omega_2, ..., \omega_\ell) \in \Omega$ such that $(\mathbb{X}_1 \approx \mathbb{X}_2 \approx ... \approx ... \mathbb{X}_\ell) \in \mathbb{X}$, then $(\omega_1, \omega_2, ..., \omega_\ell) \in \mathbb{C}$. Alternatively if there exist $(\omega_1, \omega_2, ..., \omega_\ell) \in \mathbb{C}$ then the corresponding characteristics sets are identical in the sense that. $(\mathbb{X}_1 \approx \mathbb{X}_2 \approx ... \approx ... \mathbb{X}_\ell) \in \mathbb{X}$.

## Postulate 7.4.2.3: Universal Partitioning of the Object Set

The universal object set is a partition with respect to $\mathbb{C}$-categories such that $\mathbb{C}_\ell \neq \varnothing$, $\bigcap\limits_{\ell \in \mathbb{L}^\infty} \mathbb{C}_\ell = \varnothing$, $\forall \ell$ and for any $i \neq j \in \mathbb{L}^\infty$, $\mathbb{C}_i \cap \mathbb{C}_j = \varnothing$ with $\bigcup\limits_{\ell \in \mathbb{L}^\infty} \mathbb{C}_\ell = \Omega = \mathbb{U}$ in relational continua and unity.

It may be noticed that corresponding to each element, $\omega_\ell \in \Omega$, there is a set of attributes, $\mathbb{X}_\ell$, that identifies it. The collection of all, $\omega_\ell \in \Omega$ with the set of attributes $\mathbb{X}_\ell$ constitutes a category $\mathbb{C}_\ell$, and the collection of all these categories constitutes the objective universe $\mathbb{U}$. The objective universe is organized in a way where each object can be identified by its specific characteristics set and placed in the category to which it belongs (for example, the category of humans or living things). The universal object set, $\Omega$, is simply the objective universe, $\mathbb{U}$, without the defining characteristics set, $\mathbb{X}_\ell$ that partitions, $\Omega$ into categories of entities. Thus

$$\Omega = \left\{ \mathbb{C}_\ell \mid \ell \in \mathbb{L}^\infty \right\} = \left\{ (\omega_\ell, \mathbb{X}_\ell) \mid \ell \in \mathbb{L}^\infty \right\} \qquad \text{where } \mathbb{X} = \bigcup\limits_{\ell \in \mathbb{J}^\infty} \mathbb{X}_\ell ,$$

$$\mathbb{J}^\infty = \bigcup\limits_{\ell \in \mathbb{L}^\infty} \mathbb{J}_\ell \text{ and } \#\mathbb{X}_\ell = \#\mathbb{J}_\ell.$$ Both $\mathbb{U}$ and $\Omega$ represent the collection of the categories of reality from which there is the primary category that by its internal contradictions, conflicts and dynamics gives rise to others as derived categories. In other words the universal object set is composed of primary and secondary categories. If the primary category is

assumed to be matter then all the derived categories must be shown to be reducible to matter .The study of the dynamics of the relational structure between the primary categories and the derived categories is the subject matter of the theory of categorial conversion.

Definitions (7.4.2.2 - 7.4.2.4) specify the objective existence of entities, states, processes and events that constitute the sources of a *characteristics-based information set* is the objective information. The sources present two important items of the universal object set $\Omega$ and the universal characteristics set $\mathbb{X}$. The two sets are considered as factual reality in the sense that their existence is independent of awareness of any object in the universal object set. These elements will exist whether other elements are aware or not. Thus, the characteristics-based information set is objective reality. Furthermore, the universal object set is infinitely closed under *categorial formation.* The structure in defining characteristics-based information may be represented in an epistemic geometry that shows the cognitive path as an objective information square in Figure 7.4.2.2.

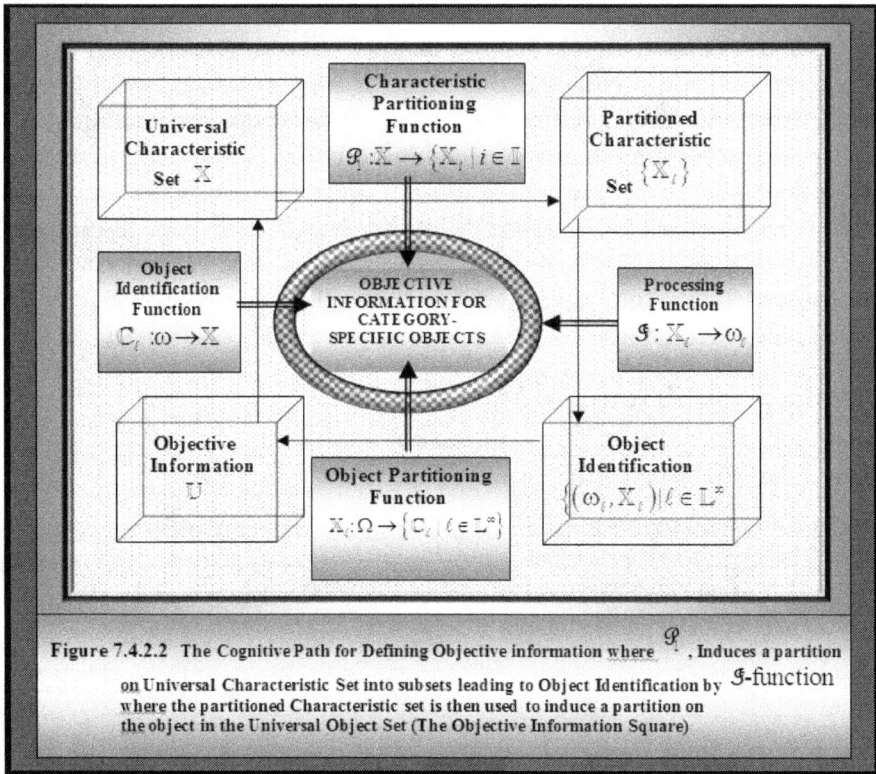

**Figure 7.4.2.2  The Cognitive Path for Defining Objective information where** $\mathscr{P}$ **, Induces a partition on Universal Characteristic Set into subsets leading to Object Identification by** $\mathscr{S}$-**function where the partitioned Characteristic set is then used to induce a partition on the object in the Universal Object Set (The Objective Information Square)**

The process of category formation on the basis of informational definition in Figure 7.4.2.2 is such that first we have a set-to-subset mapping, followed by a set-to-object mapping. This is then followed by a set-to-subset mapping which is completed by a point-to-set mapping as an illustrative path of definition of objective information and category formation.

### 7.4.3 The Subjective Information Structure

To conclude the specifications, definitions and representations of the overall information structure on the basis of which the theory of categorial conversion may be constructed, it is useful to deal with subjective aspects of information. The understanding of the subjective aspects of information is importantly relevant to the understanding of any theory including the theory of categorial conversion. It is also relevant to the claims of knowledge, demonstration of claims and the beliefs in the claims. Let us begin with the observation that every element in the universal object set sends attributive signals that correspond to the number of attributes that define its identity. The attribute signals create conditions for possible awareness by other objects, especially cognitive objects in terms of relations between the source objects and recipient objects creating source-target configuration. Every entity in the universal object set is both a source object and a recipient object in the sense that it sends and receives information through signals. In this way, the source-recipient modules establish relational continua that are defined by information flows and subjectively interpreted by the recipients. Let us put this idea in a definitional mode.

### Definition 7.4.3.1: Attribute Signal Set

If $s$ is an attribute signal that corresponds to a characteristic $x \in \mathbb{X}_\ell$ and sent from source $\omega_\ell \in \Omega$, then the collection of all such attribute signals from $\mathbb{X}_\ell$ is an *attribute signal set* $\mathbb{S}_\ell$ from the object $\omega_\ell$ that may be written as:

$$\mathbb{S}_\ell = \left\{ \left( s_{\ell 1}, s_{\ell 2}, \cdots, s_{\ell j}, \cdots \right) \mid \omega_\ell \in \Omega, \mathbb{X}_\ell \subset \mathbb{X}, j \in \mathbb{J}_\ell \subset \mathbb{J}^\infty \text{ and } \ell \text{ is fixed in } \mathbb{L}^\infty \right\},$$

where $\mathbb{J}_\ell$ is a finite index set of attribute signals from the source.

The elements of the attribute signal set pass through *cognitive filters*, and become processed and transformed into a *perception characteristics set* that establishes a *set of information relations*. Thus $\mathbb{S}_\ell$ defines conditions of *subjective information* which we define as relation-based information about an object $\omega_\ell \in \Omega$ from its information characteristics set, $x \in \mathbb{X}_\ell$. This is shown an epistemic geometry in Figure 7.4.3.1 as a subjective information square.

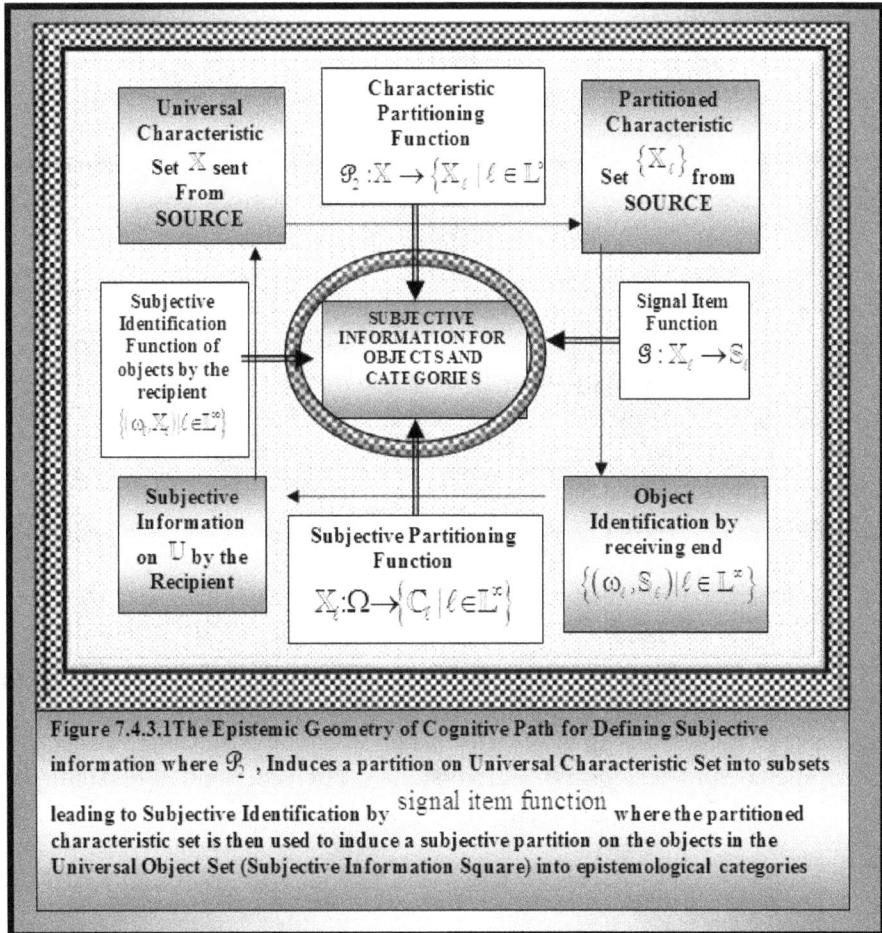

Figure 7.4.3.1 The Epistemic Geometry of Cognitive Path for Defining Subjective information where $\mathcal{P}_2$ , Induces a partition on Universal Characteristic Set into subsets leading to Subjective Identification by signal item function where the partitioned characteristic set is then used to induce a subjective partition on the objects in the Universal Object Set (Subjective Information Square) into epistemological categories

The definition of information that has been offered divides the concept of information into two interrelated sub-concepts of *characteristic-based information* and *relation-based information*. The characteristic-based information is defined by *attributes* of objects, states, processes and events that exist independently of the awareness of objects in the universal object space whose cognitive activities may attach meanings and interpretations. The relation-based information defines attribute characteristic signals between the source objects and recipient objects where naming, vocabulary, meaning and interpretations are attached to the attribute signals. The naming, meaning and interpretation require capacities for a processing mechanism of the received attribute signals. The meaning and interpretation that result from the processing mechanism of the attribute signals require awareness of recipient objects. As such, information relations among objects, states and processes are subjective and forms the basis for the development of theories including the theory of categorial conversion. For the development of any theory, the subjective information structure may be taken as an axiomatic information structure, empirical information structure or both.

The view of the organic concept of information as two interrelated sub-concepts of objective and subjective puts information into the heart of the philosophical problem of *what there is* (the objective) and how *what there is* can be known (subjective). It also forms the heart of the scientific-philosophical problem of what is a primary category, what are derived categories, and how do these categories link in the unity of understanding. This approach for defining information is a semantic theory of information that can form the foundation for the analysis and explanation of decision-choice rationality in general and specific rationalities such as classical rationality, bounded rationality, and others in particular, from the viewpoint of decision-choice environment induced by information and cognitive limitations in understanding destruction-construction processes in nature and society. It will further allow us to introduce the fuzzy paradigm with fuzzy rationality as a generalized methodology in understanding qualitative motions, subjective phenomena and categorial conversions. The conceptual system of definitions and information representations is shown in Figure 7.4.3.2.

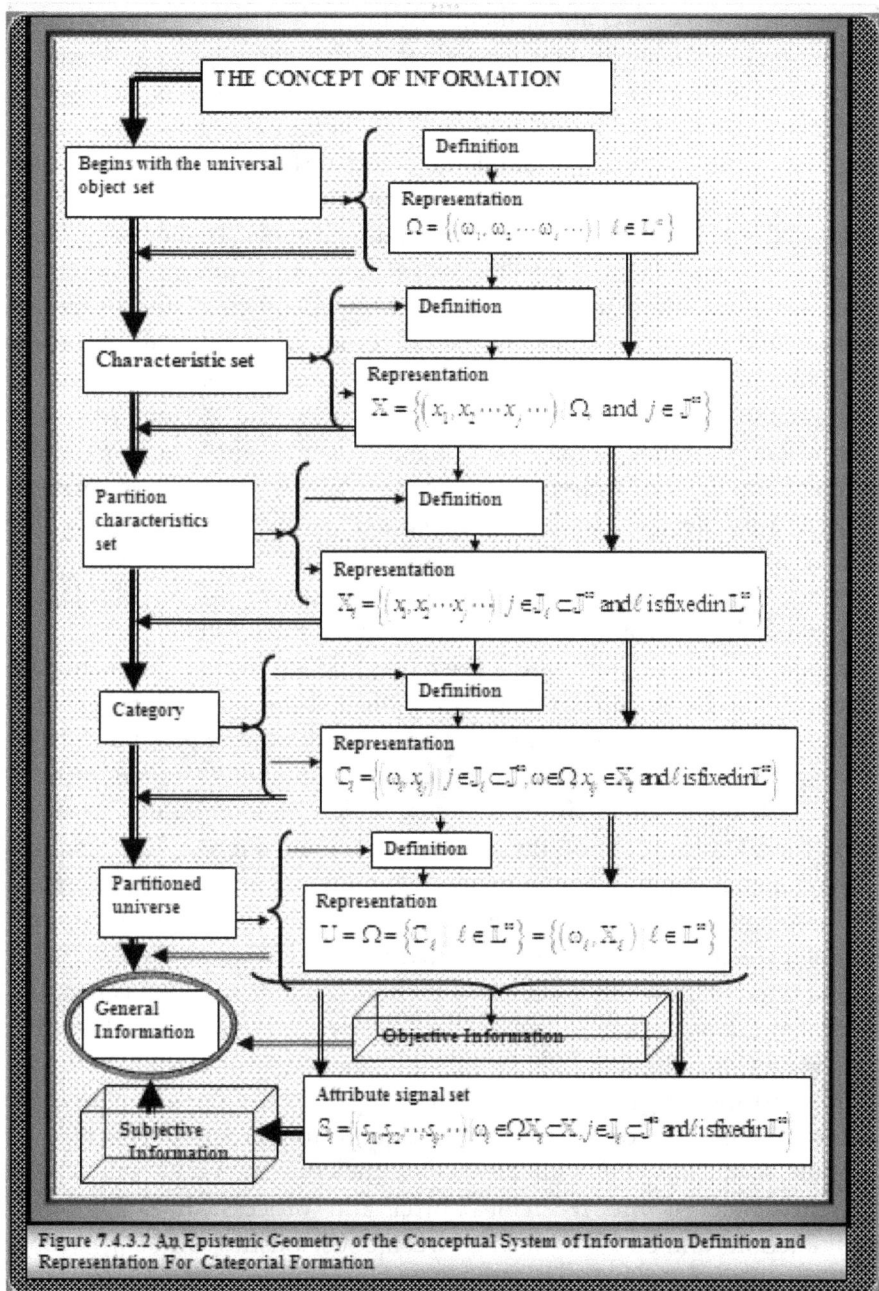

Figure 7.4.3.2 An Epistemic Geometry of the Conceptual System of Information Definition and Representation For Categorial Formation

At each state, for each object in the universe of objects, for each process in the world of processes, and for each event in the world of events, there is *something* about them that uniquely defines their individual identity and existence at their natural states that allows the identification of their existence by cognitive agents. This something is that which establishes the *varieties* in nature as well as defines channels of grouping or categorizing in terms of similarities and differences. It is the same something that gives rise to the capacity for language development composed of vocabulary, grammar and epistemic framework. The varieties thus establish the differences and similarities among objects and processes where their natural existence is independent of the awareness of living and non-living things in the universal object set. This objective-subjective quality approach to information definition allows us to separate the *philosophical problem of existence* from the *epistemic problem of knowability*. Existence of objects is objective in the ontological space and independent of the awareness of other objects, while knowability is subjective in the epistemological space and requires awareness on the part of affected objects. Alternatively, the objective-subjective quality approach to information definition allows us to separate the ontological identity of objects from epistemological identity of objects.

The awareness of the existence of objective information (characteristic-based information) is established through communication channels and sending of attribute signals among objects given a capacity of awareness. What is meant by awareness is an ability of a recipient to receive attribute signals from sources, assign meanings and interpret them for identification of the objects and their source. The attribute (characteristic) signals must be produced and transmitted by objects from the source. The signals must be received, processed, assigned meaning and interpreted by receiving objects that allow relations to be established among objects from this source and at the receiving end. This is relation-based information whose subjectivity is derived from the fact that knowability of characteristic-based information from the source in addition to meanings and interpretations of attribute signals depend on the conscious state of the receiving objects and process. The type of response to the subjective information at the conscious state will depend on conscience of the receiving object and process. It is here that contradictions and conflicts arise between consciousness and conscience.

The attribute signals are transmitted by energy and responded to through material media. Different classes of attribute signals will be transmitted and responded to through a material media with differential properties. The *degree of accuracy of transmission* of attribute signals depends on the quality of the set of properties of channels of transmission and the source. The *degree of accuracy* of reception at the receiving end will depend on the efficiency of the receptor component of the receiving object. The *degree of efficiency of the receptor* component of the receiving object will also depend on the organizational complexity of the awareness mechanism of the receiving object and its processing efficiency. The *degree of accuracy of response* by the receiving object will depend on the qualitative properties of the receiving object. The type of relationships that may be established among the sources and receiving ends will depend on the efficiencies of the receptor, processor and response mechanisms. Importantly, the *characteristic-based information* (objective information) defines and sets parameters of the environment for both the objects at the source and objects at the receiving end since all of them belong to the universal object set. The *relation-based information* (subjective information) defines the opportunity to append meanings and interpretations of the nature and type of environment in which objects exist. It is a stimuli-response process whose degrees of efficiency and interpretational accuracy depend on sending, receiving processing and interpretational mechanisms of entities with appropriate capacities. In fact, viewed as a system, it is self-exciting and self-correcting.

While all objects and processes in the universal object set produce and receive attribute signals in the *environment,* not all of them are probably equipped with the capacity to be aware of the signals that fill the environment through the information characteristics set. Thus, relationships that are formed among objects and processes may be divided into *active* and *passive*. Similarly, objects may be divided into those with an awareness capacity and those with a non-awareness capacity. Corresponding to awareness and non-awareness capacities are active and passive relations respectively. The objects with an awareness capacity to process and respond to attribute signals have active information relations while those with non-active awareness to process and respond to received attribute responses have passive information with other objects. The objects that can process and respond to received attribute signals around them can do so with varying degrees of accuracy. Such degrees of accuracy depend on the nature of the conscious state of objects and processes. Among the objects that have awareness capacity, there are some that intentionally seek information from their relevant environments for responses and adjustments to the active relations that they form with other objects. Special properties are required of objects and processes in order to be able to process, interpret, assign meaning to, and respond to attribute signals from other objects, processes and events in the general environment in which objects find themselves.

The nature of responses of objects in the environment to attribute signals and the relationships that may be established among objects in the environment depend on the complexities of internal organizations of objects for accuracies in receiving, naming, processing and interpreting such attribute signals. Just as information relations can be classified as either passive or active, so also can the objects and processes be grouped. If an object is incapable of responding to attribute signals then we shall refer to it as "info-passive". On the other hand, an object or process is said to be "info-active" if such an object or process is not only capable of the awareness of attribute signals, but is capable of responding to them. As presented, an object that is info-passive has passive relations with all other objects and hence lacks capacity of environmental awareness and learning. It, however has an internal awareness on the basis of its experiential information structure. Info-active objects and processes have active relations with all other objects and hence possess capacity of environmental awareness and learning. The info-active objects and processes while possessing the property of environmental awareness have differential capacity and accuracy to receive, recognize (names) process and respond to attribute signals. The differences are qualitative in the complexities of internal arrangements of characteristics, polyrhythmic vibrations and energies of objects.

The info-active objects and processes are those that are associated with differential development of cognitive systems. Out of this set emerges an intelligent life whose members additionally and intentionally seek information for executing a purpose. The sub-set of those members possessing intelligent life is composed of humans who are capable of creating names of categories and developing concepts from the simple meaning of attribute signals, and integrating them into a reason to understand externality and internality of environment for the decision-choice process, where accumulation of experience and revision of epistemological identity become engrained attributes and culture of behavior.

We may infer from the above decisions, that mathematical theory of information or informatics is devoted to the study of relation-based information. The measurements of the content of information is basically measurements of *quantity* of the content of attribute signals that reach the receiving objects and allow reductions of (possibilities) ignorance and improvement of awareness of the environment by info-active objects. The theories in sciences are about the studies of characteristics-based information that allow the understanding of the character and behavior of objects, as well as re-enforce the clarity of information relations and the content of attribute signals. The theories of scientific measurements are thus devoted to measurements of attributes in either aggregate or specificity that allow functional relations to be established between objective information and subjective information (that is, between characteristic-based information and relation-based information). Every object and process contains characteristic-based information whose received relation-based information will be less than or equal to that of characteristic-based information for any given receiving object or process in the epistemic activities of establishing the existence of ontological reality.

## 7.5 The Logico-Mathematical Structure of the Theory of Categorial Conversion under Fuzzy Laws of Thought

In the discussions on fuzzy paradigm relative to the classical paradigm as they may relate to the reasoning on categorial conversion in nature and society including the construction of thought, the process of knowing and the possible use of knowing to engineer transformation, we mentioned the concepts of polarity, duality, relational continuum and unity under the Africentric principle of opposites. It will be useful now to outline the essential characteristics of the concept of duality and how it presents a useful analytical tool for general understanding of the conversion of socio-natural categories, computability conditions of conversion and the design of a powerful cognitive framework for reasoning in simple and complex systems under static and dynamic conditions in various stages of qualitative dispositions. For analytical clarity, it is useful to note that the concepts of polarity, duality, relationality, relational continuum and unity, are derived from the organic principle of opposites, given the universal object set. The presence of qualitative dispositions, corresponding subjectivity, and inexactness of quantitative dispositions and defectiveness of objectivity in all elements renders the fuzzy paradigm with its laws of thought and corresponding mathematics as a powerful analytical framework for understanding the theory of categorial conversion in constructionism-reductionism duality.

## 7.5.1 The Concepts of the Actual and Potential in the Theory of Categorial Conversion

The theory of categorial conversion is about explanatory and prescriptive sciences of negation of negation in the actual-potential elements in the universal system. It starts its development from the notion that every element belongs to a category and every category resides in an actual-potential polarity just as every element and every category reside in a positive-negative duality. There are two types of universe which are made up of the actual and the potential. The actual universe is composed of the elements of *what there is*. The potential universe is composed of the elements of *what would be*. Both the actual and potential universes are infinite in structure. Every actual element has at least one corresponding potential element in conflict and competition. Similarly, every actual element has a corresponding characteristics set through its duality and every potential has also a corresponding characteristics set through its duality. The behavioral activities of the characteristics sets for both the actual and potential are defined in an *action space* $\mathbf{A}$ which is a collection of negative and positive actions under the principles of conflict. The action space, given an index set $\mathbb{L}$ may be defined as:

$$\mathbf{A} = \{\mathbf{a}_1, \mathbf{a}_2 \cdots \mathbf{a}_\lambda \mid \lambda \in \mathbb{L}\} \qquad (7.5.1.1)$$

For the purpose of verifying positive action and its effects on actual-potential polarities and positive-negative dualities, the characteristics sets may be identified as actual, $\mathfrak{A}$, and potential, $\mathfrak{P}$. The positive and negative elements have already been identified. The actual element is always a *singleton set* which will serve as the primary category in the transformation process. Its identity is revealed by the relational structure of its *negative* and *positive characteristic sets*. For any given actual element, there corresponds a set of potential elements ready to replace it. The set of potential elements relative to the replacement of the actual may be finite or infinite with elements that must be compared to the actual for initializing an internal transformation. The identity of each element in the potential is revealed by its internal relational structure of the negative and positive characteristic sets. An actual category is $\mathbb{C}_\mathfrak{A}$, and an actual characteristic set is $\mathbb{X}_\mathfrak{P}$. A potential category is $\mathbb{C}_\mathfrak{P}$ and a potential characteristic set is $\mathbb{X}_\mathfrak{P}$. In the space of the actual the characteristics set of each category may then be identified as $\mathbb{X}_{\mathbb{C}_\mathfrak{A}}$ while in the space of the potential the characteristic set of each category may be identified as $\mathbb{X}_{\mathbb{C}_{\mathfrak{p}}}$.

Let us keep in mind that each actual element and each potential element have a different characteristic set of the category to which it belongs. In this way, a completely qualitative transformation requires the establishment of differences between the identity of the actual and the identity of the potential elements, such that $\mathbb{C}_{\mathfrak{p}} \neq \mathbb{C}_{\mathfrak{a}}$, and hence the potential category is not the same as the actual category. This is called *categorial difference* which allows comparative distinction between different structures. The *action space* is a union of a set of *positive actions*, $\mathbf{A}^{\mathrm{P}} = \left\{ \mathbf{a}_1, \mathbf{a}_2 \cdots \mathbf{a}_\lambda \mid \lambda \in \mathbb{L}^{\mathrm{P}} \right\}$ and a set of *negative actions*, $\mathbf{A}^{\mathrm{N}} = \left\{ \mathbf{a}_1, \mathbf{a}_2 \cdots \mathbf{a}_\lambda \mid \lambda \in \mathbb{L}^{\mathrm{N}} \right\}$, such that $\mathbf{A} = \mathbf{A}^{\mathrm{P}} \bigcup \mathbf{A}^{\mathrm{N}}$, $\mathbf{A}^{\mathrm{P}} \bigcap \mathbf{A}^{\mathrm{N}} \neq \varnothing$ $\mathbb{L} = \mathbb{L}^{\mathrm{P}} \bigcup \mathbb{L}^{\mathrm{N}}$ and $\mathbb{L}^{\mathrm{P}} \bigcap \mathbb{L}^{\mathrm{N}} \neq \varnothing$ . The conditions of action interdependence in the action space and the avoidance of double counting are shown by the non-exclusivity of the action sets and non-exclusivity of the index sets that may be consistent with strategic complement. The negative and positive actions reside in duality under relational continuum and unity within a zero-sum power game of dominance and conversion of the actual to a potential or dominance and maintenance of the actual.

The relational structure of the actual-potential polarity and positive-negative duality with corresponding characteristic sets is shown as an epistemic geometry in Figure 7.5.1.1 in terms of zonal analysis. In Zone I the negative characteristics sets are the same for actual and potential elements, and hence there will be no qualitative conflict if the positive characteristics sets are the same. This is also the case in Zone III. The behavior of the characteristics sets in these Zones are said to be in temporary zonal qualitative equilibrium in terms of action selection and the outcome of changes in the state. The temporary qualitative equilibrium state is transferred to the respective dualities and the poles in which the dualities reside. Both Zones II and IV have differential negative and positive characteristics sets with conflicts seeking to negate each other for dominance to maintain the existing state or to actualize a potential. The behavior of these positive and negative characteristics sets are in disequilibrium that is translated to the corresponding dualities and to the poles of the actual-potential polarity in which the dualities reside.

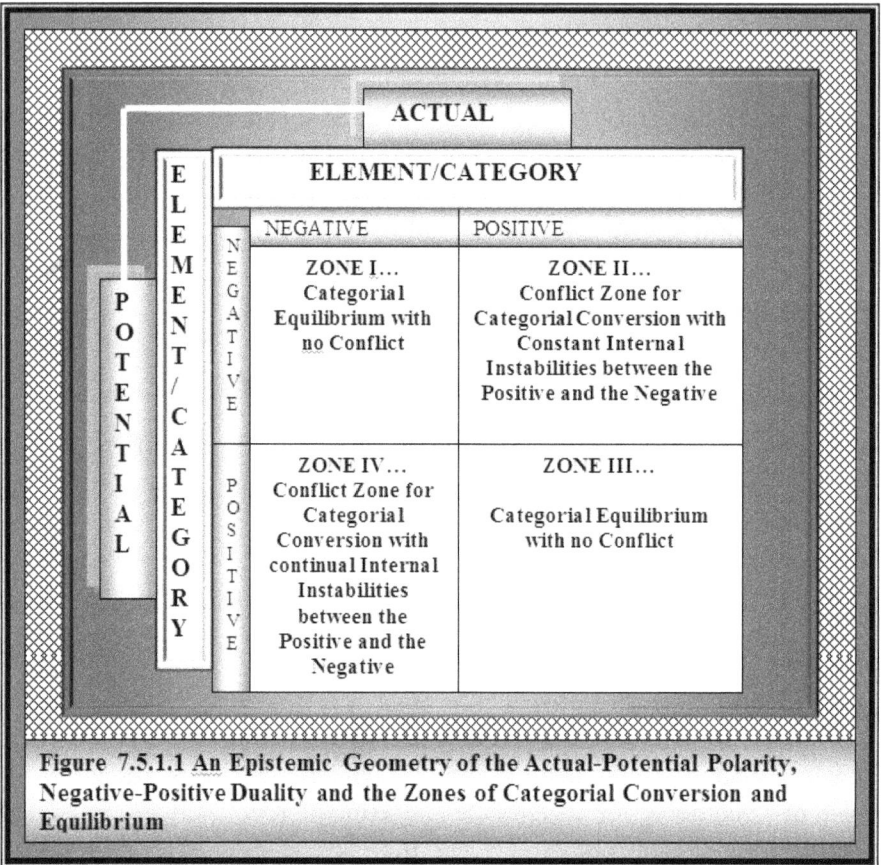

| | | NEGATIVE | POSITIVE |
|---|---|---|---|
| | | ZONE I...<br>Categorial Equilibrium with no Conflict | ZONE II...<br>Conflict Zone for Categorial Conversion with Constant Internal Instabilities between the Positive and the Negative |
| | | ZONE IV...<br>Conflict Zone for Categorial Conversion with continual Internal Instabilities between the Positive and the Negative | ZONE III...<br>Categorial Equilibrium with no Conflict |

**Figure 7.5.1.1 An Epistemic Geometry of the Actual-Potential Polarity, Negative-Positive Duality and the Zones of Categorial Conversion and Equilibrium**

The potential category emerges from the actual though the internal dynamics of the actual. It is the behavior of the internal conditions of the actual that also determines how the seed of the potential will geminate and grow as well as the kind of the potential that will replace it. In this way, a potential becomes the actual and the actual becomes a potential through the internal dynamics of the actual-potential polarity. The actual under its own dynamics to become a potential is to be derived from the actual. The actual always has a seed of self-destruction where every actual element serves as a primary category for an unknown potential. Similarly, every actual is a derivative from a previous primary category which has become a derived potential. The working mechanism in the potential-actual structure of socio-natural processes reveals itself as the disappearance of the old and the emergence of the new through the system's internal dynamics. It is under these conditions that the studies of self-organizing stochastic and self-organizing fuzzy-stochastic systems find important utility in understanding the theory of categorial conversion that may be extended to a unified theory of engineering sciences. This also holds for self-correcting and self-exited systems such as the information-knowledge production system, social systems and the universe itself. The Africentic implication drawn from these systems of categorial conversion is the postulate of never-ending process of creation where permanency of categorial elements is temporary.

The results of these processes are the complex works of synergetics of matter, energy and information. It is under these conditions that a claim can be made for the self-healing power and self-destructive power of the human body because the human body has both the seed of life and the seed of death with mutual negation. At the level of cognition, the science and art of knowing has its roots in information that works through knowledge, wisdom, and intelligence in the activities of innovation, *datamatics*, analytics and decision-choice activities under the principle of intentionality. Linguistic representations and symbolism of the information of the characteristic sets project a differential nature of these characteristic sets over the action space. Given the information structure, negative and positive strategic and tactical actions are developed to shape the decision-choice behavior in the action space. In this respect, and in relation to the transformations in the qualitative disposition of elements, one may speak of sub-theories of positive and negative actions with corresponding strategies in the action space within the theory of categorial conversion. Here, the qualitative disposition is intimately linked to quantitative disposition of the elements as they naturally exist in the actual-potential polarities.

The development of the theory of positive action, at the level of explanatory science is to create a framework, techniques and methods to explain how through the internal dynamics of an element a qualitative development of an internal force is generated to move an element in $\mathbb{C}_\mathfrak{p}$ into $\mathbb{C}_\mathfrak{a}$, and an element in the actual $\mathbb{C}_\mathfrak{a}$ into the potential $\mathbb{C}_\mathfrak{p}$. At the level of prescriptive science, the objective of the theory of categorial conversion is to develop a set of prescriptive rules which when followed will allow the transfer of the actual into the potential and the actualization of the potential. The whole process is simply the destruction of the actual and the creation of the potential. It is also *potentialization* of the actual and the *actualization* of the potential through creative destructions within the construction-destruction duality. In other words, from the space of the actual, the main concern of the theory of positive action involves the understanding of the dynamics of the behavior of the relative negative-positive characteristics sets to the destruction of the actual and the movement of it into the potential by altering its identity. This is the first negation. The main concern in the space of the potential is the understanding of the dynamics of the behavior of how the relative negative-positive characteristics sets become altered in favor of the potential and to bring it into the space of the actual with a new identity. This is the second negation. The two negations involve negation of negation with a never ending process of transformation within actual-potential dualities and within the actual-potential polarities.

When these processes take place by the system's own internal dynamics, we speak of *categorial conversion*, and the theory that explains these internal dynamics is what has been developed here as *the theory of categorial conversion* composed of intra-categorial and inter-categorial conversions. At the level of natural categorial conversion, examination of matter-energy production is to reflect the information content of natural actual-potential variables in order to understand how the natural potential is set against the natural actual. At the level of social categorial conversion, the knowledge-ideological production process is to examine the *information content* of the actual-potential variables in order to understand how to set the social potential against the social actual or to protect the actual from any potential element. From the space of ideology to the space of the actions, the main social concern shifts to the examination of the knowledge relevance to the transforming variables that reveal the goals and constraints imbedded in national interests and social vision. In the ideological plane, the test of transformation is defined on the credibility of the information content in terms of quality and quantity, while in the action space, the test of conversion is defined on the knowledge content for the acceptance of social vision, national interest and the social goal-objective set in terms of *categorial distance* defined by *categorial difference* in the actual-potential polarity. Let us keep in mind that in the theory of categorial conversion, the actual-potential polarity is the state of the process where the actual-potential relational structure defines the *state variables* of the qualitative control system. The corresponding dualities in relational continuum and unity generate the *control variables* that must be manipulated to create qualitative motion to bring about conversion where energies are generated by the duals in conflict and contradiction to define power relation for either positive or negative action in the aggregate to produce the force of motion required to move an element from one category to the other or to maintain the element in the same category.

## 7.5.2: The Concepts and Roles of Categories in the Theory of Categorial Conversion

The concept and the definition of category have been introduced and explicated in earlier chapters as well as in [R3.7] [R3.10] [R3.13] (see also [R2]). In this section, as part of the development of the mathematical representation of categorial conversion, it is useful to affirm the existence and the nature of categories to define the conditions of convertibility.

### Proposition 7.5.2.1: Existence of Categories

Every element belongs to a category and is defined by its opposites. Its identity is specified by a duality where every dual is specified by a non-empty negative characteristic set, $\mathbb{X}_C^N \neq \varnothing$, and a non-empty positive characteristic set, $\mathbb{X}_C^P \neq \varnothing$, such that its complete characteristic set, $\mathbb{X}_C = \left( \mathbb{X}_C^N \cup \mathbb{X}_C^P \right)$, with the conditions that $\left( \mathbb{X}_C^N \cap \mathbb{X}_C^P \right) \neq \varnothing$, $\left( \#\mathbb{X}_C^N \lesseqgtr \#\mathbb{X}_C^P \right)$ and $\pi^P = \left( \dfrac{\#\mathbb{X}_C^P}{\#\mathbb{X}_C} \right)$, $\pi^N = \left( \dfrac{\#\mathbb{X}_C^N}{\#\mathbb{X}_C} \right)$ where $\pi^P + \pi^N = 1$, in order to establish the identity of the element and its category in the ontological space. Its relative characteristic set is defined as $\rho_C = \left( \dfrac{\#\mathbb{X}_C^P}{\#\mathbb{X}_C^N} \right) \gtreqless 1$. The potential and the actual elements present themselves within the actual-potential polarity where the identity of every pole is defined by its residing duality and the identity of every dual is defined by its residing relative negative-positive characteristics set.

### Note: 7.5.2.1

As it has been previously pointed out, the negative and positive characteristics sets divide a unit into two opposites. They also unite them into a unity in a relational continuum. The negative and positive characteristics sets may be taken as discrete entities of attributes with nothing in common in order to define the qualitative essence of the elements and the category to which they belong. They may also be taken as residing in a continuum that expresses a smooth transition between the extremes to establish the identity of the elements through linkages. The linkages of the negative and positive characteristic sets are through relations that must be established, depending on the categorial qualitative disposition. The relative characteristic set defines the categorial quantitative disposition of each element and points to the home category. The relations are complementary, supplementary, give-and-take (reciprocity) and others such as asymmetry of force, dominance, governance and others that conditional necessity may require their defining attributes. All types of relations fall under the analytical concept of relationality which establishes the unity of the opposites in a continuum. The relations may change as time proceeds and transformations of characteristics take place to alter both the categorial qualitative and quantitative dispositions. It is this process of transformation that allows the potential to negate the actual and the actual to negate the potential in the transformation-substitution process from within the internal dynamics of relevant elements.

The essential thinking ideas in the theory of categorial conversion are: a) both negative and positive characteristics interact in fulfilling and supplying something that both the negative and the positive characteristics lack, to ensure their mutual existence and the survival of the elements in the category of their residence; and b) both negative and positive characteristics interact in fulfilling and supplying something that both the negative and the positive characteristics lack, to ensure their mutual destruction and the negation of the elements in the category of their residence. For example, the existence of neo-colonialism is such that the neo-colonialist supplies something to maintain its status as an imperialist while the neo colonial state supplies something to maintain its status as an occupied and a slave state. The qualitative reversal of this give-and-take process alters the relationship and brings about a negation. In this respect, the neo-colonialism and neo-colonial state supply each other with interdependent elements of freedom and oppression for their mutual destruction and negation through the internal dynamics of the actual pole and potential pole of the actual-potential polarity. Here, neo-colonialism with the neo-colonial state is the actual pole, and non-neocolonialism with independence is the potential pole.

It is this relationality that ensures the quality, identity and integrity of the elements for distinction from others in the relational space through informational acquaintance. The relationality of the positive and negative sub-sets of characteristics is essential in defining the concepts of dualism and duality as they relate to conflicts and tension under the principles of opposites. It further ensures the paradigms of categorial conversion with the developments of their logics and corresponding mathematics needed in the construct of revolutionary philosophy and ideology for social transformation as well as for the understanding of the dynamics of the emergence of the new and the disappearance of the old where life resides in death and death resides in life in actual-potential duality. Let us keep in focus that every element in the universal object set presents to the cognitive agents both *real cost* and *real benefit* or negative and positive characteristic sets as seen in terms of harm and usefulness whether conceptualized or not. It is this cost-benefit interdependence that defines the construction-destruction process under the principle of transformation-substitution actions for continual categorial conversion. The problem of this cost-benefit duality in socio-natural decision-choice actions in categorial conversions is described as the *asantrofi-anoma problem* in Akan linguistics where every element is composed of cost and benefit in relational continuum and unity which prevent a selection of benefit without the cost.

At this point and from Figure 5.2.2.2.1, we may specify the actual and potential negative characteristic sets as $\mathbb{X}^N_{C_\mathbf{a}}$ and $\mathbb{X}^N_{C_\mathbf{p}}$ respectively, and the actual and potential positive characteristic sets as $\mathbb{X}^P_{C_\mathbf{a}}$ and $\mathbb{X}^P_{C_\mathbf{p}}$. Since the space of the actual is given, we shall concentrate on the space of the potential. The proposition of the existence of categories combines the essential elements of qualitative and quantitative disposition of the universal elements. Using the concepts of negative and positive characteristics sets, we have specified the analytical distinction and similarity between the concepts of dualism and duality. Dualism relates to separation of opposites through non-relationality while duality relates to unity of the opposites through relationality. This idea has been extended to distinguish the classical paradigm with excluded middle from the fuzzy paradigm with a continuum and unity. The basic analytical structure of the logic of categorial conversion may be conceptualized in terms of the cancer-non-cancer disease system and a search for a cure or the management of the growth.

# CHAPTER EIGHT

## The Mathematical Problem and the
## Solution of the Categorial Conversion of Actual-Potential Polarity in Socio-Natural Systemicity

Given the mathematical foundations, symbolic representation and conditions of convertibility, the mathematical problem and the needed solution may now be presented. This require the introduction of the concept of action space and the space tactics and strategies as directly or indirectly utilized by the negative and positive characteristic set associated with any actual-potential polarity. Every actual or potential element of a category is in a state of both active internal positive and negative actions defined in an action space within the field of negation and transformations that brings about a change in polar identities leading to a new actual-potential polarity. The universe is seen in terms of a collection of actual-potential polarities expressed through categories, dualities and negative and positive characteristic sets with conflicts, energy and force that generate internal qualitative self-motion. In other words, the universe is a collection of self-exciting, self-correcting, self-learning and self-transforming elements. The resolution to one categorial conversion of actual-potential polarity leads to the rise of another conversion problem of a new actual-potential polarity. In this respect, any categorial conversion of an existing actual-potential polarity is a destruction of the existing actual. It is also a contestant of a new potential to be actualized in the new actual-potential polarity, where the nature of the information structure shapes the intensity of conflicts and contradictions of the underlying forces and the direction of resistance and conversion. This categorial-conversion process gives meaning to the statement that everything has capacity of change as well as under transformation from within.

## 8.1 The Structures of the Polarity, Duality and the Action Space in Categorial Convertions

Transformation of actual-potential polarities depends on the negative-positive polar actions from the action space $\mathbf{A}$. The action space $\mathbf{A}$ as has been specified in an above section is composed of positive and negation actions. The positive characteristics set selects a positive action to execute a program of positive conversion on behalf of the positive duality to transform the potential pole of the actual-potential polarity, while the negative characteristics set selects a negative action to execute a program of resistance to the positive conversion on behalf of the negative duality to maintain the conditions of the actual pole of the actual-potential polarity in the way it is. These are the conditions of the *actualization-potentailization process* in the conditions of a negation-of-negation phenomenon that is connected to the action space. The actualization of the potential and the potentialization of the actual will depend on the relative effectiveness of the positive and negative actions where positive action is always opposed to and in support of negative action and vice versa in any duality under the principles of continuum and unity for qualitative transformation induced by categorial conversion. Within the action space is the space of strategies, $\mathbf{S}$, and the space of tactics, $\mathbf{T}$, where a strategy may be seen as a sequence of tactical steps that belongs to the action field and is undertaken by both the positive and negative characteristic sets. The structural relation in continuum and unity is shown as an epistemic geometry in Figure 8.1.1. Let us keep in mind that the whole categorial-conversion system is such that every pole has a residing duality whose identity is revealed by either the duality is negative or positive within the conversion problem of interest. The duality is negative or positive according to the nature of the dominating dual.

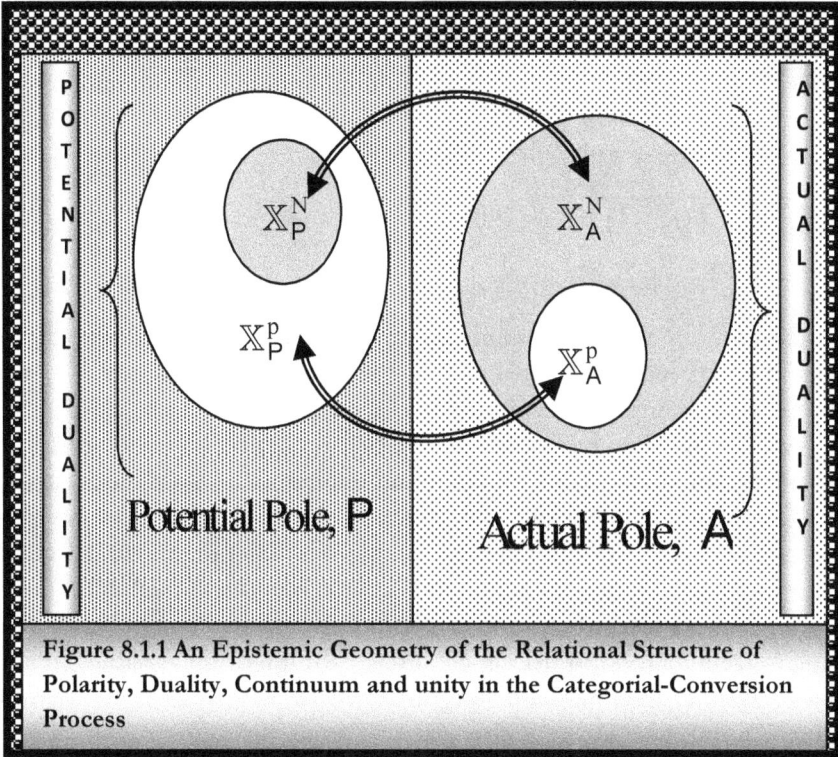

**Figure 8.1.1 An Epistemic Geometry of the Relational Structure of Polarity, Duality, Continuum and unity in the Categorial-Conversion Process**

The categorial conversion process is such that the identity of every duality, $\mathbf{D}$ may be defined as a function of the negative and positive proportions at any point of time as well as the transformation dynamics of their proportions. In this respect, we may specify the potential duality $\mathbf{D}^{\mathfrak{p}}$ that resides in the potential pole $\mathfrak{p}$ and actual duality $\mathbf{D}^{\mathfrak{a}}$ that resides in the actual pole $\mathfrak{a}$ as:

$$\left.\begin{array}{l}\mathbf{D}^{\mathfrak{p}}=\mathbf{D}^{\mathfrak{p}}\left(\pi_{\mathfrak{p}}^{\mathrm{P}}(t),\pi_{\mathfrak{p}}^{\mathrm{N}}(t)\right),\dfrac{\partial\mathbf{D}^{\mathfrak{p}}}{\partial\pi_{\mathfrak{p}}^{\mathrm{P}}}\dot{\pi}_{\mathfrak{p}}^{\mathrm{P}}>0,\dfrac{\partial\mathbf{D}^{\mathfrak{p}}}{\partial\pi_{\mathfrak{p}}^{\mathrm{N}}}\dot{\pi}_{\mathfrak{p}}^{\mathrm{N}}<0,\ \text{Potential Duality}\\[4mm]\mathbf{D}^{\mathfrak{a}}=\mathbf{D}^{\mathfrak{a}}\left(\pi_{\mathfrak{a}}^{\mathrm{P}}(t),\pi_{\mathfrak{a}}^{\mathrm{N}}(t)\right),\dfrac{\partial\mathbf{D}^{\mathfrak{a}}}{\partial\pi_{\mathfrak{a}}^{\mathrm{P}}}\dot{\pi}_{\mathfrak{a}}^{\mathrm{P}}<0,\dfrac{\partial\mathbf{D}^{\mathfrak{a}}}{\partial\pi_{\mathfrak{a}}^{\mathrm{N}}}\dot{\pi}_{\mathfrak{a}}^{\mathrm{N}}>0,\ \text{Actual Duality}\end{array}\right\}\quad(8.1.1)$$

Let us notice that $\left(\pi_{\mathfrak{p}}^{\mathrm{P}}(t)\ \text{and}\ \pi_{\mathfrak{p}}^{\mathrm{N}}(t)\right)$ constitute a description of the positive and negative duals in the potential duality with time rates of conversions $\left(\dot{\pi}_{\mathfrak{p}}^{\mathrm{P}}(t)\ \text{and}\ \dot{\pi}_{\mathfrak{p}}^{\mathrm{N}}(t)\right)$ in the potential pole, while $\left(\pi_{\mathfrak{a}}^{\mathrm{P}}(t)\ \text{and}\ \pi_{\mathfrak{a}}^{\mathrm{N}}(t)\right)$ constitute a description of the positive and negative duals with time rate of conversions $\left(\dot{\pi}_{\mathfrak{a}}^{P}(t) \text{ and } \dot{\pi}_{\mathfrak{a}}^{N}(t)\right)$ in the actual duality in the actual pole. The system is such that as the proportion of the positive (negative) characteristic set increases, the proportion of then negative (positive) falls since the conditions $\left(\pi_{\mathfrak{a}}^{P}(t)+\pi_{\mathfrak{a}}^{N}(t)=1\right)$ and $\left(\pi_{\mathfrak{p}}^{P}(t)+\pi_{\mathfrak{p}}^{N}(t)=1\right)$ must hold at all times with $\left(\left|\dot{\pi}_{\mathfrak{a}}^{P}\right|=\left|\dot{\pi}_{\mathfrak{a}}^{N}\right|\right)$ and $\left(\left|\dot{\pi}_{\mathfrak{p}}^{P}\right|=\left|\dot{\pi}_{\mathfrak{p}}^{N}\right|\right)$. The strategic action for any polarity is defined in dualities with continuum and unity where every positive (negative) action has its opposing negative (positive) action. An increase in the positive (negative) proportion is completely compensated by a fall in the negative (positive) proportion by the principle of mutual negation in the categorial conversion. The positive-negative action process is similar to the behavior of relevance-irrelevance duality of a selection of strategy to accomplish a goal. As the positive (negative) action in the actual duality increases (falls), the negative (positive) action falls (increases).

The integration of all these conditions allows one to relate the changes in increases and decreases in terms of linguistic variables that are subject to interpretation and judgment within any duality by cognitive agents. Within the duality, the positive dual defined by the positive characteristic set deals with the positive action in strategies and tactics; the negative dual defined by the negative characteristic set deals with the negative action in strategies and tactics in conflicts and qualitative changes of the duality. The linguistic variables are expressed in linguistic quantities such as big, medium and small and then related to degrees of effectiveness of actions that are undertaken by the characteristic sets in either the potential or the actual pole of the actual-potential polarity. These linguistic quantities are defines in a fuzzy space. All the internal conflicts, force and energy are generated by the actions and strategic behavior of the negative and positive characteristic sets and in terms of positive action and negative action. The direction of qualitative motion of the duality **D** depends on the changes in the relative characteristic set. The positive and negative actions are in a continuous domain and may now be specified in dualistic form as:

$$\mathbf{a}^{P}=\mathbf{a}^{P}\left(\mathbb{X}^{P},\mathbb{X}^{N}\right),\ \frac{\partial\mathbf{a}^{P}}{\partial\mathbb{X}^{P}}>0,\ \frac{\partial\mathbf{a}^{P}}{\partial\mathbb{X}^{N}}<0 \ \ \text{Pasitive Action (8.1.2a)}$$

$$\mathbf{a}^N = \mathbf{a}^N\left(\mathbb{X}^N, \mathbb{X}^P\right), \quad \frac{\partial \mathbf{a}^N}{\partial \mathbb{X}^N} > 0, \quad \frac{\partial \mathbf{a}^N}{\partial \mathbb{X}^P} < 0 \quad \text{Negative Action} \quad (8.1.2b)$$

The selected positive or negative action may either be big, medium or small in the degree of effectiveness. These various degrees of effectiveness are in continuum and may be related to the transformational activities within the actual-potential polarities through the residing dualities with the positive and negative characteristic sets.

The degrees of effectiveness of the positive action are in relation to changing the actual to allow the emergence of a new actual. In other words, the positive action is directed toward the destruction of the conditions that allow the actual to exist and be maintained. The reward to the positive characteristic set for the positive action is the destruction of the existing actual and the emergence of a new actual from the potential. The degrees of effectiveness of the negative action are in relation to maintaining the actual and preventing the emergence of a new actual. In other words, the negative action is directed toward the maintenance of the conditions that allow the actual to exist and be sustained by preventing the emergence of a new actual from the potential. The reward to the negative characteristic set for the negative action is the maintenance of the existing actual and the prevention of the emergence of a new actual from the potential. The degrees of effectiveness of both positive and negative actions may be specified in terms of fuzzy membership characteristic functions, where the effectiveness of the actions contributing to the categorial conversion are specified as fuzzy characteristic functions which also implicitly depend on the individual actions of the members in the negative and positive characteristic sets in the same duality of either the potential pole or the actual pole of the actual-potential polarity.

In specifying the actions in the categorial conversion process, let us keep in mind that one only needs to deal with the negative and positive actions of the duality in the actual pole of the polarity. There is only one actual element and there are many elements of the potential. The conflict is about a change and no-change phenomenon. The destruction of the existing actual may bring an unintended potential element that gives meaning to the phrase *unintended consequences*. The activities of the positive and negative poles, through their residing dualities, are interdependent in a relational continuum and unity. The opposite activities of the actual are relationally taking place in the potential where an element in the potential space may be actualized. The negative and positive actions in the duality of the actual pole simultaneously transform both the actual and the potential poles through their opposite relational interaction in continuum and unity. In this respect, the actions in the actual pole are constrained by, as well as dependent on the behavior of the potential duality and its pole. Similarly, the positive actions undertaken by the positive characteristic set in the actual duality are constrained by the negative actions undertaken by the negative characteristic set, as the categorial conversion process is related to the negation of the actual and the negation of the potential. The process is such that the actions in the actual pole must be such that the preferentially relevant element in the potential space must be identified to constitute the potential pole to the actual pole in forming the actual-potential polarity. It must also be kept in mind, as it has been discussed and explained in the previous chapters, that there are more than one potential element in the potential space which is pregnant with possibilities [R3.7] [R3.10] [R3.13][R10] [R10.2] [R10.11] [R10.18]. It is therefore possible, that the desired potential element may not be actualized, and instead, the non-desired potential element may be actualized in the process especially in social polarities and also in intervention practices in natural polarities.

In the actual pole, we may specify both the positive and negative actions under the conditions effectiveness relative to the values of the sizes of the actions. The sizes of effectiveness as actions increase may be specified as fuzzy sets with membership characteristic functions that are also implicitly interdependent on individual actions of the elements of the characteristic sets in the actual duality. The conditional activities are such that if $\mathbf{a}_i^P$ and $\mathbf{a}_j^N$ are the actions of the $i$-element in the positive characteristic set and j-element in the negative characteristic set, then

$$\mathbf{a}^P = \sum_{i=1}^{\#(\mathbb{X}^P)} \mathbf{a}_i^P \text{ and } \mathbf{a}^N = \sum_{i=1}^{\#(\mathbb{X}^N)} \mathbf{a}_i^N$$

respectively. Let us keep in mind that the aggregative actions may be constructed as weighted values $\gamma_i$ and $\lambda_j$ where the weights may be specified as degrees of individual participations.

Relational continuum and unity imply an inseparability and interdependence of the actual and potential dualities as well as the actual and potential poles of all polarities. The interdependence implies that the choice of positive action by the positive characteristic set in the actual duality, besides being constrained by the negative action of the negative characteristic set, also affects the simultaneous behavior of the positive and negative characteristic sets of the potential pole by inducing qualitative motion to define the disappearance of the old and the emergence of the new under the principles of continuum and unity. The process reveals itself in terms of expansions and contractions of the positive and negative characteristic sets in opposite directions for dominance.

When the positive characteristic set in the actual duality expands due to an increasing positive action in converting some elements in the negative characteristic set, the negative characteristic set in the actual duality contracts, while simultaneously, the positive characteristic set contracts and the negative characteristic set expands in the potential duality. There is an export-import process of the characteristics taking place in the actual-potential polarity changing the polar qualitative disposition through the behavior of the quantitative disposition of the characteristic sets. In this respect, four qualitative state equations are required for the descriptive notion of the states. Two equations of motion relate to the state of the actual duality and the pole, while two relate to the state of the potential duality as seen in the potential pole in the actual-potential polarity.

## 8.2 Fuzzy Rationality, Degrees of Effectiveness of the Positive and Negative Actions in Logical Dualization

The interdependent qualitative state equations may be specified for actual duality and its pole as well as the potential duality and its pole in terms of the degrees of effectiveness of the positive and negative actions under fuzzy rationality, relational continuum and unity. Let $\mathfrak{D}^A$ be the fuzzy set of degrees of effectiveness of actions in the actual duality.

### 8.2.1 Actual Duality in the Actual Pole

Degrees of Effectiveness of Positive Action

$$\mathfrak{A}^{\mathfrak{a}} = \left\{ \left( \mathbf{a}^{\mathrm{p}}, \mu_{\mathfrak{A}^{\mathfrak{a}}}\left(\mathbf{a}^{\mathrm{p}}\right)\right) \mid \mathbf{a}^{\mathrm{P}} \in \mathbf{A}^{\mathfrak{a}}, \mu_{\mathfrak{A}^{\mathfrak{a}}}\left(\mathbf{a}^{\mathrm{p}}\right) \in (0,1), \frac{d\mu_{\mathfrak{A}^{\mathfrak{a}}}\left(\mathbf{a}^{\mathrm{p}}\right)}{d\mathbf{a}^{\mathrm{p}}} > 0 \right\} \qquad (8.2.1.1a)$$

Degrees of Effectiveness of Negative Action

$$\mathscr{B}^{\mathfrak{a}} = \left\{ \left( \mathbf{a}^{N}, \mu_{\mathscr{B}^{\mathfrak{a}}} \left( \mathbf{a}^{N} \right) \right) \mid \mathbf{a}^{N} \in \mathbf{A}^{\mathfrak{a}}, \mu_{\mathscr{B}^{\mathfrak{a}}} \left( \mathbf{a}^{N} \right) \in (0,1), \frac{d\mu_{\mathscr{B}^{\mathfrak{a}}} \left( \mathbf{a}^{N} \right)}{d\mathbf{a}^{N}} < 0 \right\} \quad (8.2.1.1b)$$

In other words, as the activities of the positive characteristic set increases the degree of effectiveness of the positive action increases due to the expansion of the positive characteristic set in the actual duality. In response, the negative characteristic set is reduced in the actual duality leading to a reduction of the effectiveness of the negative action in the actual pole. In this respect, the action system is such that $\mathfrak{A}^{\mathfrak{a}} \subset \mathfrak{D}^{\mathfrak{a}}$ and $\mathscr{B}^{\mathfrak{a}} \subset \mathfrak{D}^{\mathfrak{a}}$.

## 8.2.2 Potential Duality in the Potential Pole

Similar representation may be specified for the potential duality where $\mathfrak{D}^{\mathfrak{p}}$ is the fuzzy set of degree s of effectiveness of actions in the potential duality.

Degrees of Effectiveness of Positive Action

$$\mathfrak{A}^{\mathfrak{p}} = \left\{ \left( \mathbf{a}^{P}, \mu_{\mathfrak{A}^{\mathfrak{p}}} \left( \mathbf{a}^{P} \right) \right) \mid \mathbf{a}^{P} \in \mathbf{A}^{\mathfrak{p}}, \mu_{\mathfrak{A}^{\mathfrak{p}}} \left( \mathbf{a}^{P} \right) \in (0,1), \frac{d\mu_{\mathfrak{A}^{\mathfrak{p}}} \left( \mathbf{a}^{P} \right)}{d\mathbf{a}^{P}} < 0 \right\} \quad (8.2.2.1a)$$

Degrees of Effectiveness of Negative Action

$$\mathscr{B}^{\mathfrak{p}} = \left\{ \left( \mathbf{a}^{N}, \mu_{\mathscr{B}^{\mathfrak{p}}} \left( \mathbf{a}^{N} \right) \right) \mid \mathbf{a}^{N} \in \mathbf{A}^{\mathfrak{p}}, \mu_{\mathscr{B}^{\mathfrak{p}}} \left( \mathbf{a}^{N} \right) \in (0,1), \frac{d\mu_{\mathscr{B}^{\mathfrak{p}}} \left( \mathbf{a}^{N} \right)}{d\mathbf{a}^{N}} > 0 \right\} \quad (8.2.2.1b)$$

## 8.2.3 Relational Interactions, Continuum and Unity in the Actual-Potential Polarity

As the activities of the negative and positive characteristic sets are taking place in the actual duality of the actual pole, reversal activities are simultaneously taking place in the potential duality of the potential pole. In the potential duality and in the potential pole, the size of the negative characteristic set expands as the size of the negative characteristic set in the actual duality shrinks leading to increasing negative action in the potential duality. At the same time, the size of the positive characteristic set in the potential duality shrinks, leading to a reduction in the degree of effectiveness of the negative action in the potential duality and the potential pole. In this respect, the action system in the potential duality is such that $\mathfrak{A}^{\mathfrak{p}} \subset \mathfrak{D}^{\mathfrak{p}}$ and $\mathfrak{B}^{\mathfrak{p}} \subset \mathfrak{D}^{\mathfrak{p}}$.

The complete interactive process in the actual-potential polarity is in continuum through the principle of the export-import process in the residing duality of actual pole of the actual-potential polarity as the double negation takes place. The categorial conversion process is such that as the size of the positive characteristic sub-set of the duality expands with an increasing degree of effectiveness of the positive action in the positive dual of the duality within the actual pole, while the size of the negative characteristic sub-set contracts with a decreasing degree of effectiveness of the negative action in the negative dual with the actual pole to create the dominance of the positive dual in the residing duality of the actual pole. Similarly, as the size of the positive characteristic sub-set contracts with a decreasing degree of effectiveness of the positive action in the positive dual of the duality within the potential pole, the size of the negative characteristic sub-set expands with an increasing degree of effectiveness of negative action in the negative dual of the duality within the potential pole to create the dominance of the negative dual in the duality of the potential pole. The result of these processes is to potentialize the actual and actualize a potential, where the changes in the size of the characteristic sub-sets preset conditions of quantitative disposition to generate conditions of change of qualitative disposition in the quantity-quality duality with relational continuum and unity. Here, the principle of opposites finds meaning in the opposite activities and actions of the negative and positive element that will replace the actual as a possibility that may either be unknown or at best fuzzily known. The analytics of the categorial conversion is on the qualitative dynamics of the actual duality which must be related to a desired potential element from the potential space.

The understanding of categorial conversion is through the understanding of competing actions of the negative and positive characteristic sets in the duality of the actual pole of the actual-potential polarity where such actions must be defined in terms of degrees of effectiveness of negations captured by the descriptive notion of the fuzzy membership characteristic sets on linguistic numbers of degrees of effectiveness of actions in the actual pole in terms of strategic interactions. The strategic interactive behaviors of both negative and positive characteristic sets express themselves in a series of conflicts and temporary qualitative equilibrium states through the quantitative behaviors of the characteristic sets. The conflicts and the temporary qualitative equilibrium states may be specified as fuzzy process in terms of *categorial fuzzy decision problem* through the membership characteristic functions that involve the degrees of effectiveness of the positive and negative actions by the competing positive and negative characteristic sub-sets or the duals to bring about double negation.

The process of continual negations through categorial conversions at each temporary equilibrium state may be seen through the competing strategic interactions captured through the fuzzy membership characteristic functions of the degrees of effectiveness of the positive and negative actions of the respective characteristic sets creating a fuzzy strategic decision problem, $\Delta^{\mathfrak{a}}$ where

$$\Delta^{\mathfrak{a}} = \left( \mathfrak{A}^{\mathfrak{a}} \cap \mathfrak{B}^{\mathfrak{a}} \right) = \left\{ \left( \mathbf{a}, \mu_{\Delta^{\mathfrak{a}}}(\mathbf{a}) \right) \mid \mathbf{a} \in \left( \mathfrak{A}^{\mathfrak{a}} \cap \mathfrak{B}^{\mathfrak{a}} \right) \subset \mathcal{D}^{\mathfrak{a}}, \mu_{\Delta^{\mathfrak{a}}}(\mathbf{a}) = \left( \mu_{\mathfrak{A}^{\mathfrak{a}}}(\mathbf{a}) \wedge \mu_{\mathfrak{B}^{\mathfrak{a}}}(\mathbf{a}) \right) \in [0,1] \right\}$$

(8.2.3.1a)

A similar fuzzy decision-choice problem may be specified for the potential duality and pole as:

$$\Delta^{\mathfrak{p}} = \left( \mathfrak{A}^{\mathfrak{p}} \cap \mathfrak{B}^{\mathfrak{p}} \right) = \left\{ \left( \mathbf{a}, \mu_{\Delta^{\mathfrak{p}}}(\mathbf{a}) \right) \mid \mathbf{a} \in \left( \mathfrak{A}^{\mathfrak{p}} \cap \mathfrak{B}^{\mathfrak{p}} \right) \subset \mathcal{D}^{\mathfrak{p}}, \mu_{\Delta^{\mathfrak{p}}}(\mathbf{a}) = \left( \mu_{\mathfrak{A}^{\mathfrak{p}}}(\mathbf{a}) \wedge \mu_{\mathfrak{B}^{\mathfrak{p}}}(\mathbf{a}) \right) \in [0,1] \right\}$$

(8.2.3.1b)

This is a way of representing strategic interactions in the fuzzy information space in which positive and negative characteristic sets work for negation of negation. An introduction of a probability measure over fuzzy events will be required if the system of strategic interactions is defined in a fuzzy-stochastic information space where it becomes necessary to deal with quantum fuzzy-stochastic variables. The system is complex where the breakeven point of the size of a mutual degree of effectiveness of the actions of the positive and negative characteristic sets is complicated as a point of transition for both the positive and negative categorial conversions.

The fuzzy strategic interactive decision-choice problem for the double negation must be solved to obtain the breakeven point through the use of a fuzzy optimization principle for the actual duality and pole. The disappearance of the actual implies the emergence of the new which may or may not be desired. From the breakeven point, the success of the categorial conversion of the positive action or the success of the retention of the existing state by the negative action are definable. The simultaneous strategic interactions are taking place in both the space of the actual and the space of the potential where the essential difference is that the actual is known and the potential is a set of possibilities that are not known or fuzzily anticipated since the destruction of the actual through the transformation-substitution process implies a substitution of a potential. In both spaces, the degrees of effectiveness of the negative actions are constraints on the degrees of effectiveness of the negative actions and vice versa. This is the goal-constraint process in all decision-choice processes with the implications defined within the cost-benefit duality. The fuzzy categorial-conversion decision-choice problem may be specified as maximization of the degree of effectiveness of the positive action subject to the degree of effectiveness of the negative action in the fuzzy space or fuzzy-stochastic space where $\mu_{\Delta^{\mathfrak{a}}}(\mathbf{a}) = \left(\mu_{\mathfrak{A}^{\mathfrak{a}}}(\mathbf{a}) \wedge \mu_{\mathfrak{B}^{\mathfrak{a}}}(\mathbf{a})\right)$ is optimized. The fuzzy categorial-conversion decision problem needs some modification if the goal is the maintenance of the actual. In this respect, the problem is to maximize the degree of effectiveness of the negative action subject to the degree of effectiveness of the positive action to create the negative dominance in the fuzzy space

The optimization of the fuzzy categorial-conversion decision-choice problem is simply to compute the breakeven point that defines equal degrees of effectiveness of the positive and negative actions to provide information that will allow for the analysis of zones of conversion.

$$\underset{\mathbf{a} \in \left(\mathfrak{A}^{\mathfrak{a}} \cap \mathfrak{B}^{A\mathfrak{a}}\right)}{\text{Opt}} \mu_{\Delta^{\mathfrak{a}}}(\mathbf{a}) = \underset{\mathbf{a} \in \left(\mathfrak{A}^{\mathfrak{a}} \cap \mathfrak{B}^{\mathfrak{a}}\right)}{\text{Opt}} \left(\mu_{\mathfrak{A}^{\mathfrak{a}}}(\mathbf{a}) \square \mu_{\mathfrak{B}^{\mathfrak{a}}}(\mathbf{a})\right) \quad (8.2.3.2)$$

The solution may be defined as an equivalent fuzzy computable system of maximization and minimization with the following equivalent theorem

Theorem 8.2.3.1

The problem $\underset{\mathbf{a} \in \left(\mathfrak{A}^{\mathfrak{a}} \cap \mathfrak{B}^{\mathfrak{a}}\right)}{\text{Opt}} \mu_{\Delta^{\mathfrak{a}}}(\mathbf{a}) = \underset{\mathbf{a} \in \left(\mathfrak{A}^{\mathfrak{a}} \cap \mathfrak{B}^{\mathfrak{a}}\right)}{\text{Opt}} \left(\mu_{\mathfrak{A}^{\mathfrak{a}}}(\mathbf{a}) \square \mu_{\mathfrak{B}^{\mathfrak{a}}}(\mathbf{a})\right)$

is equivalent to

$$\left.\begin{array}{l} \displaystyle\max_{\mathbf{a}\in\left(\mathfrak{A}^{\mathfrak{a}}\cap\mathfrak{B}^{A\mathfrak{a}}\right)}\mu_{\mathfrak{A}^{\mathfrak{a}}}(\mathbf{a}) \\[3ex] \text{s.t}\left[\mu_{\mathfrak{A}^{\mathfrak{a}}}(\mathbf{a})-\mu_{\mathfrak{B}^{\mathfrak{a}}}(\mathbf{a})\right]\le 0 \\[3ex] \displaystyle\lim_{\mathbf{a}\to\infty}\mu_{\mathfrak{A}^{\mathfrak{a}}}(\mathbf{a})\to 1 \text{ and } \lim_{\mathbf{a}\to\infty}\mu_{\mathfrak{B}^{\mathfrak{a}}}(\mathbf{a})\to 0 \\[2ex] \displaystyle\lim_{\mathbf{a}\to 0}\mu_{\mathfrak{A}^{\mathfrak{a}}}(\mathbf{a})\to 0 \text{ and } \lim_{\mathbf{a}\to 0}\mu_{\mathfrak{B}^{\mathfrak{a}}}(\mathbf{a})\to 1 \end{array}\right\}=\mathbf{a}^{*}$$

$$\left.\begin{array}{l} \displaystyle\min_{\mathbf{a}\in\left(\mathfrak{A}^{\mathfrak{a}}\cap\mathfrak{B}^{\mathfrak{a}}\right)}\mu_{\mathfrak{A}^{\mathfrak{a}}}(\mathbf{a}) \\[3ex] \text{s.t}\left[\mu_{\mathfrak{A}^{\mathfrak{a}}}(\mathbf{a})-\mu_{\mathfrak{B}^{\mathfrak{a}}}(\mathbf{a})\right]\le 0 \\[3ex] \displaystyle\lim_{\mathbf{a}\to\infty}\mu_{\mathfrak{A}^{\mathfrak{a}}}(\mathbf{a})\to 1 \text{ and } \lim_{\mathbf{a}\to\infty}\mu_{\mathfrak{B}^{\mathfrak{a}}}(\mathbf{a})\to 0 \\[2ex] \displaystyle\lim_{\mathbf{a}\to 0}\mu_{\mathfrak{A}^{\mathfrak{a}}}(\mathbf{a})\to 0 \text{ and } \lim_{\mathbf{a}\to 0}\mu_{\mathfrak{B}^{\mathfrak{a}}}(\mathbf{a})\to 1 \end{array}\right\}=\mathbf{a}^{*}$$

Similarly,

The problem $\displaystyle\mathop{\mathrm{Opt}}_{\mathbf{a}\in\left(\mathfrak{A}^{\mathfrak{a}}\cap\mathfrak{B}^{\mathfrak{a}}\right)}\mu_{\Delta^{\mathfrak{a}}}(\mathbf{a})=\mathop{\mathrm{Opt}}_{\mathbf{a}\in\left(\mathfrak{A}^{\mathfrak{a}}\cap\mathfrak{B}^{\mathfrak{a}}\right)}\left(\mu_{\mathfrak{A}^{\mathfrak{a}}}(\mathbf{a})\square\mu_{\mathfrak{B}^{\mathfrak{a}}}(\mathbf{a})\right)$ is

equivalent to

$$\left.\begin{array}{l} \min_{\mathbf{a}\in\left(\mathfrak{A}^{\mathfrak{a}}\cap\mathfrak{B}^{\mathfrak{a}}\right)}\mu_{\mathfrak{B}^{\mathfrak{a}}}\left(\mathbf{a}\right) \\[2em] \text{s.t } \left[\mu_{\mathfrak{A}^{\mathfrak{a}}}\left(\mathbf{a}\right)-\mu_{\mathfrak{B}^{\mathfrak{a}}}\left(\mathbf{a}\right)\right]\le 0 \\[2em] \lim_{\mathbf{a}\to\infty}\mu_{\mathfrak{A}^{\mathfrak{a}}}\left(\mathbf{a}\right)\to 1 \text{ and } \lim_{\mathbf{a}\to\infty}\mu_{\mathfrak{B}^{\mathfrak{a}}}\left(\mathbf{a}\right)\to 0 \\[1em] \lim_{\mathbf{a}\to 0}\mu_{\mathfrak{A}^{\mathfrak{a}}}\left(\mathbf{a}\right)\to 0 \text{ and } \lim_{\mathbf{a}\to 0}\mu_{\mathfrak{B}^{\mathfrak{a}}}\left(\mathbf{a}\right)\to 1 \end{array}\right\} = \mathbf{a}^{*}$$

$$\left.\begin{array}{l} \max_{\mathbf{a}\in\left(\mathfrak{A}^{\mathfrak{a}}\cap\mathfrak{B}^{\mathfrak{a}}\right)}\mu_{\mathfrak{B}^{\mathfrak{a}}}\left(\mathbf{a}\right) \\[2em] \text{s.t } \left[\mu_{\mathfrak{A}^{\mathfrak{a}}}\left(\mathbf{a}\right)-\mu_{\mathfrak{B}^{\mathfrak{a}}}\left(\mathbf{a}\right)\right]\le 0 \\[2em] \lim_{\mathbf{a}\to\infty}\mu_{\mathfrak{A}^{\mathfrak{a}}}\left(\mathbf{a}\right)\to 1 \text{ and } \lim_{\mathbf{a}\to\infty}\mu_{\mathfrak{B}^{\mathfrak{a}}}\left(\mathbf{a}\right)\to 0 \\[1em] \lim_{\mathbf{a}\to 0}\mu_{\mathfrak{A}^{\mathfrak{a}}}\left(\mathbf{a}\right)\to 0 \text{ and } \lim_{\mathbf{a}\to 0}\mu_{\mathfrak{B}^{\mathfrak{a}}}\left(\mathbf{a}\right)\to 1 \end{array}\right\} = \mathbf{a}^{*}\mathfrak{a}$$

The solutions to these stated equivalent fuzzy optimization problems provide the needed breakeven point where $\mu_{\Delta^{\mathfrak{a}}}\left(\mathbf{a}^{*}\right)=\mu_{\mathfrak{A}^{\mathfrak{a}}}\left(\mathbf{a}^{*}\right)=\mu_{\mathfrak{B}^{\mathfrak{a}}}\left(\mathbf{a}^{*}\right)$. For the proof of this theorem see [R4.17] [R4.41][R4.48] [R5][R5.3] [R5.9]. This may be illustrated geometrically as:

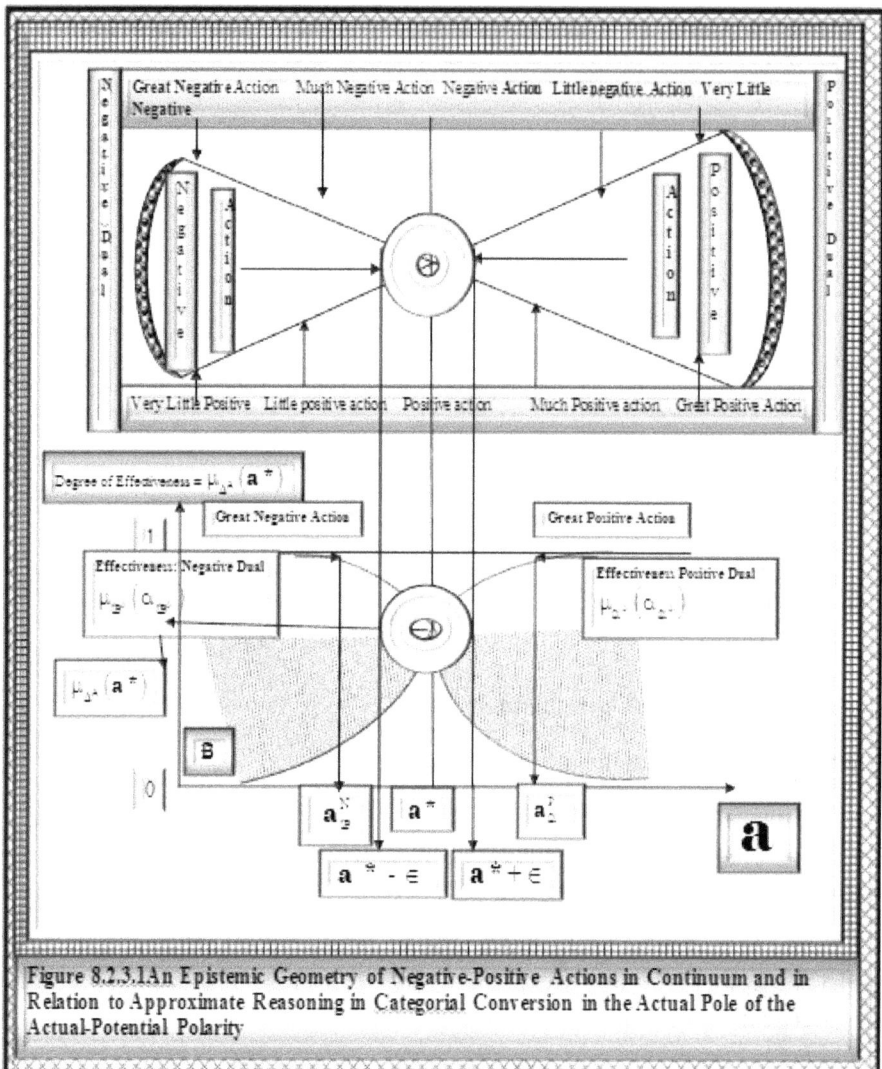

Figure 8.2.3.1 An Epistemic Geometry of Negative-Positive Actions in Continuum and in Relation to Approximate Reasoning in Categorial Conversion in the Actual Pole of the Actual-Potential Polarity

Note $\alpha_{\mathfrak{B}}a = \mathbf{a}_{\mathfrak{B}}^{N}$ and $\alpha_{\mathfrak{A}}a = \mathbf{a}_{\mathfrak{A}}^{P}$ for symbolic convenience.

## 8.3 Catergorial-Conversion Analytics in System Dynamics

A question arises as to the role as to the role that the characteristic sets play to induce the conversion of categories. The answer requires some interesting analytics. At the fuzzy optimal point of degree of action effectiveness, there is a breakeven point of the categorial conversion such that $\mathbf{a}^{P*} = \mathbf{a}^P\left(\mathbb{X}^P, \mathbb{X}^N\right) = \mathbf{a}* = \mathbf{a}^N\left(\mathbb{X}^N, \mathbb{X}^P\right) = \mathbf{a}^{N*}$. The categorial-conversion system is sitting at a point of transition between the conditions of actual to potential or conditions to maintain and affirm the integrity of the actual. With the use of the optimal transitional point the categorial-conversion system may be partitioned into fuzzy zones where: a) $\left(\mathbf{a}*\text{-}\epsilon, \mathbf{a}*\text{+}\epsilon\right)$ defines a zone of indifference, b) $\left(\mathbf{a}^N < (\mathbf{a}*-\epsilon)\right)$ defines the negative dominance zone and c) $\left(\mathbf{a}^P > (\mathbf{a}*+\epsilon)\right)$ defines the positive dominance zone as they are shown on Figure 7.6.2.1. In these zones, two operative categorial moments are identifiable. They are the general positive categorial moment which is associated with the activities of the positive characteristic sub-set and the negative general categorial moment which is associated with the actions of the negative characteristic sub-set.

At the breakeven point, the positive action is increasing in the positive characteristic set such that $\left(\mathbf{a}^P > (\mathbf{a}*+\epsilon)\right)$ implies that $\left(\pi^P > \pi^N\right)$ with $\mu_{\mathfrak{A}^{\mathbf{a}}}(\mathbf{a}) \to 1$ and $\mu_{\mathfrak{B}^{\mathbf{a}}}(\mathbf{a}) \to 0$ as $(\mathbf{a}\text{-}\mathbf{a}*)\uparrow$. Similarly, $\left(\mathbf{a}^N < (\mathbf{a}*-\epsilon)\right)$ implies that $\left(\pi^N > \pi^P\right)$ with $\mu_{\mathfrak{A}^{\mathbf{a}}}(\mathbf{a}) \to 1$ and $\mu_{\mathfrak{B}^{\mathbf{a}}}(\mathbf{a}) \to 0$ as $(\mathbf{a}\text{-}\mathbf{a}*)\uparrow$ Furthermore, there exists an $\alpha$-threshhold zone beyond which a conversion to the new actual can be claimed with fuzzy conditionality. There also exists a $\alpha$-threshhold zone beyond which the maintenance of the integrity of the existing actual can be claimed with fuzzy conditionality. These threshold zones are called the zones of *complete convertibility* with fuzzy conditionality. These zones and other corresponding zones may be specified by the use of a fuzzy decomposition theorem in terms of $\alpha$-level sets as shown below.

$$\left. \begin{array}{l} \mathbf{a}^P > \alpha_{\mathfrak{A}} = \text{the zone of complete positive dominance} \\[2em] \mathbf{a}^P \in \left((\mathbf{a}*+\epsilon), \alpha_{\mathfrak{A}}\right) = \text{the zone of partial positive dominance} \\[2em] \mathbf{a}^P \in \left(\mathbf{a}*, (\mathbf{a}*+\epsilon)\right) = \text{the zone of } \epsilon \text{ - positive indifference} \end{array} \right\} \quad (8.3.1)$$

317

$$\mathbf{a}^{N} < \alpha_{\mathfrak{B}} = \text{the zone of complete negetive dominance}$$

$$\mathbf{a}^{N} \in \left(\alpha_{\mathfrak{A}}, (\mathbf{a}^{*} - \varepsilon)\right) = \text{the zone of partial positive dominance} \qquad (8.3.2)$$

$$\mathbf{a}^{N} \in \left((\mathbf{a}^{*} - \varepsilon), \mathbf{a}^{*}\right) = \text{the zone of } \varepsilon\text{-negative indifference}$$

A question arises as to how one relates the zones of dominance and indifference to the categorial-conversion process of the actual-potential polarity. As conceptualized, the universal object set may be partitioned into the set of the actual elements, $\mathfrak{A}$, and the set of potential elements $\mathfrak{P}$. The set of the actual elements may be partitioned into $\mathfrak{A}_0, \mathfrak{A}_I$ and $\mathfrak{A}_n$ such that $\mathfrak{A} = \left(\mathfrak{A}_0 \cup \mathfrak{A}_I \cup \mathfrak{A}_n\right)$ at any moment of time. The sets define three zones of categorial-conversion activities. The symbol $\mathfrak{A}_0$ defines the existing (old) actual, $\mathfrak{A}_I$ defines the transitional zone of the old and new actual, and $\mathfrak{A}_n$ defines the new actual that has just emerged from the potential. It is useful to always keep in mind the concepts of duality, polarity with relational continuum and unity. Additionally, every duality is composed of positive and negative duals which are defined by positive and negative characteristic sub-sets, where the duality is said to be positive (negative) if positive (negative) dual relationally dominates the negative (positive) dual. Every pole of the actual-potential polarity has a residing duality, where the actual (potential) pole is said to be a negative (positive) pole if the residing duality is that of negative (positive). An example of the old, the transitional and the new in a socio-political system would be $\mathfrak{A}_0$ is a colonial state, $\mathfrak{A}_I$ is a transitional state between colonialism and independence, and $\mathfrak{A}_n$ defines an independent state in a socio-political arrangement under colonialism-independence polarity. In a natural system, one may consider cancer-non-cancer polarity where $\mathfrak{A}_0$ is a healthy cell state, $\mathfrak{A}_I$ is the transition between a healthy cell state and a cancer-cell state, and $\mathfrak{A}_n$ is cancer-cell state. It may be kept in mind that each of $\mathfrak{A}_0$, $\mathfrak{A}_I$, $\mathfrak{A}_n$ contains cancer characteristic sub-set and non-cancer characteristic sub-set in different relative proportions. The partition is undertaken with the application of the fuzzy decomposition theorem [R4.17] [R4.37] [R4.39] [R4.41][R4.48] [R4.73].

As has been discussed, each element in the set $\{\mathfrak{A}_0, \mathfrak{A}_1, \mathfrak{A}_n\}$ constitutes a theoretical category of qualitative zones of transformation in a relational continuum and unity with continual categorial dynamics, where every state sits in a temporary qualitative equilibrium in the categorial-conversion process. Each stage in any zone is defined by a duality. If $\omega \in \Omega$, then

$$\omega \in \mathfrak{A}_0 \Leftrightarrow \forall \mathbf{a}^N, \mathbf{a}^P \in \mathbf{A} \quad \mathbf{a}^N > \mathbf{a}^P \ldots\ldots\ldots(a)$$

$$\omega \in \mathfrak{A}_1 \Leftrightarrow \forall \mathbf{a}^N, \mathbf{a}^P \in \mathbf{A} \quad \mathbf{a}^N \cong \mathbf{a}^P \ldots\ldots\ldots(b) \qquad (8.3.3)$$

$$\omega \in \mathfrak{A}_n \Leftrightarrow \forall \mathbf{a}^N, \mathbf{a}^P \in \mathbf{A} \quad \mathbf{a}^P > \mathbf{a}^N \ldots\ldots\ldots(c)$$

Each element $\omega \in \Omega$ is passing through transitional stages under a internally defined qualitative motion that supports the corresponding state equation for categorial conversion. The qualitative motions are of two types of forward progressive and backward retrogressive motions through the dynamics of the relative characteristic sets. The motions require operations of the internal forces generated by the internal energies to move the element from one qualitative stage to another by negating the relative relationship of negative and positive actions through their relative characteristic sets. The negative action generates a general negative categorial-conversion moment $\mathbf{C}^N$ that defines the factors required for negative action $\mathbf{a}^N$ to dominate the positive action $\mathbf{a}^P$. In this respect, the measure of the negative categorial conversion moment is the measure of the degree of the effectiveness $\delta^N$ of the negative action. Similarly, the positive action generates a general positive categorial-conversion moment $\mathbf{C}^P$ that defines the factors required for positive action $\mathbf{a}^P$ to dominate the negative action $\mathbf{a}^N$. In this respect, the measure of the positive categorial conversion moment is the measure of the degree of the effectiveness $\delta^P \in \mathfrak{D}^{\mathfrak{a}}$ of the positive action. Negation of negation allows a categorial conversion equation of the form that shows forward and backward motions.

## Theorem 8.3.1 Categorial Belonging

$$\omega \in \mathfrak{A}_0 \Leftrightarrow \forall \mathbf{a}^N, \mathbf{a}^P \in \mathbf{A}^P,\ \mathbf{a}^N > \mathbf{a}^P \text{ and } \not\exists \delta^P \ni \delta^P \left(\mathbf{a}^N > \mathbf{a}^P\right)\omega \Rightarrow \left(\mathbf{a}^P > \mathbf{a}^N\right)\omega \dots\dots\dots\text{(a)}$$

$$\omega \in \mathfrak{A}_0 \Leftrightarrow \forall \mathbf{a}^N, \mathbf{a}^P \in \mathbf{A}\ \mathbf{a}^N \cong \mathbf{a}^P \text{ and } \not\exists \left(\delta^N, \delta^P\right), \ni \left(\delta^N, \delta^P\right)\left(\mathbf{a}^P \cong \mathbf{a}^N\right)\omega \Rightarrow \left(\delta^N, \delta^P\right)\left(\mathbf{a}^N \geq \mathbf{a}^P\right)\omega..\text{(b)}$$

$$\omega \in \mathfrak{A}_h \Leftrightarrow \forall \mathbf{a}^N, \mathbf{a}^P \in \mathbf{A}^P,\ \mathbf{a}^P > \mathbf{a}^N \text{ and } \not\exists \delta^N \ni \delta^N \left(\mathbf{a}^P > \mathbf{a}^N\right)\omega \Rightarrow \left(\mathbf{a}^N > \mathbf{a}^P\right)\omega \dots\dots\dots\text{(c)}$$

The theorem of categorial belonging acknowledges two important conditions of convertibility. The types of conditions involve general *categorial conversion moments* generated by intra-categorial conversion, and the *measure of degrees of effectiveness* of the conversion moments as generated by intra-categorial conversion and expressed by a *categorial transfer function* relative to *categorial transversality conditions*. The categorial conversion is in a continuum as well as in unity and is always present with a distribution of measures of degrees of effectiveness over the spectrum of the categorial conversion in a never-ending process leading to *internal crisis*, where the result of the conversion process is seen in a discrete domain for the emergence of the new with a new power structure and enhanced force of dominance, and the disappearance of the old with a diminished old power structure and a weakened force of conversion. The categorial-conversion process is to establish a new regime of negative-positive dominance for categorial power and force defined in the space of conflicts and contradictions. The process is analog while the result is digital (discrete) in form. Within the set of degrees of effectiveness $\mathfrak{D}^{\mathfrak{a}}$ there is one $\delta^P \in \mathfrak{D}^{\mathfrak{a}}$ which defines the threshold of positive categorial conversion, and there is $\delta^N \in \mathfrak{D}^{\mathfrak{a}}$ which also establishes the threshold of negative categorial conversion. Let $\delta^{P*} \in \mathfrak{D}^{\mathfrak{a}}$ define the threshold measure of effectiveness of the positive conversion and $\delta^{N*} \in \mathfrak{D}^{\mathfrak{a}}$ define the threshold measure of effectiveness of the negative conversion. The threshold measure of effectiveness for either positive or negative action will be called *Nkrumah Delta*. The categorial moment, categorial transfer function and

## Definition 8.3.1: Nkrumah Delta

Nkrumah Delta $\delta^* \in \mathfrak{D}^{\mathfrak{a}}$ is the measure of a threshold categorial moment beyond which categorial conversion occurs to potentialize the existing actual and actualize an element in the potential in the qualitative dynamics of the actual-potential polarity.

The categorial moment, the categorial transfer function and Nkrumah Delta are necessary conditions for transformation of any categorial element. The Nkrumah delta will vary different actual-potential polarities, and will vary over time and categorial generations for any given actual-potential polarity. From the conditions of categorial belonging, one must examine the conditions of categorial convertibility. The conditions of categorial convertibility are associated with conditions of categorial belonging and specified in terms of existence of a measure of degree of effectiveness of the categorial-conversion moment, categorial transfer function and the limiting processes in the qualitative space regarding changes in the positive and negative actions that are supported by motions in the quantitative space regarding the cardinality of the positive and negative characteristic sets to satisfy the conditions categorial transversality. The conditions may be stated as a theorem of existence.

## Theorem 8.3.2 Categorial Conversion

Given any general categorial-conversion moment $\mathbf{C}$ if $\mathbf{C}$ with $\left(\mathbf{a}^{N}, \mathbf{a}^{P} \in \mathbf{A}\right)$ then

$$\mathbf{C}\left(\mathbf{a}^{N} > \mathbf{a}^{P}\right) \omega \ni \lim \mu_{\mathfrak{A}^{\mathfrak{a}}}(\mathbf{a}) \to 1 \text{ and } \mu_{\mathfrak{B}^{\mathfrak{a}}}(\mathbf{a}) \to 0 \text{ as } \mathbf{a} \uparrow, \text{ then } \exists\, \delta^{P} \in \mathfrak{D}^{\mathfrak{a}}$$

such that $\delta^{P}\left(\mathbf{a}^{N} > \mathbf{a}^{P}\right) \omega \in \mathfrak{A}_{0} \Rightarrow \left(\mathbf{a}^{P} > \mathbf{a}^{N}\right) \omega \in \mathfrak{A}_{n}$.

Similarly,

If $\mathbf{C}\left(\mathbf{a}^{P} > \mathbf{a}^{N}\right) \omega \ni \lim \mu_{\mathfrak{B}^{N}}(\mathbf{a}) \to 1$ and $\mu_{\mathfrak{A}^{P}}(\mathbf{a}) \to 0$ as $\mathbf{a} \downarrow$ then $\exists\, \delta^{N} \in \mathfrak{D}^{\mathfrak{a}}$

such that $\delta^{N}\left(\mathbf{a}^{P} > \mathbf{a}^{N}\right) \omega \in \mathfrak{A}_{N} \Rightarrow \left(\mathbf{a}^{N} > \mathbf{a}^{P}\right) \omega \in \mathfrak{A}_{0}$ where $\omega \in \Omega$ the universal characteristic set.

It may be kept in mind that $\omega \in \mathfrak{A}_{0}$ implies an existing actual and $\omega \in \mathfrak{A}_{n}$ implies a realization of a new actual that has been converted from the potential. The theorem of existence involves a construction-reduction process in forward-backward motions in the sense that all elements belonging to $\mathfrak{A}_n$ constitute categorial derivatives from the primary category of elements in the old actual. In other words, the new actual is reducible to the old actual and there is a process of categorial reductionism as well as there is a process of categorial constructionism. The general categorial moment is always present with any actual-potential polarity with distribution of degrees of effectiveness. The threshold measures of degree of categorial moments (that is the Nkrumah delta) for either negative $\delta^{N^*} \in \mathfrak{D}^{\mathfrak{A}}$ or positive $\delta^{P^*} \in \mathfrak{D}^{\mathfrak{A}}$ action must be internally manufactured where their production requires changes and organizational rearrangements of the relative structure of the negative and positive characteristic sets in the quantitative space. One may reason from the set and action relativities where:

$$\frac{\#\mathbb{X}^P}{\#\mathbb{X}^N} = \frac{\mathbf{a}^P}{\mathbf{a}^N} = \frac{\sum_{i=1}^{\#\mathbb{X}^P} \mathbb{X}_i^P}{\sum_{j=1}^{\#\mathbb{X}^N} \mathbb{X}_j^N} \tag{8.3.4}$$

Given equal degree of action effectiveness of each element one can rearrange from eqn. (8.3.4) and write:

$$\frac{\mathbf{a}^P}{\#\mathbb{X}^P} = \frac{\mathbf{a}^N}{\#\mathbb{X}^N} \tag{8.3.5}$$

It may be noted that as $\mathbf{a}_{\mathfrak{A}}^P \uparrow \Rightarrow \mathbf{a}_{\mathfrak{A}}^N \downarrow$ and for the equality principle $\#\mathbb{X}_{\mathfrak{A}}^P \uparrow \Rightarrow \#\mathbb{X}_{\mathfrak{A}}^N \downarrow$ in the actual pole through the actual duality, $\mathbf{D}^{\mathfrak{A}}$ A reverse process is also taking place in the potential pole in that as $\left( \mathbf{a}_{\mathfrak{A}}^P \uparrow \mathbf{a}_{\mathfrak{p}}^P \downarrow \right) \Rightarrow \left( \mathbf{a}_{\mathfrak{A}}^N \downarrow \mathbf{a}_{\mathfrak{P}\mathfrak{p}}^N \uparrow \right)$ under the principle of equality and actual-potential qualitative dynamics $\left( \#\mathbb{X}_{\mathfrak{A}}^P \uparrow \#\mathbb{X}_{\mathfrak{p}}^P \downarrow \right) \Rightarrow \left( \#\mathbb{X}_{\mathfrak{A}}^N \downarrow \#\mathbb{X}_{\mathfrak{p}}^N \uparrow \right)$.

The implication is that the number of elements contributing to positive aggregate action must rise while the number of elements contributing to the negative action must fall in the actual pole. Similarly, the number of elements contributing to positive aggregate action must fall while the number of elements contributing to the negative action must rise in the potential pole. The result is that $\left( \dfrac{\mathbf{a}^P}{\mathbf{a}^N} \uparrow \Rightarrow \dfrac{\#\mathbb{X}^P}{\#\mathbb{X}^N} \uparrow \right)_{\mathfrak{A}}$ while

$$\left( \frac{\mathbf{a}^{\mathrm{P}}}{\mathbf{a}^{\mathrm{N}}} \downarrow \ \Rightarrow \ \frac{\#\mathbb{X}^{\mathrm{P}}}{\#\mathbb{X}^{\mathrm{N}}} \downarrow \right)_{\mathfrak{P}}$$ to affect the needed degree of effective

measure of categorial conversion. The paths of categorial conversion are presented in an epistemic Geometry in Figure 8.3.1 where the infinity ∞ defines a never-ending process of categorial conversions in all socio-natural elements in the ontological space. These relative numbers may be seen in terms of how Nkrumah conceptualized them in symbolic representation in the Appendix to his book Consciencism when the theory of categorial conversion is given an application to social transformations. It may be emphasized that the theory of categorial conversion establishes the necessary conditions for the general theory of socio-natural transformations. These are the conditions of categorial convertibility. The conditions are *categorial moment*, *categorial transfer function* and *categorial transversality conditions*. The construct of the sufficient conditions comes under the theory of Philosophical Consciencism which will be taken up in a separate monograph.

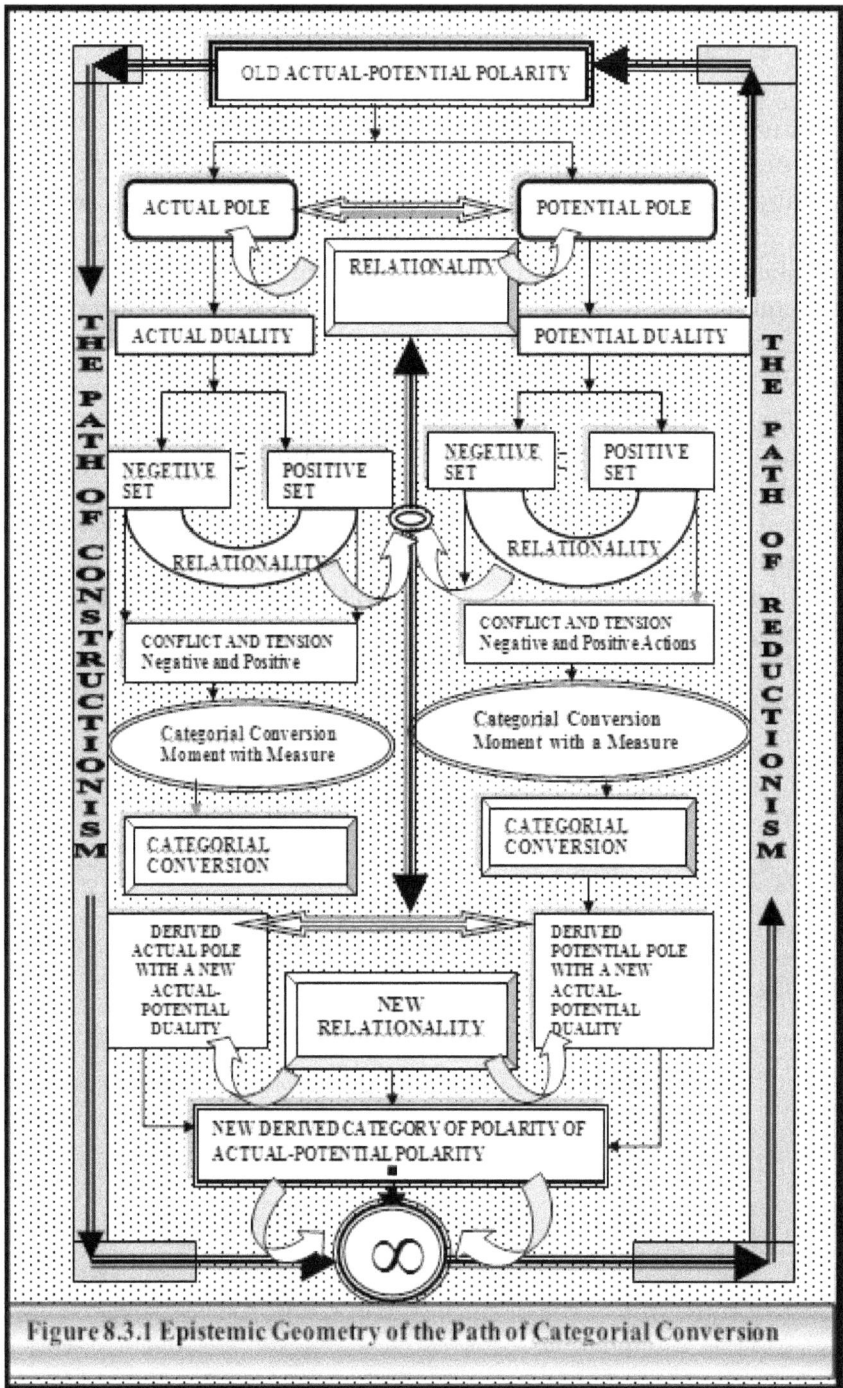

Figure 8.3.1 Epistemic Geometry of the Path of Categorial Conversion

## 8.4 The Nature of the Categorial Conversion Moment and Its Measure

It has been discussed in the previous sections that the categorial-conversion moment $\mathbf{C}$ and the corresponding distribution of degrees of measures of effectiveness in bringing about categorial conversion has an element $\delta^{*} \in \mathfrak{D}^{\mathfrak{A}}$ (Nkrumah delta) that defines the threshold of degree of effectiveness whether one considers positive or negative categorial conversion. The categorial conversion moment is a general transformation function that depends on the behavior of negative action $\mathbf{a}^{N}\left(\mathbb{X}_{\mathfrak{A}}^{N}\right) \in \mathbf{A}$, positive action $\mathbf{a}^{P}\left(\mathbb{X}_{\mathfrak{A}}^{P}\right) \in \mathbf{A}$, and the measures of degrees of effectiveness of the negative and positive categorial conversion moments $\delta^{N}, \delta^{P} \in \mathfrak{D}^{\mathfrak{A}}$ respectively. It may be written as $\mathbf{C} = \mathbf{C}\left(\mathbf{a}^{P}, \mathbf{a}^{N}, \delta^{P}, \delta^{N}\right)$ where the variables depend on their respective characteristic sets. It may also be noted that the degree of effectiveness of the negative categorial moment depends on the negative characteristics set through the negative action and hence $\delta^{N} = \delta^{N}\left(\mathbf{a}^{N}\left(\mathbb{X}^{N}\right)\right) \in \mathfrak{D}^{\mathfrak{A}}$. Similarly, the degree of effectiveness of the positive categorial moment depends on the positive characteristics set through the positive action and hence $\delta^{P} = \delta^{P}\left(\mathbf{a}^{P}\left(\mathbb{X}^{P}\right)\right) \in \mathfrak{D}^{\mathfrak{A}}$. In all categorial conversions, the measure of the degree of effectiveness must be manufactured from within the category of interest. There are many categories in the space of the actual and hence each category has a corresponding categorial conversion process taking place. Let $\mathbf{C}^{\mathfrak{A}}$ be the set of categorial conversion moments with an index set $\mathfrak{I}$ such that $\mathbf{C}_{i} \in \left(\mathbf{C}^{\mathfrak{A}} \mid i \in \mathfrak{I}\right)$ that are generated by intra-categorial conversion activities as a necessary condition. Each categorial conversion moment has its own uniqueness in terms of form, content and the resource requirement to bring about the threshold-degree of effectiveness that will induce the conversion. Conversion means a movement from one category to another distinguished by qualitative characteristics. The movement is carried by a transfer function which is manufactured by intra-categorial conversion activities as a necessary condition. Thus, corresponding to $\mathbf{C}_{i} \in \left(\mathbf{C}^{\mathfrak{A}} \mid i \in \mathfrak{I}\right)$ there is a set of transfer functions

$\mathbf{F}_i \in \left( \mathfrak{F}^{\mathfrak{a}} \mid i \in \mathfrak{J} \right)$ that together must satisfy the set of categorial transversality conditions $\mathbf{T}_i \in \left( \mathfrak{T}^{\mathfrak{a}} \mid i \in \mathfrak{J} \right)$ which will vary in content and form as the categories vary structure and form. The sets $\mathbf{C}_i \in \left( \mathfrak{C}^{\mathfrak{a}} \mid i \in \mathfrak{J} \right)$ and $\mathbf{F}_i \in \left( \mathfrak{F}^{\mathfrak{a}} \mid i \in \mathfrak{J} \right)$ constitute the controlled variables while the state of the system is defined by a category and $\mathbf{T}_i \in \left( \mathfrak{T}^{\mathfrak{a}} \mid i \in \mathfrak{I} \right)$ provides the visible conditions of the success of the categorial conversion. The implication here is that a transformation-substitution process works with cost-benefit rationality through the elasticity of categorial conversion in the creation of the threshold values of effectiveness. The sufficiency conditions of the convertibility are expressed through internal decision-choice activities under antagonistic preferences of dualistic dominance and power.

From eqns.(8.3.4) the equation of motion governing the categorial conversion may be developed, where the dynamics of qualitative and quantitative dispositions are linked. In this respect, let $Z = \dfrac{\#\mathbb{X}^{\mathrm{P}}}{\#\mathbb{X}^{\mathrm{N}}}$ defines the value of relative characteristic set, then it may be specified at each qualitative state that:

$$\begin{cases} Z = 1 \Rightarrow \text{negative - positive power balance} \\ Z > 1 \Rightarrow \text{positive power dominance and idntity} \\ Z < 1 \Rightarrow \text{negative power dominance and identity} \end{cases} \qquad (8.3.6)$$

The relative characteristic set may be specified as a function of negative and positive actions where in the process of the struggle for dominance the number of the negative qualitative characteristics can be changed by the activities or in-activities of the negative action as time progresses. Similarly, the number of the positive qualitative characteristics can be changed by the activities or in-activities of the positive action as time progresses.

$$Z = \varphi \left( \#\mathbb{X}^{\mathrm{P}} \left( \mathbf{a}^{\mathrm{P}} (t) \right), \#\mathbb{X}^{\mathrm{N}} \left( \mathbf{a}^{\mathrm{N}} (t) \right) \right), \qquad (8.3.6)$$

The function $\varphi(\bullet)$ is complex, continuous an infinitely differentiable in its entire domain. For simplicity, we may write it as the function of negative and positive actions in the form:

$$Z = \varphi\left(\mathbf{a}^{\mathrm{P}}(t), \mathbf{a}^{\mathrm{N}}(t)\right) \qquad (8.3.7)$$

Equations (8.3.6) and (8.3.7) define the path of categorial enveloping where the equation of motion governing the system is made to depend on the internal negative and positive actions in the form

$$\dot{Z} = = \frac{\partial \varphi}{\partial \mathbf{a}^{\mathrm{P}}} \dot{\mathbf{a}}^{\mathrm{P}} + \frac{\partial \varphi}{\partial \mathbf{a}^{\mathrm{N}}} \dot{\mathbf{a}}^{\mathrm{N}}$$

(8.3.8)

From equation (8.3.7), we may write

$$\left.\begin{array}{l} \dot{Z} = = \dfrac{\partial \varphi}{\partial \mathbf{a}^{\mathrm{P}}} \dot{\mathbf{a}}^{\mathrm{P}} + \dfrac{\partial \varphi}{\partial \mathbf{a}^{\mathrm{N}}} \dot{\mathbf{a}}^{\mathrm{N}} \gtreqless 0 \\[3mm] \dot{Z}{=}0 \Rightarrow \dfrac{\partial \varphi}{\partial \mathbf{a}^{\mathrm{P}}} \dot{\mathbf{a}}^{\mathrm{P}} = \dfrac{\partial \varphi}{\partial \mathbf{a}^{\mathrm{N}}} \dot{\mathbf{a}}^{\mathrm{N}} \Rightarrow \text{no change in power relation} \\[3mm] \dot{Z}{>}0 \Rightarrow \dfrac{\partial \varphi}{\partial \mathbf{a}^{\mathrm{P}}} \dot{\mathbf{a}}^{\mathrm{P}} > \dfrac{\partial \varphi}{\partial \mathbf{a}^{\mathrm{N}}} \dot{\mathbf{a}}^{\mathrm{N}} \Rightarrow \text{change in power relation in favor of positive characteric set} \\[3mm] \dot{Z}{<}0 \Rightarrow \dfrac{\partial \varphi}{\partial \mathbf{a}^{\mathrm{P}}} \dot{\mathbf{a}}^{\mathrm{P}} < \dfrac{\partial \varphi}{\partial \mathbf{a}^{\mathrm{N}}} \dot{\mathbf{a}}^{\mathrm{N}} \Rightarrow \text{change in power relation in favor of negative characteric set} \end{array}\right\}$$

(8.3.9)

These equations of motion may be linked to the decision-choice system of the Philosophical Consciencism that generates socio-natural effort and degrees of participation effort where the strategies and counter strategies may be formulated as a control dynamic game with fuzzy-stochastic information structure.

## Definition 8.4.1

The elasticity of positive categorial conversion is the proportionate change of the positive action for a given proportionate change of the cardinality of the positive characteristic set. It may be represented as:

$$\mathscr{E}^{\mathrm{P}} = \frac{\left(\Delta \mathbf{a}^{\mathrm{P}} \middle/ \mathbf{a}^{\mathrm{P}}\right)}{\left(\Delta \# \mathbb{X}^{\mathrm{P}} \middle/ \# \mathbb{X}^{\mathrm{P}}\right)}$$

Similarly, the elasticity of negative categorial conversion may be defined and computed as:

$$\mathcal{E}^N = \frac{\left(\Delta\mathbf{a}^N \middle/ \mathbf{a}^N\right)}{\left(\Delta\#\mathbb{X}^N \middle/ \#\mathbb{X}^N\right)}$$

## Definition 8.4.2

The cross-elasticity of positive (negative) categorial conversion is the proportionate change of the positive (negative) action for a given proportionate change of the negative (positive) action that may be computed as:

$$\mathcal{E}^{PN} = \frac{\left(\Delta\mathbf{a}^P \middle/ \mathbf{a}^P\right)}{\left(\Delta\mathbf{a}^N \middle/ \mathbf{a}^N\right)} \text{ , and } \quad \mathcal{E}^{NP} = \frac{\left(\Delta\mathbf{a}^N \middle/ \mathbf{a}^N\right)}{\left(\Delta\mathbf{a}^P \middle/ \mathbf{a}^P\right)} \quad \text{which may also}$$

be expressed in terms of set cardinalities as:

$$\mathcal{E}^{PN} = \frac{\left(\Delta\#\mathbb{X}^P \middle/ \#\mathbb{X}^P\right)}{\left(\Delta\#\mathbb{X}^N \middle/ \#\mathbb{X}^N\right)} \text{ , and } \quad \mathcal{E}^{NP} = \frac{\left(\Delta\#\mathbb{X}^N \middle/ \#\mathbb{X}^N\right)}{\left(\Delta\#\mathbb{X}^P \middle/ \#\mathbb{X}^P\right)}$$

The own positive and negative elasticities of categorial conversion show the degree of effectiveness of the positive and negative actions by the positive and negative characteristic sets while the cross-elasticities shows the measure of the relative impact on the relative actions of the characteristic sets. The effectiveness of the positive action is seen in terms of the degree to which the negative action is reduced, the cardinality of the positive characteristic set is increased and the cardinality of the negative characteristic set is reduced. In this respect, the elasticities of the positive and negative actions and characteristic sets work in opposite directions that can be seen from the cross elasticities

$$\frac{\left(\Delta\mathbf{a}^P \middle/ \mathbf{a}^P\right)}{\left(\Delta\mathbf{a}^N \middle/ \mathbf{a}^N\right)} = \frac{\left(\Delta\#\mathbb{X}^P \middle/ \#\mathbb{X}^P\right)}{\left(\Delta\#\mathbb{X}^N \middle/ \#\mathbb{X}^N\right)} \Rightarrow \frac{\left(\Delta\mathbf{a}^P \middle/ \mathbf{a}^P\right)}{\left(\Delta\#\mathbb{X}^P \middle/ \#\mathbb{X}^P\right)} = \frac{\left(\Delta\mathbf{a}^N \middle/ \mathbf{a}^N\right)}{\left(\Delta\#\mathbb{X}^N \middle/ \#\mathbb{X}^N\right)}$$

The values of the elasticities of categorial conversion and the measure of categorial moment will depend on the resource availability and commitment, and the relative strengths of the internal organizations of the positive and negative characteristic sets that mutually operate as constraints on each other's actions in the categorial conversion process. The resource availability, resource commitment and relative strengths of the internal organizations will vary over different categories, stages, generations, conditions and time that provide information inputs for strategic action.

The essential conceptual elements in the theory of categorial conversion are categories define by quality, actual-potential polarity, duality, negative characteristic set, positive characteristic set, negative action, positive action, experiential information and payoffs that are defined in terms of final outcomes. The payoffs are expressed in terms dominance and power that lead to categorial conversion or the maintenance of the actual. The toolbox of reasoning includes relational continuum, relational unity, fuzzy paradigm with its logic and mathematics, degrees of effectiveness of the actions, categorial moment, the measures of categorial moments and the threshold measure defining point of conversion when the categorial transversality conditions are satisfied. The appropriateness of the use of fuzzy paradigm of thought lies in the fact that none only is the experiential information vague and inexact but that the categories are qualitatively defined. The categorial conversion process is such that:

$$\delta^{P*} = \delta^{P*}\left(\mathbf{a}^{P}\left(\mathbb{X}^{P}\right)\right) \in \mathfrak{D}^{\mathfrak{a}} \quad \delta^{N*} = \delta^{N*}\left(\mathbf{a}^{N}\left(\mathbb{X}^{N}\right)\right) \in \mathfrak{D}^{\mathfrak{a}}$$

The $\delta$'s must be internally created from the organization and reorganization of the over-all characteristics set into the relative composition. External help presented as intervention can only be successful if such a help is internally accepted and incorporated into the internal basic characteristics. This is how, for example, medical intervention works or fails. It is also how external intervention in different social set-ups to change value structure works or fails. Examples may be found in all areas that one may conceive.

## 8.5 Game Theory and the Theory of Categorial Conversion

The theory of categorial conversion is about transformations due to the internal dynamics of elements under the principles of opposite with conflicts characterized by fuzzily defined opposite characteristic sets of negative and positive sets expressed respectively as negative and positive duals which constitute a duality. All the transformation processes pass through penumbral regions of tactical and strategic decision-choice actions defined in terms of actions and counter action within the duals and dualistic preference relations in search of power and dominance for categorial existence. The basic characteristics of the theory of categorial conversion are polarity, duality under the principles of opposites, relational continuum and unity. There are opposites which work under strategies and counter strategies for any given information structure. The internal conflicts are between the positive and negative characteristic sets of any given entity in a fuzzily defined category. The energy require to produce qualitative motion and then categorial conversion is generated by the internal contradictions and conflicts from the strategic and tactical actions by the negative and positive characteristic sets for negation and negation of negation. The strategic actions come as positive and negative action and are realized as power games by the duals in each duality in the penumbral regions defined by fuzzy information structure reflected in the experiential information structure where each dual seeks dominance by converting some members of the characteristic sub-set of the opposing dual to enhance its size for increasing power, force and dominance.

## 8.5.1 Types of the Categorial-Conversion Game, Information and Negation of Negation

The negation of negation is a mutual struggle that may be viewed as the development of tactics and strategies to accomplish the direction of mutual negation from the basic structure of the contradictory behavior by the positive and negative characteristic subsets (duals of the duality). In this respect, the actions and counter actions by the negative and positive characteristic sub-sets may be studied not only as strategic games but as an optimal controlled games which are played by the opposite duals for power and dominance. The theory of categorial conversion may then be seen from the viewpoint of an internal controlled dynamic game theory in conflict resolution between the duals to bring about conversion of the actual-potential polarity where the game may be viewed as evolutionary game for the explanation of disappearance of the old and the emergence of the new. A question arises as to what type of game theory will be useful to allow good intellectual insights into the understanding of transformational dynamics under the principles of categorial conversion. The question of what type of game theory implies that there are different types of game theory dealing with conflicts and strategic interactions in polarity and duality under relational continuum and unity. The search for an answer to this question brings us to the classificatory science of classification and sub-classification of games theory under defined criteria. Generally, the criteria include 1) the number of players, 2) the nature of payoffs, 3) individuals or coalitions and 4) the information structure. When game theory is seen in the framework of decision theory, then the information structure imposes the types of games that are involved in the conflicts. The relevant classification is presented as an epistemic geometry in Figure 8.5.1.1.

The assumed information structure, its interpretation and representation drive all decision-choice actions including strategic and non-strategic games. In the process of knowing and decision-choice actions, the information structure, from which knowledge is constructed, comes to us as empirical or axiomatic information or both [R3.7][R3.10] [R3.13] R8.44] [R8.47] [R8.55] [R8.66]. All decision-choice actions including strategic games and knowledge production in all areas of interest are not only information dependent but information determined in the energy field of socio-natural activities. Such information include preferences which are captured by some utility, payoffs, possible risks, general incentive structure and the established and non-communicated rules of the relevant game. The elements of the information structure define the environment and the overall field of the game and decision-choice actions. The assumed information structure in turn defines the nature of the paradigm required to formulate the game on the principle of opposite where conflicts are the foundations of relational continuum and the unity of the opposite. It, further imposes the structure on the theory that may emerge as well as dictates the explanatory and prescriptive structures for understanding and action [R3.13].

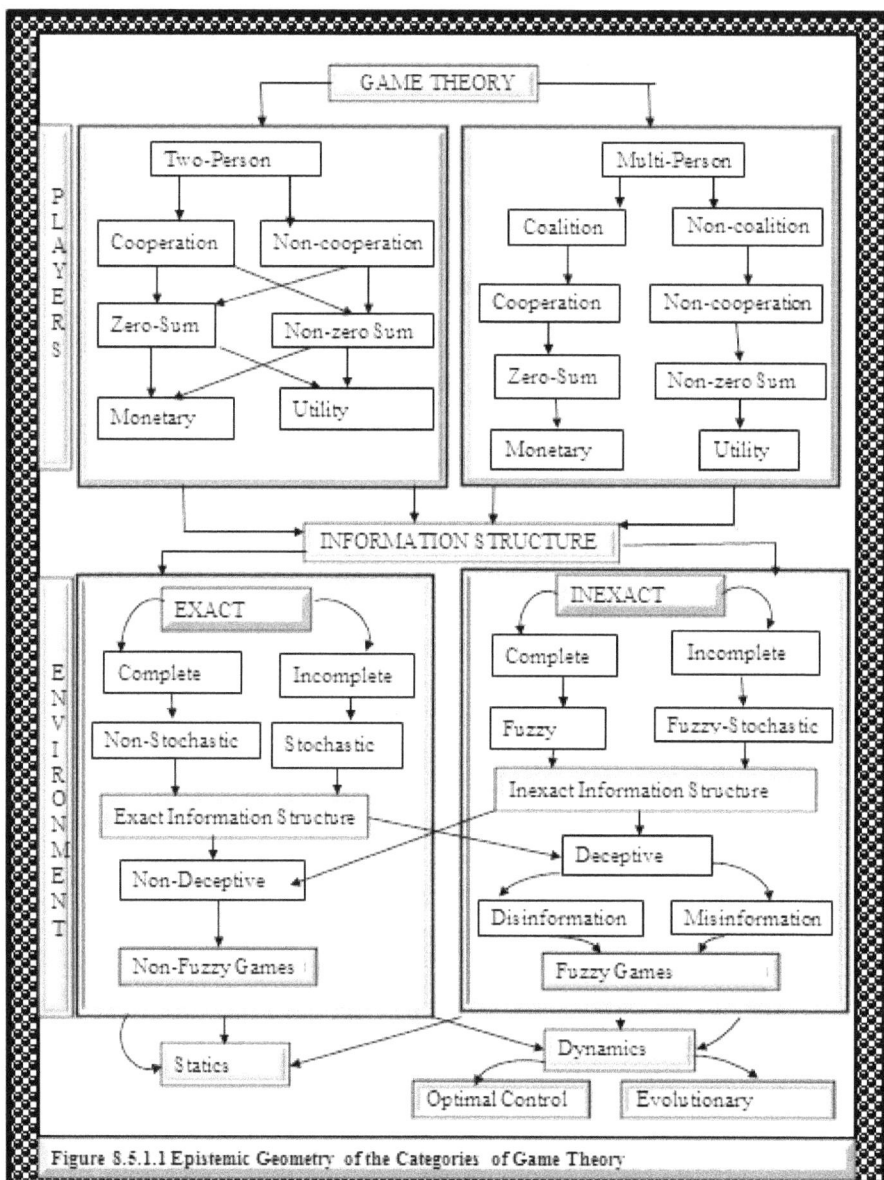

Figure 8.5.1.1 Epistemic Geometry of the Categories of Game Theory

## 8.5.2 Information, Paradigm of Thought and Types of Game Theory

The information on the payoffs may be defined over the qualitative or quantitative space. When the payoffs of the game are completely defined over the quantitative space, then one is dealing with quantity-time phenomena where time may be fixed (static case) or moving (dynamic case). Alternatively, when the payoffs are completely defined over qualitative space then one deals with quality-time phenomena which may fit in either a static case or a dynamic case. The meaning and the representation of the concepts of statics and dynamics will vary from category to category, from phenomenon to phenomenon and over categorial generations. The environment established by the exact information structure, whether complete or incomplete, is referred to as the *classical decision-choice environment*, where the information structure may be either exact and complete with an exact-variable representation or exact and incomplete with a random-variable representation. All the classical game theories deal with quantity-time phenomena whether they are stochastic or non-stochastic combined with static or dynamic conditions. In the structure of the classical environment, the problems of game theory are formulated, solved and analyzed under the classical paradigm of thinking, where information is exact and the laws of thought follow those of the Aristotelian system with the admittance of non-contradiction under the principle of excluded middle. The laws of thought used to process exact information, formulate decision-choice problems, solve and analyze them are the *classical laws of thought*, where contradictions are not allowed and the phenomena of qualitative disposition are difficult to handle. The laws of thought that do not admit contradiction, as a truth-value, with corresponding mathematics and logic of reasoning used to process exact information representation in all areas of knowledge is the *classical paradigm*. The class of games that are formulated, solved and analyzed under the classical paradigm in both static and dynamic spaces is referred to in this monograph as *classical game theory*, otherwise it falls into the non-classical game theory. The classical variable representation, while it can accommodate information incompleteness, cannot meaningfully accommodate information when it contains vagueness, quality, inexactness, deceptions and subjectivity. The information structure with these characteristics is called inexact information structure and defined in fuzzy spaces.

The decision-choice environment established by the inexact, subjective, deceptive and qualitative information structure, whether complete or incomplete, is referred to as the *non-classical decision-choice environment* in which the non-classical game theories are defined. This information structure comes under the general name fuzzy information structure with the corresponding fuzzy decision-choice environment.

The non-classical game theories deal with quantity-time, quality-time and quantity-quality-time phenomena in a fuzzy decision-choice environment whether the games are stochastic or non-stochastic combined with static or dynamic conditions. In the structure of the non-classical environment, the problems of game theory are formulated, solved and analyzed under the non-classical paradigm of thinking, where information may be qualitative, deceptive, subjective and inexact, and the laws of thought follow the fuzzy system or approximate system of reasoning with the admittance of contradiction as one of the truth values under the principle of relational continuum and unity. The laws of thought used to process inexact information, formulate decision-choice problems, solve and analyze them are the *fuzzy laws of thought* which admit contradiction. The laws of thought with corresponding mathematics and logic of reasoning for processing fuzzy information structure is the *fuzzy paradigm* that can handle subjectivity, qualitative disposition and deception. The class of games that are formulated, solved and analyzed under the fuzzy paradigm in fuzzy decision-choice environment is referred to in this monograph as *fuzzy game theory* as encompassing the non-classical game theory.

## Definition 8.5.2.1

The classical game theory is a class of all game theories that are characterized by exact information structure whether such information structure is complete or incomplete and uses the toolbox of the classical paradigm composed of its laws of thought and mathematics in its information representation, problem formulation, abstraction of solution, and analyses of relative strategies and outcomes.

## Definition 8.5.2.1

The fuzzy game theory is a class of all game theories that are characterized by inexact information structure whether such information structure is complete or incomplete, and uses the toolbox of the fuzzy paradigm composed of its laws of thought and mathematics in its information representation, problem formulation, abstraction of solution, and analyses of relative strategies and outcomes.

There are some important ideas that must be noted about the two definitions. The first idea to note is that these definitions are presented in the information space in terms of the assumptions that may be imposed. The second idea is that the assumed structure of the information space

334

imposes a special constraint set on the acceptable paradigm for information representation, methods of reasoning and the mathematics for information processing. The choice of the information structure and the paradigm of thought for information processing will affect both the explanatory and prescriptive paths of the selected game strategies and the possible outcomes. If deceptive information and vague information are accepted as part of the formation of strategies, then, it is not clear how the classical paradigm incorporates these elements into the formulation and analysis of conflicts, contradictions and strategies. These deficiencies in the information structure cannot be meaningfully treated in the same way as incomplete information with probabilistic reasoning, where the thought process deals with quantity of information in terms of variables and parameters that are defined in the exact space (for discussions on exact and inexact information spaces see [R3.10] [R3.13]. To understand these difficulties, one must separate the quantity of information measure specified in some meaningful volume from the quality of information defined in terms of vagueness, quality and subjectivity. In our current general theory of knowledge, the degree of completeness of information is measured in the probability space through the representation of random variable; while the degrees of vagueness, quality and subjectivity are measured in the possibility space through the representation of fuzzy variable. The simultaneous consideration of the degrees of completeness, vagueness and subjectivity are measured over the interactive phenomena of the possibility-probability spaces through the representation of *fuzzy-random variable* or *random-fuzzy variable* [R3.9] [R3.13] [R4.37] [R6.12] [R6.25] [R6.28] [R6.36] [R6.41] [R6.42] [R6.47].

Deceptive and vague information structures are defined in the inexact space that deals with quality of information with subjective interpretation whether complete or incomplete. Here, the information processing requires a logical reasoning different from probabilistic reasoning in the domain of the classical paradigm devoted to the exact quantitative information space. The logical process for dealing with vague and deceptive information structures in all decision-choice activities is possibilistic reasoning where lines of thought allow one to deal with both qualitative and quantitative information in an inexact space. In this respect, when the information of the game is defined in an inexact space, subjectivity is required to clarify the degrees of exactness, correctness and acceptance that may be attached to the information which must be processed for the selection of a strategy in the game space. The understanding of the deceptive information structure is extremely important in all decisions, such as negotiations, trading, war and political activities that present themselves as games. The power of the deceptive information structure cannot be underestimated. It forms an integral part of national ideological construct, choice of strategy and counter strategy in war games, diplomatic games, economic games, political games, pursuit-avoidance games and many more. It is made up of *disinformation* and *misinformation* as is shown in figure 8.5.1.1 (for full discussion on deceptive information structure see [R3.10] [R3.13] [R13.8] [R13.9]). In true social games as opposed to natural games, an inclusion of deceptive information structure to capture propaganda will change the nature of the game formulation. It is the presence of this deceptive information structure that calls for the principle of transparency in negotiations and governmental activities as well as opposed to governmental information classification.

### 8.5.3    Polarity, Duality, Relationality and Game Theory

The information conditions which generate game types have been discussed. Any game may be intentional or unintentional which may be played for simple entertainment or for changing the conditions of social or natural existence or both. All games entail conditions of opposites where such opposites may be either artificially or naturally created to establish either natural or social opposites. In general, social opposites are created for spectator games that are designed for entertainment. Some social opposites are created for power game to subjugate and control resources and income for economic advantage. Examples of social opposites are imperialism-colonialism polarity, occupier-occupied polarity and others. Natural opposites are part of the creative process and arise for transformation of conditions and structures of socio-natural existence. Examples include life-death polarity, deceased-healthy polarity and others. The poles of all polarities contain dualities under internal relational contradiction conflicts and tensions. Polarity involves negation of negation in the conditions of the game within the actual-potential polarity for categorial conversion. Dualities are the instruments of the negations of the poles, and the behaviors of the negative and positive characteristic (subsets duals) define the action activities of strategies and counter strategies in the game space for qualitative transformation of elements and categories. In fact, with a given information structure, the whole of democratic decision-choice system is based on the principle of opposites and defined in vote-casting game space wherever, whatever and whenever a collective decision-choice action is required. These opposites are structured as dualities with conflicts and contradictions under relational continuum and unity for any given social actual-potential polarity,

In terms of quality-time space, the game theory must be viewed in terms of transformational actions in the struggles between the actual and the potential where such actions are executed through the residing dualities with the corresponding relative negative-positive characteristic sub-sets (duals). Here, the game theory is to help to explain and forecast the result of the transformation action for categorial conversion, for example, evolutionary games in fuzzy-stochastic spaces. The game space may be defined by information that may or may not be correctly anticipated by the opposite characteristic sub-sets or the duals of the duality. Additionally, there are players unsure of the negative and positive poles of the polarity, where the payoffs at each stage is the degree of success of categorial conversion and the final payoff is the creation of the threshold delta for complete negation. The payoff of the game is increasing degree of power and dominance of the duals seeking to negate the actual or preserve the actual. The information for categorial conversion is composed of experiential information, possible strategies of the opposites and the progress of the reward of power or dominance that may be accrued. In general, the information is incomplete and vague and hence defined in the fuzzy-stochastic space [R3.10]. The payoff is qualitatively defined by the success of the categorial conversion of the actual to the potential by the *positive duality* through the *positive actions* of the *positive characteristic sub-set* (positive dual) or the retention of the actual by the negative duality through the *negative actions* of the negative characteristic sub-set (negative dual).

The game is about continual internal self-transformation in nature and society where categorial conversion implies interdependence of change and permanence, new and old, beginning and end, all of which are defined and established in terms of the actual and potential poles of the actual-potential polarity and may be expressed in terms of nothingness-somethingness polarity [R1.92]. Here, categorial conversion and the categorial-conversion game involving tactical and strategic actions find meaning and expression in existence and non-existence which contain the elements and the power of categorial conversion which defines the never-ending process of transformation from within

337

elements in the poles of the actual-potential polarity. The extensive discussions on the theory of categorial conversion and game theory are left to be developed and written. One thing that stands out is that every actual-potential polarity is a contest where the reward is an incremental gain to attain the Nkrumah Delta $\left[ \delta^{P*} = \delta^{P*}\left( \mathbf{a}^{P}\left( \mathbb{X}^{P} \right) \right) \in \mathfrak{I}^{a}, \delta^{P*} = \delta^{P*}\left( \mathbf{a}^{P}\left( \mathbb{X}^{P} \right) \right) \in \mathfrak{I}^{a} \right]$ to either maintain the existing actual or to actualize a potential by slow and continual accumulation of power for dominance. Every contest is a zero-sum game where $\left( \delta^{P*} + \delta^{N*} = 0 \right)$ which simply implies a destruction-construction game or substitution-transformation process of actual-potential poles of ontological elements. The categorial-conversion game is fuzzy-stochastic dynamic with fuzzy-random variable or stochastic-fuzzy dynamic with random-fuzzy variable. As a dynamic game, it is either optimal control game or evolutionary game where the important variables are defined in the quality-quantity-time space under real fuzzy-cost-benefit rationality defined in fuzzy-cost-benefit time space [R4.17] [R4.18]. By understanding the nature of ontological evolutionary dynamics in terms of qualitative disposition, one can meaningfully connect such an understanding to control dynamics of qualitative disposition in the quality-quantity-time space. Such an understanding may also be extended to social game space of principal-agent problems, pursue-avoidance problems and fringe-incumbency problems. The pursue-avoidance problems include independence-neo-colonialist games, principal-agent problems include governance-people games while fringe-incumbency problems include colonialism-imperialist games. The conceptual structures of these problems and corresponding games have been discussed in [R13.8][R13.9].

# Multidisciplinary References

## R1: On Africa and Foundations of African Thought System

**[R1.1]** Abraham, W.E., *The Mind of Africa*, London, Weidenfeld and Nicholson, 1962.

**[R1.2]** Ahuma, Attoh S.R.B., *The Gold Cost Nation and National Consciousness*, in Langley, Ayo J. (ed.), *Ideologies of Liberation in Black Africa: 1856-1970*, London, Rex Collings, 1979. pp. 161-172.

**[R1.3]** Ajala, Adekunle, *Pan-Africanism: Evolution, Progress and Prospects*, London, Andre Deutsch, 1973.

**[R1.4]** Allen, J. *et al., Without Sanctuary: The Lynching Photography in America*, Santa Fe, N.M., Tween Palms Publishers, 2000.

**[R1.5]** Allen, James P., *Genesis in Egypt: The Philosophy of Ancient Egyptian Creation* Accounts (Yale Egyptological Studies, #2), New Haven, Yale University, Department of Near Eastern Languages and Civilizations, 1988.

**[R1.6]** Alston, W. and R.B. Brandt (eds.), *The Problems of Philosophy*, Boston, MA, Allyn andBacon, 1967.

**[R1.7]** Amate, C.O.C., *Inside OAU: Pan-Africanism in Practice*, New York, St. Martin's Press, 1986.

**[R1.8]** Amen, Ra Un Nefer, *Metu Neter* Vol. 1, Bronx, New York, Khamit Co. Pub., 1977.

**[R1.9]** Amin, Samir, *Neo-Colonialism in West Africa*, New York, Monthly Review Press, 1973.

**[R1.10]** Amin, Samir, *Eurocentrism*, New York, Monthly Review Press, 1989.

**[R1.11]** Amo Afer, A.G., *The Absence of Sensation and the Faculty of Sense in the Human Mind and Their Presence in our Organic and Living Body, PhD Dissertation and other Essays1727-1749*, Halle, Wittenberg, Jena, Martin Luther University Translation, 1968.

**[R1.12]** Ani, M., *Yurugu: An African-Centered Critique of European Cultural Thought and Behavior*, Trenton, N.J., Africa World Press, 1994.

**[R1.13]** Apostel, Leo, *African Philosophy: Myth or Reality*, Ghent, Belgium, Scientific Publishers, 1981.

**[R1.14]** Appolus, Emil (ed.), *The Resurgence of Pan-Africanism*, London, Freedman Brothers, 1974.

**[R1.15]** Aristote, N., *Politique*, Books I and II, Paris, Les Belles Lettres, 1960.

**[R1.16]** Armah, Ayi Kwei, *Two Thousand Seasons*, Oxford, Heineman, 1973.

**[R1.17]** Armah, Ayi Kwei, *Osiris Rising*, Dakar, Senegal, Africa Per Ankh, 1995.

**[R1.18]** Asante, Molefi K. *et al.* (eds.), *African Culture: The Rhythms of Unity*, Trenton, New Jersey, African World Press Inc., 1990.

**[R1.19]** Asante, Molefi K., *Afrocentricity*, Trenton, New Jersey, Africa World Press, Inc., 1989.

**[R1.20]** Asante, Molefi K. *et al.* (eds.), *African Intellectual Heritage: Book of Sources*, Philadelphia, PA, Temple University Press, 19

**[R1.21]** Asante, Molefi K., *Kement, Afrocenticity and Knowledge*, Trenton, New Jersey, Africa World Press, Inc., 1990.

**[R1.22]** Asante, Molefi K and Ama Mazama (Eds.), Encyclopedia of African Religion Thousand Oaks, California, Sage, 2009

**[R1.23]** Asante, S.K.B., *Pan-African Protest: West Africa and the Italo-Ethiopian Crisis*, 1934-1941, Legon History Series, London, Longmans, 1977.

**[R1.24]** Ashby, M.A., *The African Origins of Civilization: Book I, Mystical Religion and Yoga Philosophy*, Miami, FL, Cruzan Mystic Books, 1995.

**[R1.25]** Ashby, M.A., *The African Origins of Civilization, Book II: African Origins of Western Civilization, Religion and Philosophy*, Miami, FL, Cruzan Mystic Books, 2001.

**[R1.26]** Ashby, M.A., *The African Origins of Civilization Book III: The African Origins of Eastern Civilization, Religion, Yoga Mysticism and Philosophy*, Miami, FL, Cruzan Mystic Books, 2001

**[R1.27]** Ashby, M.A., Egyptian *Yoga Vol. I,: The Philosophy of Enlightenment*, Miami, FL, Cruzan Mystic Books, 1997

**[R1.28a]** Auma-Osolo, A., *Cause-Effects of Modern African Nationalism on the World Market*, University Press of America, 1983.

**[R1.28b]** Austen, R. A., *In Search of Sunjata: The Mande Oral Epic as History, Literature and Performance*, Bloomington, Indiana University Press, 1999

**[R1.29]** Axinn, Sidney, "Kant, Logic and Concept of Mankind," *Ethics*, Vol. 48 (XLVIII), 1958, pp. 286-291.

**[R1.30]** Bakewell, Charles M., *Source Book in Ancient Philosophy*, New York, Charles Scribner's Sons, 1909.

**[R1.30a]** Bangura, Abdul K., *African Mathematics: From Bones to Computers*, Lanhan, MD, USA, University Press of America, 20012.

**[R1.31]** Bascom, William, *African Art in Cultural Perspective: Introduction*, New York, Norton, 1973        .

**[R1.32]** Bebey, Francis, *African Music: A People's Art*, Westport, CT, Lawrence Hill and Co., 1980.

**[R1.33]** Bell, Richard H., *Understanding African Philosophy: A Cross-cultural Approach to Classical and Contemporary Issues*, New York, Routledge, 2002.

**[R1.34]** Ben-Jochannan, Joseph A.A., *Cultural Genocide in the Black and African Studies Curriculum*, New York, River Nile Universal Books, 1972.

**[R1.35]** Ben-Jochannan, Joseph A.A., *Africa: Mother of Western Civilization*, Baltimore, MD, Black Classic Press, 1988.

**[R1.36]** Ben-Jochannan, Joseph A.A. *et al.*, *African Origins of Major Western Religions*, New York, Alkebulan Books, 1970.

**[R1.37]** Ben-Jochannan, Joseph A.A., *Black Man of the Nile*, Baltimore, MD, Black Classic Press, 1989.

**[R1.38]** Ben-Jochannan, Joseph A.A., *We the Black Jews*, Baltimore, MD, Black Classic Press, 1983.

**[R1.39]** Ben-Jochannan, Joseph A.A., Hugh Brooks and Kempton Webb, *Africa: Land, People and Cultures of the World*, New York, W.H. Sadlier, 1971.

**[R1.40]** Berkeley, George, "Material Things are Experiences of Men or God" in [R1.5], 1967, pp. 658-668.

**[R1.41]** Blyden, E.W., "African Life and Customs" in Langley, Ayo J. (ed.), *Ideologies of Liberation in Black Africa:1856-1970*, London, Rex Collings, 1979 pp. 78-87..

**[R1.42]** Blyden, Edward W., *Christianity, Islam and the Negro Race*, Baltimore, MD, Black Classic Press, 1994.

**[R1.43]** Boahen, A. Adu, *African Perspectives on Colonialism*, Baltimore, Johns Hopkins University Press, 1987.

**[R1.44]** Bonnefoy, Yves, *Mythologies*, Vols 1-2, Chicago, University of Chicago Press, 1991.

**[R1.45]** Bovill, E.W., *The Golden Trade of the Moors*, London, Oxford University Press, 1958.

**[R1.46]** Bovill, E.W., *Caravans of the Old Sahara*, London, Oxford University Press, 1933.

**[R1.47]** Breasted, James H., *Ancient Records of Egypt*, Vols. 1-5, Chicago, The University of Chicago Press, 1906-1907.

**[R1.48]** Breasted, James Henry, *Development of Religion and Thought in Ancient Egypt*, New York, Charles Scribner's Sons, 1912.

**[R1.49]** Browdes, Anthony T., *Exploiting the Myths, Vol. 1: Nile Valley Contribution to Civilization*, Washington, D.C., The Institute of Karmic Guidance, 1992.

**[R1.50]** Browdes, Anthony T., *Egypt on the Potomac*, Washington, D.C., IKG Publishers, 2004.

**[R1.51]** Brown, Lee (ed.), *African Philosophy: New and Traditional Perspectives*, New York, Oxford University Press, 2004.

**[R1.52]** Budge, Willis E.A., *Osiris and the Egyptian Resurrection*, Vols. 1 and 2, New York, Dover, 1973.

**[R1.53]** Budge, Willis E.A., *The Gods of Egyptians*, Vols. 1 and 2, New York, Dover, 1969.

**[R1.54]** Budge, Willis E.A., *Amulets and Talismans*, New Hyde Park, University Books, 1961.

**[R1.55]** Budge, Willis E.A., *The Bandlet of Righteousness, an Ethiopian Book of the Dead*, London, Luzac and Co., 1929.

**[R1.56]** Budge, Willis E.A., *The Egyptian Book of the Dead*, New York, Dover, 1967.

**[R1.57]** Budge, Willis E.A., *The Papyrus of Ani*, Vols. 1, 2 and 3, New York, G. P. Putman's sons, 1913.

**[R1.58]** Budge, Willis E.A., *The Negative Confession*, New York, Bell Publishing Co., 1960.

**[R1.59]** Budge, Wallis, *The Egyptian Sudan.* Vols. I and II, London, Kegan, Trench & Co. 1907.

**[R1.60a]** Cabral, A., *Return to the Source*, New York, Monthly Review Press, 1973.

**[R1.60b]** Calame-GRIAULE, G., *Words and the Dogon World*, Trans D. LaPin, Philadelphia, Institute for the Study of Human Issues, 1986.

**[R1.61]** Cameron, J., *The African Revolution*, New York, Random House, 1961.

**[R1.62]** Chinweizu, The *West and the Rest of Us: White Predators, Black Slaves and the African Elite*, New York, Vintage Books, 1975.

**[R1.63]** Chomsky, Noam, *Pirates and Emperors: International Terrorism in the Real World*, New York, Claremont Research Publication, 1986.

**[R1.64]** Chomsky, Noam, *Profit Over People*, New York, Seven Stories Press, 1999.

**[R1.65]** Chomsky, Noam and E. S. Herman, *The Washington Connection and Third World Fascism*, New York, South End Press, 1979.

**[R1.66]** Clark, Gordon Haddon, *Thales to Dewey: a History of Philosophy*, Boston, MA, Houghton Mifflin, 1957.

**[R1.67]** Clark, Rundle R.T., *Myth and Symbol in Ancient Egypt,* New York, Thames and Hudson, 1978.

**[R1.68]** Clarke, John H., *Notes on African World Revolution: Africa at the Crossroads,* Trenton, N.J. African World Press, 1991.

**[R1.69a]** Clarke, John H., *Marcus Garvey and the Vision of Africa*, New York, Random House, 1974.

**[R1.69b]** Comaroff, J. and J. Comaroff, *Of Revelation and Revolution: Christianity, Colonization and Consciousness in South Africa*, Chicago, University of Chicago Press, 1991

**[R1.70a]** Cowan, L.G., *The Dilemmas of African Independence*, New York, Warker and Co., 1964.

**[R1.70b]** Craton, Michael, *Sinews of Empire*, New York, Anchor Press, 1974.

**[R1.71]** Cromwell, Adelaide M. (ed.), *Dynamics of the African Afro-American Connection: From Dependency to Self-Reliance*, Washington, D.C., Howard University Press, 1987.

**[R1.72]** Cruse, Harold, *The Crisis of the Negro Intellectual: A History and Analysis of the Failure of Black Leadership*, New York, Quill Press, 1967.

**[R1.73]** Danquah, J.B., *Friendship and Empire*, London, Fabian Colonial Bureau, 1949.

**[R1.74]** Danquah J.B., *The Akan Doctrine of God: A Fragment of Gold Coast Ethics and Religion*, London, Frank Cass and Co., 1968.

**[R1.75]** Danquah, J.B., *Ancestors, Heroes and God*, Kibi, Ghana, George Boakie Pub. Co., 1938.

**[R1.76]** Davidson, Basil, *The Lost Cities of Africa*, Boston, Little, Brown & Co., 1959.

**[R1.77]** Davis, Kortright, *Emancipation Still Comin': Explorations in Caribbean Emancipatory Theology*, New York, Orbis, 1990.

**[R1.78]** Dawson, Christopher, *The Making of Europe, Part 1: The Foundations*, New York, The World Pub. Co., 1956.

**[R1.79]** De Buck, Adriaan and Alan H. Gardiner (eds.), *The Egyptian Coffin Texts*, Vols. 1-7, Chicago, University of Chicago Press, 1935-1961.

343

**[R1.80]** Descartes, René, *The Philosophical Works of Descartes*, Cambridge, Cambridge University Press, 1931.

**[R1.81]** Descartes, René, "Man as Two Substances" in [R1.5], 1962, pp. 386-402.

**[R1.82]** Descartes, René, *Meditations on First Philosophy*, New York, Boobs-Merrill and Co, 1960.

**[R1.83a]** Diop, Cheikh A., *Towards the African Renaissance: Essays in Culture and Development, 1946-1960*, Berwick upon Tweed .Great Britain, Martins the Printers, *1996*

**[R1.83b]** Diop, Cheikh A., *The African Origins of Civilization: Myth or Reality*, Brooklyn, New York, Lawrence Hill, 1974

**[R1.84]** Diop, Cheikh A., *The Cultural Unity of Black Africa*, Chicago, Third World Press, 1978

**[R1.85]** Diop, Cheikh A., *Pre-colonial Black Africa*, Brooklyn, New York, Lawrence Hill, 1987.

**[R1.86]** Diop, Cheikh A., *Civilization or Barbarism*, Brooklyn, New York, Lawrence Hill, 1991.

**[R1.87]** Diop, Cheikh A., *Black Africa: The Economic and Cultural Base for a Federated State*, Brooklyn, New York, Lawrence Hill, 197.

**[R1.88]** Doane, Thomas W., *Bible Myths and Their Parallels in Other Religions*, New Hyde Park, New York, University Book, 1971.

**[R1.89]** Dodson, H. *et al., Jubilee: The Emergence of African-American Culture*, Schomburg Center for Research in Black Culture, New York, New York Public Library, 2002.

**[R1.90a]** Dompere, Kofi K., *Africentricity and African Nationalism*, Langley Park, MD, IAAS Publishers, 1992.

**[R1.90b]** Dompere, Kofi K., *The Theory of Philosophical Consciencism*, Working Monograph, Washington, D.C., Howard University Economics Department, 2013

**[R1.91]** Dompere, Kofi K., *African Union: Pan-African Analytical Foundations*, London, Adonis- Abbey Pub., 2006.

**[R1.92]** Dompere, Kofi K., *Polyrhythmicity: Foundations of African Philosophy*, London, Adonis- Abbey Pub., 2006.

**[R1.93]** Douglass, Frederick, "Fourth of July Oration," in Asante, Molefi K. *et al.* (eds.), *African Intellectual Heritage: Book of Sources*, Philadelphia, PA, Temple University Press, 19, pp. 637-640.

**[R1.94]** Dray, W.H., "Historical Understanding as Re-thinking," in Baruch Brody (ed.) Readings in the Philosophy of Science, Englewood Cliffs, NJ, Prentice-Hall Inc, 1970, pp. 167-179.

**[R1.95]** DuBois, W.E.B., *W. E. B. Du Bois: Reader* [edited by Eric J. Sundquist], New York, Oxford University Press 1996.

**[R1.96]** DuBois, W.E.B., *Dust of Dawn: An Essay Toward an Autobiography of a Race Concept*, New York, Harcourt, Brace

**[R1.97]** DuBois, W.E.B., *The World and Africa*, New York, International Publishers, 1987.

**[R1.98]** DuBois, W.E.B., *On Sociology and Black Community*, Chicago, The University of Chicago Press, 1978.

**[R1.99]** DuBois, W.E.B., *W.E.B. DuBois Speaks: Speeches and Addresses 1890-1919*, New York, Pathfinder, 1970.

**[R1.100]** DuBois, W.E.B., *The Education of Black People; Ten Critiques, 1906-1960* (Ed. H. Aptherker), New York, Monthly Review Press, 1975.

**[R1.101]** Duchein, R.N., *The Pan-African Manifesto*, Accra, Guinea Press Ltd., 1957.

**[R1.102]** Emerson, Rupert, *From Empire to Nation*, Boston, MA, Beacon Press, 1963.

**[R1.103]** Engels, Frederick, *Dialectics of Nature*, New York, International Pub., 1940.

**[R1.104]** Esedebe, P.O., *Pan-Africanism: The Idea and the Movement 1963-1976*, Washington, D.C., Howard University Press, 1982.

**[R1.105]** Fagg, William B., *Nigerian Images, The Splendor of African Sculpture*, New York, Praeger, 1963.

**[R1.106]** Fanon, Frantz, *The Wretched of the Earth*, New York, Grove Press Inc., 1963.

**[R1.107]** Fanon, Frantz, *Toward the African Revolution*, New York, Grove Press Inc., 1964.

**[R1.108]** Fanon, Frantz, *Black Skin, White Masks*, New York, Dover Press Inc., 1967.

**[R1.109]** Faulkner, R.C., *The Ancient Egyptian Coffin Texts*, Warminster, England, Aris and Philips, 1973.

**[R1.110]** Faulkner, R.C., *The Ancient Egyptian Pyramid Texts*, Oxford, Clarendon, 1969.

**[R1.111a]** Felder, C.H., *Troubling Biblical Waters: Race. Class and Family*, New York, Orbis Press, 1989.

**[R1.111b]** Finch, Charles S. III, *The Star of Deep Beginnings: The Genesis of African Science and Technology,* Khenti Inc, Decatur, Georgia, *USA,* 2001.

**[R1.112a]** Foote, George W. and W.P. Bell (eds.), *The Bible Handbook for Freethinkers and Inquiring Christians,* London, The Pioneer Press, 1921.

**[R1.112b]** Forde, D. (ed.) *African Worlds: Studies in the Cosmological Ideas and Social Values of African Peoples,* Oxford, Oxford University Press, 1954.

**[R1.113]** Frankford, Henri, *The Intellectual Adventure of Ancient Man,* Chicago, University of Chicago Press, 1957.

**[R1.114]** Frankford, Henri, *Ancient Egyptian Religion: An Interpretation,* New York, Columbia University Press, 1948.

**[R1.115]** Fraser, Douglas (ed.), *African Art as Philosophy,* New York, Interbook, 1974.

**[R1.116]** Frazer, James G., *The Golden Bough; a Study in Magic and Religion,* Vols. 1 – 13, London, Macmillan, 1911-1936.

**[R1.117]** Frobenius, Leo, *The Voice of Africa,* Vols. 1 and 2, London, Hutchinson and Company, 1913.

**[R1.118]** Gabre-Medhim, Tsegaye, "The Origin of the Trinity in Art and Religion," **in** Ben-Jochannan, Joseph A.A. *et al., African Origins of Major Western Religions,* New York, Alkebulan Books, 1970, **pp**. 99-120.

**[R1.119]** Gadamer, Hans Georg, *The Beginning of Knowledge* (Translated by Rod Coltman), New York, Continuum, 2001.

**[R1.120]** Garvey, Marcus, *The Philosophy and Opinions of Marcus Garvey,* Dover, MA, The Majority Press, 1986.

**[R1.121]** Geiss, I., *The Pan-African Movement,* London, Methuen Press, 1974.

**[R1.122]** Ghana Ministry of Information and Broadcasting, *Nkrumah's Subversion in Africa,* Accra, Ghana, Government Printing Press, 1966.

**[R1.123]** Gillings, Richard J., *Mathematics in the Times of the Pharaohs,* London, M.I.T. Press, 1972.

**[R1.124]** Glover, Ablade E. (ed.), *Adinkra Symbolism,* Accra, Liberty Press Ltd., 1971.

**[R1.125]** Glover, Ablade E., *Linguist Staff Symbolism,* Accra, Liberty Press Ltd., 1971.

**[R1.126]** Glover, Ablade E., *Stools Symbolism,* Accra, Liberty Press, 1971.

346

**[R1.127]** Goma, L. K. H., *The Hard Road to the Transformation of Africa*, Aggrey-Fraser Guggisberg Memorial Lecture 1991, University of Ghana, Legon-Accra, Communication Studies Press, 1991.

**[R1.128]** Graves, Robert and Raphael Patai, *Hebrew Myths: The Book of Genesis*, New York, Doubleday and Company, 1964.

**[R1.129]** Green, R.H. *et al., Unity or Poverty: the Economics of Pan-Africanism*, Baltimore, MD, Penguin Books, 1968.

**[R1.130]** Griaule, M., *Conversations with Ogotemmele*, London, Oxford University Press, 1969.

**[R1.131]** Groves, Charles P., *Planting of Christianity in Africa*, Vols.1-4, London, Lutterworth Press, 1948-1958.

**[R1.132]** Guyer, David, *Ghana and the Ivory Coast*, New York, An Exploration Press, 1970.

**[R1.133]** Gyekye, K., *An Essay on African Philosophical Thought: The Akan Conceptual Schemes*, New York, Cambridge University Press, 1987.

**[R1.134]** Gyekye, Kwame, *Tradition and Modernity: Philosophical Reflections on African Experience*, New York, Oxford University Press, 1997.

**[R1.135]** Hargreaves, J. D., *Prelude to the Partition of West Africa*, London, Macmillan, 1963.

**[R1.136]** Harris, Joseph E. (ed.), *Global Dimensions of the African Diaspora*, Washington, D.C., \ Howard University Press, 1982.

**[R1.137]** Hayford, Casely J.E., "African Nationality," in Langley, Ayo J. (ed.), *Ideologies of Liberation in Black Africa:1856-1970*, London, Rex Collings, 1979, pp. 203-219.

**[R1.138]** Herman, Edward N. and N. Chomsky, *Manufacturing Consent*, New York, Pantheon Books, 1988.

**[R1.139]** Hess, R., "Travels of Benjamin of Tudela," *Journal of African History*, Vol. 6, 1965, p. 17 (also in [R1.35] p. 5).

**[R1.140]** Hill, Cromwell A., and Martin Kilson (eds.), *Apropos of Africa: Sentiments of Negro American Leaders on Africa From 1800 to the 1950*, London, Frank Cass and Co., 1969.

**[R1.141]** Hillard, Asa G. (eds.), *The Teachings of Ptahhotop: The Oldest Book in the World*, Atlanta, Blackwood, 1987.

**[R1.142]** Hirschman, A. O., *National Power and the Structure of Foreign Trade*, Berkeley, University of California Press, 1945

**[R1.143]** Hochschild, Adam, *King Leopold's Ghost*, New York, Houghton Mifflin Co., 1998.

**[R1.144]** Hodgkin, Thomas, "National Movements in West Africa", *The Highway*, February, 1952, pp. 169-175.

**[R1.145]** Hoskins, L Halford (ed.) *Aiding Underdeveloped Areas Abroad, The Annals*, Vol. 268, March 1950.

**[R1.146]** Hountondji, Paulin J., *African Philosophy: Myth and Reality*, Bloomington, Indiana, Indiana University Press, 1983.

**[R1.147]** Hufbauer, G.C. *et al., Economic Sanctions in Support of Foreign Policy Goals*, Washington, D.C., Institute for international Economics, 1983.

**[R1.148]** Hughes, L. and M. Meltzer, *A Pictorial History of the Negro (African American)*, New York, Crown, 1963.

**[R1.149]** Ilyenkov, E.V., *Dialectical Logic: Essays on its History and Theory*, Moscow, Progress Pub. 1977

**[R1.150]** Jahn, Janheinz, *Muntu, The New African Culture* (Trans. Marjorie Grene), New York, Grove, 1961.

**[R1.151]** James, C.L.R., "Kwame Nkrumah: Founder of African Emancipation", *Black World*, Vol. XXI (9), July, 1972.

**[R1.152]** James, George G.M., *Stolen Legacy*, Newport News, Virginia, United Brothers Communication System, 1989.

**[R1.153]** Jeffreys, M.D.W., "The Negro Enigma," *West African Review*, September, 1951.

**[R1.154]** Johnson, De Graft J.C., *African Glory*, New York, Praeger, 1955.

**[R1.155]** July, R.W., *The Origins of Modern African Thought*, London, Faber, 1967.

**[R1.156]** Kant, Immanuel, *On History*, New York, Bobb-Merrill and Co. Inc, 1963.

**[R1.157]** Karenga, N., *Maat: The moral Ideal in Ancient Egypt; A Study in Classical African Ethics*, Los Angeles, University of Sancore Press, 2006.

**[R1.158a]** Karenga, N., *Kwanzaa: Origin, Concepts, Practices*, Los Angeles, Kawaida Publishers, 1988.

**[R1.158b]** Karenga, N., *The African American Holiday of Kwanzaa*, Los Angeles, University of Sancore Press, 1988.

**[R1.159]** Kaunda, K., "Ideology and Humanism", *Pan-African Journal*, Vol.1 (1) 1968, pp. 5-6.

**[R1.160]** Kaunda, K., *Humanism in Africa*, London, Longmans, 1966.

**[R1.161]** Keita, L., " African Philosophical Systems: A Rational Reconstruction," *The Philosophical Forum*, Vol. 9 (23) Winter-Spring, 1977.

**[R1.162]** Keita, L., " Two Philosophies of African History: Hegel and Diop," *Présence* Africaine, No. 91, Third Quarter, 1974.

**[R1.163]**Keltie, J.S., *The Partition of Africa*, London, E. Stanford Press, 1893.

**[R1.164]**Kenyatta, Jomo, *Harambee!: The Prime Minister of Kenya's Speeches 1963-1964*, New York, Oxford University Press, 1964.

**[R1.165]**Kenyatta, Jomo, *Facing Mount Kenya*, New York, Vintage Books, 1965.

**[R1.166]**Knight, Richard P., *The Symbolic Language of Ancient Art and Mythology*, New York, J.W. Bouton, 1876.

**[R1.167]**Kohn, H., *African Nationalism in the Twentieth Century,* Princeton, New Jersey, Van Nostrand, 1965.

**[R1.168]**Krafona, Kwesi (ed.), *Organization of African Unity: Essays in Honour of Kwame Nkrumah*, London, Afroword Publishing Co., 1988.

**[R1.169a]**Lane, Ambrose I. Sr., *For Whites Only? How and Why America Became a RacistNation,* Washington, DC., 21st Century Pub. Inc., 1999

**[R1.169b]**Langley, J. Ayodele, *Pan-Africanism and Nationalism in West Africa 1900-1945: A study in Ideology and Social Classes*, Oxford, Clarendon Press, 1973.

**[R1.170]**Langley, J.A., "Garveyism and African Nationalism," *Race*, Vol. 11 (2), pp. 157-172.

**[R1.171]**Langley, Ayo J. (ed.), *Ideologies of Liberation in Black Africa: 1856-1970*, London, Rex Collings, 1979.

**[R1.172a]***Leach, E. R.] Culture and Communication: The Logic by which Symbols are Connected*, Cambridge, Cambridge University Press, 1976

**[R1.172b]** Legum, C., *Pan-Africanism*, New York, Praeger Pub., 1962.

**[R1.173]**Lynch, H.R., *Edward Wilmot Blyden: Pan-Negro Patriot*, London, Oxford University Press, 1967.

**[R1.174]**MacEwan, A., *International Economic Instability and U.S. Imperial Decline*, New York, Monthly Review Press, 1990.

**[R1.175]**Mandela, Nelson, *The Struggle for My Life*, New York, Pathfinder Press, 1986.

**[R1.176a]**Martin, D., Maat and Order in African Cosmology: A Conceptual Tool for Understanding Indigenous Knowledge, *Jour of Black Studies,* Vol. 38, #6, Jul 2008, pp.951-967.

**[R1.176b]**Martin, Tony, *The Pan-African Connection: From Slavery to Garvey and Beyond*, Dover, MA, The Majority Press, 1983.

**[R1.177]**Mark Cohen, S. *et al.* (eds.), *Readings in Ancient Greek Philosophy: From Thales to Aristotle*, Indianapolis, Indiana, Hackett, 2000.

**[R1.178]**Massey, Gerald, *Ancient Egypt: The Light of the World*, Baltimore, MD, Black Classic Press, 1992.

**[R1.179]**Massey, Gerald, *A Book of the Beginnings*, Vols. 1-2, London, William and Norgate, 1881.

**[R1.180]**Massey, Gerald, *Pyramid Text*, Vols. 1-4, New York, Longmans Green, 1952.

**[R1.181]**Massey, Gerald, *The Natural Genesis*, Vols. 1 and 2, Baltimore, MD, Black Classic Press, 1998 (First published 1883).

**[R1.182]**Mathews, Wendell, *Basic Symbols and Terms of the Church*, New York, Fortress Press, 1971.

**[R1.183]**Mazrui, Ali A.A., *Nationalism and New States in Africa from About 1935*, Nairobi, Kenya, Heinemann, 1984

**[R1.184]**Mazrui, Ali A.A., toward *a Pax Africana*, Chicago, The University of Chicago Press, 1967.

**[R1.185]**Mbiti, J.S., *African Religions and Philosophy*, New York, Anchor, 1970.

**[R1.186]**M'buyinga, E., *Pan-Africanism or Neo-Colonialism: The Bankruptcy of the O.A.U.*, London, Zed Press, 1975

**[R1.187]**McCray, W.A., *The Black Presence in the Bible*, vols. 1 and 2, Black Light Fellowship, 1990.

**[R1.188]**Mendoza, M.G. *et al.* (eds), *Trade Rules in the Making: Challenges in Regional and Multilateral Negotiations*, Washington, D.C., Brookings Institutions Press, 1999.

**[R1.189]**Meyerowitz, Eva L.R., *The Divine Kingship in Ghana and Ancient Egypt*, London, Faber and Faber, 1960.

**[R1.190a]**Moody, R.A., *Life After Life*, New York, Bantam, 1976.

**[R1.190b]**Mudimbe, V. Y., *The Inventions of Africa; Gnosis, Philosophy, and Other Knowledge*, Womington, Indiana University Press, 1988.

**[R1.191]**Murapa, R., "Nkrumah and Beyond: Osagyefo, Pan-Africanist Leader," *Black World*, Vol. XXI (9), July, 1972.

**[R1.192]**Niane, D.T., *Sundiata: An Epic of Old Mali in Transactions*, G. D. Pickett (ed.), London, Longmans, 1965

**[R1.193]**Nketia, J.H., *Ethnomusicology in Ghana*, Legon, Ghana, Ghana University Press, 1969.

**[R1.194]**Nketia, J.H., *Ethnomusicology and African Music: Collected Papers Vol.1: Models of Inquiry and Interpretation*, Accra, Afram Publication, 2005

**[R1.195]**Nketia, J.H., *The Music of African*, New York, Norton, 1974

**[R1.196]** Nketia, J.H. and J. C. Dje Dje, (eds.) *Selected Reports in Ethnomusicology Vol. V Studies in African Music,* Los Angeles, Univ. of California Press 1984 .

**[R1.197]** Nkrumah, Kwame, *Toward Colonial Freedom,* London, Heinemann, 1962 (First Published 1946).

**[R1.198]** Nkrumah, Kwame, *Ghana, The Autobiography of Kwame Nkrumah,* London, Thomas Nelson and Sons Ltd., 1957.

**[R1.199]** Nkrumah, Kwame, Speech at the Conference of Independent African States, Accra, April 15, 1958.

**[R1.200]** Nkrumah, Kwame, *I Speak of Freedom,* London, PANAF Press, 1965.

**[R1.201]** Nkrumah, Kwame, Speech at the Closing Session of Casablanca Conference, Casablanca, January 7, 1961.

**[R1.202]** Nkrumah, Kwame, *Africa Must Unite*, New York, International Publishers, 1963.

**[R1.203]** Nkrumah, Kwame, *Consciencism,* London, Heinemann, 1964.

**[R1.204]** Nkrumah, Kwame, *Neo-Colonialism*, New York,

**[R1.205]** Nkrumah, Kwame, Speech at the Fourth Afro-Asian Solidarity Conference, Winneba, Ghana, May 10, 1965.

**[R1.206]** Nkrumah, Kwame, Challenge of the Congo, London, PANAF Press, 1967.

**[R1.207]** Nkrumah, Kwame, *Axioms*, New York, International Publishers, 1967.

**[R1.208]** Nkrumah, Kwame, Dark Days of Ghana, London, PA-NAF Press, 1968.

**[R1.209]** Nkrumah, Kwame, *Handbook of Revolutionary Warfare,* London, PANAF Press, 1968.

**[R1.210]** Nkrumah, Kwame, *Class Struggle in Africa,* London, PANAF Press, 1970.

**[R1.211]** Nkrumah, Kwame, *Revolutionary Path*, London, PANAF Press, 1973.

**[R1.212]** Nkrumah, Kwame, *Rhodesian File*, London, PANAF Press, 1976.

**[R1.213]** Nkrumah, Kwame, "Principles of African Studies," Address Delivered at the Time of Official Opening of the Institute for African Studies, University of Ghana, Legon, *Voice of Africa*, Vol. 3, (3) December, 1963.

**[R1.214]** Nkrumah, Kwame, *Selected Speeches, Vols. 1-6,* Compiled by Samuel Obeng, Accra, Afram Pub. 1997

**[R1.215]**Nsanze, Terence, "In Search of an African Ideology," *Pan-African Journal,* Vol.1 (1), 1968, pp. 27-30.

**[R1.216]**Nyerere, Julius K., *Freedom and Unity ( Uhuru Na Umoja),* Nairobi, Oxford University Press, 1966.

**[R1.217]**Nyerere, Julius K., *Freedom and Socialism (Uhuru Na Ujamaa),* Nairobi, Oxford University Press, 1968.

**[R1.218a]**Obenga,Theophile T, *African Philosophy During The Period Of The Pharaohs, 2800-300BC,* Popenguine, Senegal, W.A., Per Ankh Pub., 2006

**[R1.218b]**Obenga, Theophile T, *African Philosophy in World History,* Popenguine, Senegal, W.A., Per Ankh Pub., 1998

**[R1.218c]**Obenga, Theophile T, *Ancient Egypt and Black Africa,* London, Karnak Books, Africa World Press, 1992

**[R1.218d]**Odinga, Oginga, *Not Yet Uhuru,* New York, Hill and Wang, 1967.

**[R1.219]**Ofori-Ansah, Kwaku P., *Symbols of Adinkra Cloth,* Washington, D.C., 1999.

**[R1.220]**Ofosu-Appiah, L.H., *Encyclopaedia Africana* Vols. 1 and 2, New York, Reference Publication, Inc., 1977.

**[R1.221]**Okwonko, R. L., "The Garvey Movement in British West Africa," *Journal of African History,* Vol. 21, 1980, pp. 105-117.

**[R1.222]**Olugboji, D., *The United States of Africa and Realpolitik,* Lagos, Nigeria, CMS Press, 1959.

**[R1.223]**Omari, Peter T., *Kwame Nkrumah: The Anatomy of African Dictatorship,* New York, Africana Publishers, 1970.

**[R1.224]**Owusu-Ansah, J.V., *New Versions of the Traditional Motifs,* Kumasi, Ghana, Degraft Graphics and Publications, 1992.

**[R1.225]**Organization of African Unity, *Lagos Plan of Action for Economic Development of Africa, 1980-2000,* Geneva, IILS, 1981.

**[R1.226]**Ovason, David, *The Secret Architecture of Our National Capital,* New York, HarperCollins, 2000.

**[R1.227]**Padmore, George, *Pan-Africanism or Communism,* London, Dennis Dobson, 1956.

**[R1.228]**Padmore, George, *Africa: Britain's Third Empire,* London, Dennis Dobson, 1949.

**[R1.229]**Paterson, Thomas G. (ed.), *Major Problems in American Foreign Policy: Documents and Essays,* Lexington, MA, Heath and Co., 1978.

**[R1.230]**Payer, Cheryl, *The Debt Trap: The International Monetary Fund and the Third World,* New York, Monthly Review Press, 1974.

**[R1.231]**Perham, M., "Psychology of African Nationalism," *Optima*, Vol. 10, 1960, pp. 27-36.

**[R1.232]**Perkins, John, *Confessions of Economic Hit Man*, San Francisco, Berrett-Koehler Pub. Inc., 2004.

**[R1.233]**Peters, Jonathan A., *A Dance of Masks: Senghor, Achebe, Soyinka*, Washington, D.C., Three Continents Press, 1978.

**[R1.234]**Petrie, William M.F., *The Pyramids and Temples of Gizeh*, London, Field and Tuer, 1885.

**[R1.2351]**Rattray, Robert S., *Religion and Art in Ashanti*, New York, AMS Press, 1979.

**[R1.236]**Resnick, Idrian N., "The University and Development in Africa," *Pan-African Journal*, Vol. 1 (1), 1968, pp. 30-34.

**[R1.237]**Robeson, P., "Power of Negro Action," in [R1.20], pp. 522-532.

**[R1.238]**Robinson, E.A.G. (ed.), *Economic Consequences of the Size of Nations*, New York, Macmillan, 1963.

**[R1.239]**Rodney, Walter, *How Europe Underdeveloped Africa*, London, Bogle-L'Ouvertune Pub., 1972.

**[R1.240]**Roger, Joel A., *Africa's Gift to America: The Afro-American in the Making and Saving of the United States*, New York, J.A. Rogers Publications, 1959.

**[R1.241]**Rotberg, R.I., *The Rise of Nationalism in Central Africa: The Making of Malawi and Zambia, 1873-1964*, Cambridge, MA, Harvard University Press, 1965.

**[R1.242]**Sampson, G.P. (ed.), *The Role of the World Trade Organization in Global Governance*, New York, United Nations University Press, 2001.

**[R1.243]**Schwaller de Lubicz, R.A., *The Egyptian Miracle: An Introduction to the Wisdom of the Temple*, Rochester (Vermont) Inner Traditions International, 1985

**[R1.244]**Schwaller de Lubicz, R.A., *A Study of Numbers: A Guide to The constant creation of The Universe*, Rochester, (Vermont) Inner Traditions International, 1986.

**[R1.245]**Schwaller de Lubicz, R.A., *The Temple in Man: The Secrets of Ancient Egypt* Brookline, (Massachusetts), Autumn Press, 1977.

**[R1.246]**Schaller de Lubicz, R.A., *The Temple In Man: Sacred Architecture and The Perfect Man.*, Rochester, (Vermont), Inner Traditions Published, 1981

**[R1.247]**Schwaller de Lubicz, R.A., *Symbol and the Symbolic: Egypt, Science, and The Evolution of Consciousness*, Brookline, (Massachusetts) Autumn Press, 1978.

**[R1.248]**Schwaller de Lubicz, R.A., *The Temple of Man: Apet of The South at Luxor*, Rochester, (Vermont) Inner Traditions, 1998.

**[R1.249]**Schwaller de. Lubicz, R.A., *Sacred Science: The King of Pharaonic Theocracy*, New York, Inner Traditions International, 1982.

**[R1.250]**Schwaller de Lubicz, R.A., *The Temples of Karnak*, London, Thames & Hudson, 1999).

**[R1.251a]**Scranton, Laird, *The Science of the Dogon: Decoding the African Mystery Tradition*, New York, Inner Traditions,

**[R1.251b]**Seligman, Charles G., *Egypt and Negro Africa: A Study in Divine Kingship*, London, George Routledge and Sons, 1934.

**[R1.252]**Seme, Pixley Isaka, "The Regeneration of Africa," in Langley, Ayo J. (ed.), *Ideologies of Liberation in Black Africa: 1856-1970*, London, Rex Collings, 1979, pp. 261-265.

**[R1.253]**Serequeberhan, Tsenay, *The Hermeneutics of African Philosophy: Horizon and Discourse*, New York, Routledge, 1994.

**[R1.254]**Shorter, A., *African Christian Theology—Adaptation or Incarnation*, New York, Orbis, 1977.

**[R1.255]**Shorter, Aylward, W. ., *African Culture and Christian Church: An Introduction to Social and Pastoral Anthropology*, New York, Orbis Books, 1974.

**[R1.256]**Sithole, N., *African Nationalism*, New York, Oxford University Press, 1969.

**[R1.257]**Smertin, Y., *Kwame Nkrumah*, New York, International Publishers, 1987.

**[R1.258]**Smith, A., *The Geopolitics of Information, How Western Culture Dominates the World*, New York, Oxford University Press, 1980.

**[R1.259]**Snowden Jr., Frank M., *Blacks in Antiquity*, Cambridge, MA, Harvard University Press, 1970.

**[R1.260]**Steindorff, G. and K.C. Seele, *When Egypt Ruled the East*, Chicago, Chicago University Press, 1957.

**[R1.261]**Snowden Jr., Frank M., *Blacks in Antiquity: Ethiopians in the Greco-Roman Experience*, Cambridge, MA, The Belknap Press of Harvard University Press, 1970.

**[R1.262]**Sundkler, B.G.M., *Bantu Prophets in South Africa*, London, Oxford University Press, 1961.

**[R1.263]**Tambo, Oliver, *Preparing for Power: Oliver Tambo Speaks*, New York, George Braziller Inc., 1988.

**[R1.264]**Tekyi, K., "Racial Unity," in Langley, Ayo J. (ed.), *Ideologies of Liberation in Black Africa:1856-1970*, London, Rex Collings, 1979 pp. 402-404.

**[R1.265a]**Temples, Father Placide, *Bantu Philosophy*, Paris, Présence Africaine, 1959.

**[R1.265b]**Temple, R. G., *The Sirius Mystry*, London, Sidwick and Jackson, 1976.

**[R1.266]**Tennemann, Wilhelm G., *A Manual of the History of Philosophy*, London, H. G. Bohn, 1852.

**[R1.267]**Tetteh, M.N., The *Ghana Young Pioneer Movement*, The Institute of African Studies, University of Ghana, Tema, Optimum Design and Publishing Service 1985.

**[R1.268]**The Editors, "One Afrikan-Centric Continental Afrikan Government Now", *The Afrikan Truth Magazine* Vol. 1 (2) 1995.

**[R1.269]***The Lost Books of the Bible and Forgotten Book of Eden*, New York, The World Pub. Co., 1963.

**[R1.270]***The Spark* Editor, *Some Essential Features of Nkrumaism*, London, PANAF Press, 1975.

**[R1.271]***The Torah: The Five Books of Moses*, Philadelphia, The Jewish Publication of America, 1962.

**[R1.272]**Thompson, V. Bakpetu, *Africa and Unity*, London, Longman, 1969.

**[R1.273]**Thompson, W. S., *Ghana's Foreign Policy 1957-1966*, Princeton, New Jersey, Princeton University Press, 1969.

**[R1.274]**Toure, A. Sekou, *Africa on the Move*, London, PANAF Press, 1977.

[R1.275]Tuafo, Kofi Y., *Kwame Nkrumah*, Accra, Elaine Book, 2012.

**[R1.276]**United States National Security Council, *Memorandum 200 Study: Implications of World Wide Population Growth for U.S. Security and Overseas Interest*, December 10, 1974, Classified by h.c. Blaney III, Declassified July 3, 1989, Executive Order 12358.

**[R1.277]**United States of America, Department of States, *Foreign Relations of the United States, Vol. XXIV: 190*, Africa, 1964-1968.

**[R1.278]**Van Sertima, I., *African Presence in Early Europe*, New Brunswick, New Jersey, Transaction, 1986.

**[R1.279]**Van Sertima, I., *They Came Before Columbus: The African Presence in Ancient America*, New York, Random House, 1977.

**[R1.280]**Van Sertima, I. (ed.), *Great African Thinkers, Vol. 1, Cheikh Anta Diop*, New Brunswick, New Jersey, Transaction, 1987.

**[R1.281]**Van Sertima, I. (ed.), *Blacks in Science*, New Brunswick, N.J., Transaction Publishers, 1998.

**[R1.282]**Wallerstein, I., *Africa: The Politics of Unity*, New York, Random House, 1967.

**[R1.283]**Wallerstein, I., *Africa: The Politics of Independence*, New York, Random House, 1961.

**[R1.284]**Walters, Ronald W., "The Afrocentic Concept at Howard University", *New Directions*, Fall, 1990.

**[R1.285]**Walters, Ronald W., *Pan Africanism in the African Diaspora: An Analysis of Modern Afrocentric Political Movements*, Detroit, Michigan, Wayne State University Press, 1997.

**[R1.286]**Watson, R. L., *The Slave Question: Liberty and Property in South Africa*, Hanover, New England, University of New England Press, 1990.

**[R1.287]**Weber, Alfred, *History of Philosophy*, New York, Scribner's Sons, 1896.

**[R1.288]**Welsing, Cress F., *The Isis Papers: The Keys to the Colors*, Chicago, Third World Press, 1991.

**[R1.289]**Whitman, Daniel, "The Dual Soul and Double in Africa and West," *Chrysalis: Aspects of African Spirit*, Vol. 3 (1) Spring, 1988, pp. 22-27.

**[R1.290]**Williams, Chancellor, *The Destruction of Black Civilization: Great Issues of Race From 4500BC- 2000AD*, Detroit, Harlo Press, 1974

**[R1.291]**Williamson, J., *The Lending Policies of the International Monetary Fund*, Washington, D.C., Institute for International Economics, 1982.

**[R1.292]**Wilson, H. S. (ed.), *Origins of West African Nationalism*, London, Macmillan, 1969.

**[R1.293]**Witt, R. E., *Isis in the Graeco-Roman World*, Ithaca, N.Y., Cornell University Press, 1971

**[R1.294]**Woodson, Carter G., *Mis-Education of the Negro*, Washington, D.C., The Associated Publication, 1933.

**[R1.295]**Woodson, Carter G., *The African Background*, Washington, D.C., The Associated Press, 1936.

**[R1.296]**World Health Organization, *Interrelationships between Health Programs and Socio-economic Development*, Public Health Paper #49, Geneva, WHO, 1973.

**[R1.297]**Wright, Richard A. (ed.), *African Philosophy: An Introduction*, New York, University Press of America, 1984

**[R1.298]**Yesufu, T.M. (ed.), *Creating the African University: Emerging Issues of the 1970s*, Ibadan, Oxford University Press, 1973.

**[R1.299]**Zartman, William I., *International Relations in the New Africa*, Englewood Cliffs, New Jersey, Prentice-Hall, 1966.

**[R1.300]**Zeller, Edward, *Outline of the History of Greek Philosophy*, London, Longmans, Green and Co., 1914.

**[R1.301]**Zeller, Edward, *A History of Greek Philosophy From the Earliest Period to the Time of Socrates*, Vols 1 – 2, London, Longmans, Green and Co., 1881.

## R 2.    Category Theory in Mathematics, Logic and Sciences

**[R2.1]**Awodey, S., "Structure in Mathematics and Logic: A Categorical Perspective," *Philosophia Mathematica*, Vol. 3. 1996, pp209-237.

**[R2.2]**Bell, J. L., "Category Theory and the Foundations of Mathematics," British Journal *of Science*, Vol. 32, 1981, pp349-358.

**[R2.3]**Bell, J. L., "Categories, Toposes and Sets," *Synthese*, Vol. 51, 1982, pp. 393-337.

**[R2.4]**Black, M., *The Nature of Mathematics*, Totowa, N.J., Littlefield, Adams and Co., 1965.

**[R2.5]**Blass, A., "The Interaction between Category and Set Theory," *Mathematical Applications of Category Theory*, Vol. 30, 1984, pp. 5-29.

**[R2.6]**Brown, B. and J Woods (eds.), *Logical Consequence; Rival Approaches and New Studies in exact Philosophy: Logic, Mathematics and Science*, Vol. II Oxford, Hermes, 2000.

**[R2.8]**Domany, J. L.,et al., *Models of Neural Networks III*, New York Springer, 1996

**[R2.12]**Feferman, S., "Categorical Foundations and Foundations of Category Theory," in R. Butts (ed.), Logic, *Foundations of Mathematics and Computability*, Boston, Mass., Reidel, 1977, pp149-169.

**[R2.13]**Gray, J.W. (ed.) *Mathematical Applications of Category Theory* (American Mathematical Society Meeting 89<sup>th</sup> Denver Colo. 1983)., Providence, R.I., American Mathematical Society, 1984.

**[R.2.14]**Johansson, Ingvar, *Ontological Investigations: An Inquiry into the Categories of Nature, Man, and Society*, New York, Routledge, 1989.

**[R2.15]** Kamps, K. H., D. Pumplun, and W. Tholen (eds.) *Category Theory: Proceedings of the International Conference*, Gummersbach, July 6-10, New York, Springer, 1982.

**[R2.18]** Landry, E., Category Theory: the Language of Mathematics," *Philosophy of Science,* Vol. 66, (Supplement), S14-S27.

**[R2.19]** Landry E. and J.P Marquis, "Categories in Context: Historical, Foundational and Philosophical," *Philiosophia Mathematica*, Vol. 13, 2005, pp. 1-43.

**[R2.20]** Marquis, J. –P., "Three Kinds of Universals in Mathematics," in B. Brown, and J. Woods (eds.), *Logical Consequence; Rival Approaches and New Studies in exact Philosophy: Logic, Mathematics and Science*, Vol. II Oxford, Hermes, 2000, pp 191-212.

**[R2.21]** McLarty, C., "Category Theory in Real Time," *Philosophia Mathematica,* Vol.2, 1994, pp. 36-44.

**[R2.22]** McLarty, C., "Learning from Questions on Categorical Foundations," *Philosophia Mathematica,* Vol.13, 2005, pp44-60.

**[R2.23]** Rodabaugh, S. et. al., (eds.), *Application of Category Theory to Fuzzy Subsets*, Boston, Mass., Kluwer 1992.

**[R2.24]** Taylor, J.G. (ed.), *Mathematical Approaches to Neural Networks*, New York North-Holland, 1993.

**[R2.25]** Van Benthem, J. et al.(eds.), *The Age of Alternative Logics: Assessing Philosophy of Logic and Mathematics Today,* New York, Springer, 2006.

## R3.　Fuzzy Logic in Knowledge Production

**[R3.1]** Baldwin, J.F., "A New Approach to Approximate Reasoning Using a Fuzzy Logic," *Fuzzy Sets and Systems*, Vol. 2, #4, 1979, pp. 309-325.

**[R3.2]** Baldwin, J.F., "Fuzzy Logic and Fuzzy Reasoning," *Intern. J. Man-Machine Stud.*, Vol. 11, 1979, pp. 465-480.

**[R3.3]** Baldwin, J.F., "Fuzzy Logic and Its Application to Fuzzy Reasoning," in. M. M. Gupta et al. (eds.), *Advances in Fuzzy Set Theory and Applications*, New York, North-Holland, 1979, pp.96-115.

**[R3.4]** Baldwin, J.F. et al., "Fuzzy Relational Inference Language," *Fuzzy Sets and Systems*, Vol. 14, #2, 1984, pp. 155-174.

**[R3.5]** Baldsin, J. and B.W. Pilsworth, "Axiomatic Approach to Implication For Approximate Reasoning With Fuzzy Logic," *Fuzzy Sets and Systems*, Vol. 3, #2, 1980, pp. 193-219.

**[R3.6]** Baldwin, J.F. et. al., "The Resolution of Two Paradoxes by Approximate Reasoning Using A Fuzzy Logic," *Synthese*, Vol. 44, 1980, pp. 397-420.

**[R3.7]** Dompere, K. K., *Fuzzy Rationality: Methodological Critique and Unity of Classical, Bounded and Other Rationalities*, (Studies in Fuzziness and Soft Computing, vol.235) New York, Springer, 2009.

**[R3.8]** Dompere Kofi K., *Epistemic Foundations of Fuzziness*, (Studies in Fuzziness and Soft Computing, vol. 236) New York, Springer, 2009.

**[R3.9]** Dompere Kofi K., *Fuzziness and Approximate Reasoning: Epistemics on Uncertainty, Expectation and Risk in Rational Behavior*, (Studies in Fuzziness and Soft Computing, vol. 237) New York, Springer, 2009.

**[R3.10]** Dompere, Kofi K., *The Theory of the Knowledge Square: The Fuzzy Rational Foundations of Knowledge-Production Systems*, New York, Springer, 2013.

[R3.11] Dompere, Kofi K., "Cost-Benefit Analysis, Benefit Accounting and Fuzzy Decisions: Part I, Theory", *Fuzzy Sets and Systems*, Vol. 92, 1997, pp. 275-287.

[R3.12] Dompere, Kofi K., "The Theory of Social Cost and Costing For Cost-Benefit Analysis in a Fuzzy Decision Space", Fuzzy Sets and Systems. Vol. 76, 1995, pp. 1-24.

**[R3.13]** Dompere, Kofi K., *Fuzzy Rational Foundations of Exact and Inexact Sciences*, New York, Springer, 2013.

**[R3.14]** Gaines, B.R., "Foundations of Fuzzy Reasoning," *Inter. Jour. of Man-Machine Studies*, Vol. 8, 1976, pp. 623-668.

**[R3.15]** Gaines, B.R., "Foundations of Fuzzy Reasoning," in Gupta, M.M. et. al. (eds.), *Fuzzy Information and Decision Processes*, New York North-Holland, 1982, pp. 19-75.

**[R3.16]** Gaines, B. R., "Precise Past, Fuzzy Future," *International Journal. Of Man-Machine Studies.*, Vol. 19, #1, 1983, pp.117-134.

**[R3.17]** Giles, R., "Lukasiewics Logic and Fuzzy Set Theory," *Intern. J. Man-Machine Stud.*, Vol. 8, 1976, pp. 313-327.

**[R3.18]** Giles, R., "Formal System for Fuzzy Reasoning," *Fuzzy Sets and Systems*, Vol. 2, #3, 1979, pp. 233-257.

**[R3.19]** Ginsberg, M. L.(ed.), *Readings in Non-monotonic Reason*, Los Altos, Ca., Morgan Kaufman, 1987.

**[R3.20]** Goguen, J.A., "The Logic of Inexact Concepts," *Synthese*, Vol. 19, 1969, pp. 325-373.

**[R3.21]** Gottinger, H.W., "Towards a Fuzzy Reasoning in the Behavioral Science," *Cybernetica*, Vol. 16, #2, 1973, pp. 113-135.

**[R3.22]** Gottinger, H.W., "Some Basic Issues Connected With Fuzzy Analysis", in H. Klaczro and N. Muller (eds.), *Systems Theory in Social Sciences*, Birkhauser Verlag, Basel, 1976, pp. 323-325.

**[R3.23]** Gupta, M.M. et. al., (eds.), *Approximate Reasoning In Decision Analysis,* North Holland, New York, 1982.

**[R3.24]** Höhle Ulrich . and E.P. Klement , *Non-Clasical Logics and their Applications to Fuzzy Subsets: A Handbook of the Mathematical Foundations of Fuzzy Set Theory*, Boston, Mass. Kluwer, 1995.

**[R3.25]** Kaipov, V., Kh. et. al., "Classification in Fuzzy Environments," in M. M. Gupta et al. (eds.), *Advances in Fuzzy Set Theory and Applications,* New York, North-Holland, 1979, pp. 119-124,

**[R3.26]** Kaufman, A., "Progress in Modeling of Human Reasoning of Fuzzy Logic" in M. M. Gupta et al. (eds.), *Fuzzy Information and Decision Process*, New York, North-Holland, 1982, pp. 11-17.

**[R3.27]** Lakoff, G., "Hedges: A Study in Meaning Criteria and the Logic of Fuzzy Concepts," *Jour. Philos. Logic,* Vol. 2, 1973, pp. 458-508.

**[R3.28]** Lee, R.C.T., "Fuzzy Logic and the Resolution Principle," *Jour. of Assoc. Comput*. Mach., Vol. 19, 1972, pp. 109-119.

**[R3.29]** LeFaivre, R.A., "The Representation of Fuzzy Knowledge", *Jour. of Cybernetics*, Vol. 4, 1974, pp. 57-66.

**[R3.30]** Negoita, C.V., "Representation Theorems for Fuzzy Concepts," *Kybernetes*, Vol. 4, 1975, pp. 169-174.

**[R3.31]** Nowakowska, M., "Methodological Problems of Measurements of Fuzzy Concepts in Social Sciences", *Behavioral Sciences*, Vol. 22, #2, 1977, pp. 107-115.

**[R3.32]** Skala, H.J., *Non-Archimedean Utility Theory*, D. Reidel Dordrecht, 1975.

**[R3.33]** Skala, H.J., "On Many-Valued Logics, Fuzzy Sets, Fuzzy Logics and Their Applications," *Fuzzy Sets and Systems*, Vol. 1, #2, 1978, pp. 129-149.

**[R3.34]** Tan, S K.et al., "Fuzzy Inference Relation Based on the Theory of Falling Shadows," *Fuzzy Sets and Systems*, Vol. 53, #2, 1993, pp. 179-188.

**[R3.35]** Van Fraassen, B.C., "Comments: Lakoff's Fuzzy Propositional Logic," in D. Hockney et. al., (Eds.), *Contemporary Research in Philosophical Logic and Linguistic Semantics* Holland, Reild, 1975, pp. 273-277.

**[R3.36]** Yager, R.R. et. al. (eds.), An Introduction to Fuzzy Logic Applications in *Intelligent Systems*, Boston, Mass., Kluwer 1992.

**[R3.37]** Ying, M.S., "Some Notes on Multidimensional Fuzzy Reasoning," *Cybernetics and Systems*, Vol. 19, #4, 1988, pp. 281-293.

**[R3.38]** Zadeh, L.A., "Quantitative Fuzzy Semantics," *Inform. Science*, Vol. 3, 1971, pp. 159-176.

**[R3.39]** Zadeh, L.A., "A Fuzzy Set Interpretation of Linguistic Hedges," *Jour. Cybernetics*, Vol. 2, 1972, pp. 4-34.

**[R3.40]** Zadeh, L.A., "Fuzzy Logic and Its Application to Approximate Reasoning," *Information Processing 74, Proc. IFIP Congress 74*, #3, North Holland, New York, pp. 591-594, 1974.

**[R3.41]** Zadeh, L.A, "The Concept of a Linguistic Variable and Its Application to Approximate Reasoning," in K.S. Fu et. al. (eds.), *Learning Systems and Intelligent Robots*, Plenum Press, New York, 1974, pp. 1-10.

**[R3.42]** Zadeh, L.A., et. al., (eds.), *Fuzzy Sets and Their Applications to Cognitive and Decision Processes*, New York, Academic Press, 1974.

**[R3.43]** Zadeh, L.A., "The Birth and Evolution of Fuzzy Logic," *Intern. Jour. of General Systems*, Vol. 17, #(2-3) 1990, pp. 95-105.

## R4. Fuzzy Mathematics in Approximate Reasoning under Conditions of Inexactness and Vagueness

**[R4.1]** Bellman, R.E., "Mathematics and Human Sciences," in J. Wilkinson et. al. (eds.), *The Dynamic Programming of Human Systems*, New York, MSS Information Corp., 1973, pp. 11-18.

**[R4.2]** Bellman, R.E and Glertz, M., "On the Analytic Formalism of the Theory of Fuzzy Sets," Information *Science*, Vol. 5, 1973, pp. 149-156.

**[R4.3]** Butnariu, D., Fixed Points for Fuzzy Mapping," *Fuzzy Sets and Systems*, Vol. 7, #2, pp. 191-207, 1982.

**[R4.4]** Butnariu, D., "Decompositions and Range For Additive Fuzzy Measures", *Fuzzy Sets and Syst ems*, Vol. 10, #2, pp. 135-155, 1983.

**[R4.5]** Cerruti, U., "Graphs and Fuzzy Graphs" in *Fuzzy Information and Decision Processes*, New York, North-Holland, 1982, pp.123-131.

**[R4.6]** Chakraborty, M. K. et al., "Studies in Fuzzy Relations Over Fuzzy Subsets," *Fuzzy Sets and Systems*, Vol. 9, #1, pp. 79-89, 1983.

**[R4.8]** Chang, C.L., "Fuzzy Topological Spaces," *J. Math. Anal. and Applications*, Vol. 24, 1968,pp. 182-190.

**[R4.9]** Chang, S.S.L., "Fuzzy Mathematics, Man and His Environment", *IEEE Transactions on Systems, Man and Cybernetics*, SMC-2 1972, pp. 92-93,

**[R4.10]** Chang, S.S.L. et. al., "On Fuzzy Mathematics and Control," *IEEE Transactions, System, Man and Cybernetics*, SMC-2, 1972, pp. 30-34.

**[R4.11]** Chang, S.S., "Fixed Point Theorems for Fuzzy Mappings," Fuzzy *Sets and Systems*, Vol. 17, 1985, pp. 181-187.

**[R4.12]** Chapin, E.W., "An Axiomatization of the Set Theory of Zadeh," *Notices, American Math. Society*, 687-02-4 754, 1971.

**[R4.13]** Chaudhury, A. K. and P. Das, "Some Results on Fuzzy Topology on Fuzzy Sets," *Fuzzy Sets and Systems*, Vol. 56, 1993, pp. 331-336.

**[R4.14]** Chitra, H., and P.V. Subrahmanyam, "Fuzzy Sets and Fixed Points," *Jour. of Mathematical Analysis and Application*, Vol. 124, 1987, pp. 584-590.

**[R4.15]** Czogala, J. et. al., Fuzzy Relation Equations On a Finite Set," *Fuzzy Sets and Systems*, Vol. 7, #1,1982. pp. 89-101.

**[R4.16]** DiNola, A. et. al., (eds.), *The Mathematics of Fuzzy Systems*, Koln, Verlag TUV Rheinland, 1986.

**[R4.17]** Dompere, Kofi K., *Cost-Benefit Analysis and the Theory of Fuzzy Decisions: Identification and Measurement Theory* (Series: Studies in Fuzziness and Soft Computing, Vol. 158), Berlin, Heidelberg, Springer, 2004.

**[R4.18]** Dompere, Kofi K., *Cost-Benefit Analysis and the Theory of Fuzzy Decisions: Fuzzy Value Theory* (Series: Studies in Fuzziness and Soft Computing, Vol.160), Berling, Heidelberg, Springer, 2004.

**[R4.19]** Dubois, D. and H. Prade, *Fuzzy Sets and Systems*, New York, Academic Press, 1980.

**[R4.20]** Dubois, "Fuzzy Real Algebra: Some Results," *Fuzzy Sets and Systems*, Vol. 2, #4, pp. 327-348, 1979.

**[R4.21]** Dubois, D. and H. Prade , "Gradual Inference Rules in Approximate Reasoning," *Information Sciences*, Vol. 61(1-2), 1992, pp. 103-122.

**[R4.22]** Dubois, D. and H. Prade, "On the Combination of Evidence in various Mathematical Frameworks,." In: Flamm. J. and T. Luisi, (eds.), *Reliability Data Collection and Analysis*. Kluwer, Boston, 1992, pp. 213-241.

**[R4.23]** Dubois, D. and H. Prade, "Fuzzy Sets and Probability: Misunderstanding, Bridges and Gaps." *Proc. Second IEEE Intern. Conf. on Fuzzy Systems*, San Francisco, 1993, pp. 1059-1068.

**[R4.24]** Dubois, D. and H. Prade [1994], "A Survey of Belief Revision and Updating Rules in Various Uncertainty Models," *Intern. J. of Intelligent Systems*, Vol. 9, #1, pp. 61-100.

**[R4.25]** Filev, D.P. et. al., "A Generalized Defuzzification Method via Bag Distributions," *Intern. Jour. of Intelligent Systems*, Vol. 6, #7, 1991, pp. 687-697.

**[R4.26]** Goetschel, R. Jr., et. al., "Topological Properties of Fuzzy Number," *Fuzzy Sets and Systems*, Vol. 10, #1, pp. 87-99, 1983.

**[R4.27]** Goodman, I.R., "Fuzzy Sets As Equivalence Classes of Random Sets" in Yager, R.R. (ed.), *Fuzzy Set and Possibility Theory: Recent Development*, New York, Pergamon Press 1992. pp. 327-343.

**[R4.28]** Gupta, M.M. et. al., (eds), *Fuzzy Antomata and Decision Processes*, New York, North-Holland, 1977.

**[R4.29]** Gupta, M.M. and E. Sanchez (eds.), *Fuzzy Information and Decision Processes*, New York, North-Holland, 1982.

**[R4.30]** Higashi, M. and G.J. Klir, "On measure of fuzziness and fuzzy complements," Intern. *J. of General Systems*, Vol. 8 #3, 1982, pp. 169-180.

**[R4.31]** Higashi, M. and G.J. Klir, "Measures of uncertainty and information based on possibility distributions," *International Journal of General Systems*, Vol. 9 #1, 1983, pp. 43-58.

**[R4.32]** Higashi, M. and G.J. Klir, "On the notion of distance representing information closeness: Possibility and probability distributions," *Intern. J. of General Systems*, Vol. 9 #2, 1983, pp. 103-115.

**[R4.33]** Higashi, M. and G.J. Klir , "Resolution of finite fuzzy relation equations," *Fuzzy Sets and Systems*, Vol. 13, #1,1984, pp. 65-82.

**[R4.34]** Higashi, M. and G.J. Klir , "Identification of fuzzy relation systems," *IEEE Trans. on Systems, Man, and Cybernetics*, Vol.14 #2, 1984, pp. 349-355.

**[R4.35]** Jin-wen, Z., "A Unified Treatment of Fuzzy Set Theory and Boolean Valued Set theory: Fuzzy Set Structures and Normal Fuzzy Set Structures," *Jour. Math. Anal. and Applications*, Vol. 76, #1, 1980, pp. 197-301.

**[R4.36]** Kandel, A. and W.J. Byatt, "Fuzzy Processes," *Fuzzy Sets and Systems*, Vol. 4, #2, 1980, pp. 117-152.

**[R4.37]** Kaufmann, A. and M.M. Gupta , *Introduction to fuzzy arithmetic: Theory and applications*, New York Van Nostrand Rheinhold,. 1991.

**[R4.38]** Kaufmann, A., *Introduction to the Theory of Fuzzy Subsets*, Vol. 1, New York, Academic Press, 1975.

**[R4.39]** Kaufmann,A., *Theory of Fuzzy Sets*, Paris, Merson Press, 1972.

**[R4.40]** Klement, E.P. and W. Schwyhla, "Correspondence Between Fuzzy Measures and Classical Measures," *Fuzzy Sets and Systems*, Vol. 7, #1, 1982.pp. 57-70.

**[R4.41]** Klir, George and Bo Yuan, *Fuzzy Sets and Fuzzy Logic*, Upper Saddle River, NJ Prentice Hall, 1995.

**[R4.42]** Kokawa, M. et. al. "Fuzzy-Theoretical Dimensionality Reduction Method of Multi-Dimensional Quality," in Gupta, M.M. and E. Sanchez (eds.), *Fuzzy Information and Decision Processes,* New York, North-Holland, 1982, pp. 235-250.

**[R4.43]** Kruse, R. et al., *Foundations of Fuzzy Systems*, New York, John Wiley and Sons, 1994.

**[R4.44]** Lasker, G.E. (ed.), *Applied Systems and Cybernetics, Vol. VI: Fuzzy Sets and Systems*, Pergamon Press, New York, 1981.

**[R4.45]** Lientz, B.P., "On Time Dependent Fuzzy Sets", *Inform, Science*, Vol. 4, 1972, pp. 367-376.

**[R4.46]** Lowen, R., "Fuzzy Uniform Spaces," *Jour. Math. Anal. Appl.*, Vol. 82, #21981,pp. 367-376.

**[R4.47]** Michalek, J., "Fuzzy Topologies," *Kybernetika*, Vol. 11, 1975, pp. 345-354.

**[R4.48]** Negoita, C.V. et. al., *Applications of Fuzzy Sets to Systems Analysis*, Wiley and Sons, New York, 1975.

**[R4.49]** Negoita, C.V., "Representation Theorems for Fuzzy Concepts," *Kybernetes*, Vol. 4, 1975, pp. 169-174.

**[R4.50]** Negoita, C.V. et. al., "On the State Equation of Fuzzy Systems," *Kybernetes*, Vol. 4, 1975, pp. 231-214.

**[R4.51]** Netto, A.B., "Fuzzy Classes," *Notices, American Mathematical Society,* Vol.68T-H28, 1968, pp.945.

**[R4.52]** Pedrycz, W., "Fuzzy Relational Equations with Generalized Connectives and Their Applications," *Fuzzy Sets and Systems*, Vol. 10, #2, 1983, pp. 185-201.

**[R4.53]** Raha, S. et. al., "Analogy Between Approximate Reasoning and the Method of Interpolation," *Fuzzy Sets and Systems*, Vol. 51, #3, 1992, pp. 259-266.

**[R4.54]** Ralescu, D., "Toward a General Theory of Fuzzy Variables," *Jour. of Math. Analysis and Applications*, Vol. 86, #1, 1982, pp. 176-193.

**[R4.55]** Rodabaugh, S.E., "Fuzzy Arithmetic and Fuzzy Topology," in G.E. Lasker, (ed.), *Applied Systems and Cybernetics, Vol. VI: Fuzzy Sets and Systems*, Pergamon Press, New York, 1981 pp. 2803-2807.

**[R4.56]** Rosenfeld, A., "Fuzzy Groups," *Jour. Math. Anal. Appln.*, Vol. 35, 1971, pp. 512-517.

**[R4.57]** Ruspini, E.H., "Recent Developments In Mathematical Classification Using Fuzzy Sets," in G.E. Lasker, (ed.), *Applied Systems and Cybernetics, Vol. VI: Fuzzy Sets and Systems*, Pergamon Press, New York, 1981. pp. 2785-2790.

**[R4.58]** Santos, E.S., "Fuzzy Algorithms," *Inform. and Control*, Vol. 17, 1970, pp. 326-339.

**[R4.59]** Stein, N.E. and K Talaki, "Convex Fuzzy Random Variables," *Fuzzy Sets and Systems*, Vol. 6, #3, 1981, pp. 271-284.

**[R4.60]** Triantaphyllon, E. et. al., "The Problem of Determining Membership Values in Fuzzy Sets in Real World Situations," in D.E. Brown et. al. (eds), *Operations Research and Artificial Intelligence: The Integration of Problem-solving Strategies*, Boston, Mass., Kluwer, 1990, pp. 197-214.

**[R4.61]** Tsichritzis, D., "Participation Measures," *Jour. Math. Anal. and Appln.*, Vol. 36, 1971, pp. 60-72.

**[R4.62]** Turksens, I.B., "Four Methods of Approximate Reasoning with Interval-Valued Fuzzy Sets," *Intern. Journ. of Approximate Reasoning*, Vol. 3, #2, 1989, pp. 121-142.

**[R4.63]** Turksen, I.B., "Measurement of Membership Functions and Their Acquisition," *Fuzzy Sets and Systems*, Vol. 40, #1, 1991, pp. 5-38.

**[R4.64]** Wang, P.P. (ed.), *Advances in Fuzzy Sets, Possibility Theory, and Applications*, New York, Plenum Press, 1983.

**[R4.65]** Wang, Zhenyuan, and George Klir, *Fuzzy Measure Theory*, New York, Plenum Press, 1992.

**[R4.66]** Wang, P.Z. et. al. (eds.), *Between Mind and Computer: Fuzzy Science and Engineering*, Singapore, World Scientific Press, 1993.

**[R4.67]** Wang, S., "Generating Fuzzy Membership Functions: A Monotonic Neural Network Model," *Fuzzy Sets and Systems*, Vol. 61, #1, 1994, pp. 71-82.

**[R4.68]** Wong, C.K., "Fuzzy Points and Local Properties of Fuzzy Topology,"*Jour. Math. Anal. and Appln.*, Vol. 46, 19874, pp. 316-328.

**[R4.69]** Wong, C.K., "Categories of Fuzzy Sets and Fuzzy Topological Spaces," *Jour. Math. Anal. and Appln.*, Vol. 53, 1976, pp. 704-714.

**[R4.70]** Yager, Ronald R., and Dimitor P Filver, *Essentials of Fuzzy Modeling and Control*, New York, John Wiley and Sons, 1994.

**[R4.71]** Yager, R.R. et. al., (Eds.), *Fuzzy Sets, Neural Networks, and Soft Computing*, New York, Nostrand Reinhold, 1994.

**[R4.72]** Yager, R.R., "On the Theory of Fuzzy Bags," *Intern. Jour. of General Systems*, Vol. 13, #1, 1986, pp. 23-37.

**[R4.73]** Zadeh, L.A., "A Computational Theory of Decompositions," *Intern. Jour. of Intelligent Systems*, Vol. 2, #1, 1987, pp. 39-63.

**[R4.74]** Zimmerman, H.J., *Fuzzy Set Theory and Its Applications*, Boston, Mass, Kluwer, 1985.

**R5.  Fuzzy Optimization, Decision-Choices and Approximate Reasoning in Sciences**

**[R5.1]** Bose, R.K. and Sahani D, "Fuzzy Mappings and Fixed Point Theorems," *Fuzzy Sets and Systems*, Vol. 21, 1987, pp. 53-58.

**[R5.2]** Butnariu D. "Fixed Points for Fuzzy Mappings," *Fuzzy Sets and Systems*, Vol. 7, 1982, pp.191-207.

**[R5.3]** Dompere, Kofi K., "Fuzziness, Rationality, Optimality and Equilibrium in Decision and Economic Theories" in Weldon A. Lodwick and Janusz Kacprzyk (Eds.), *Fuzzy Optimization: Recent Advances and Applications* (Series: Studies in Fuzziness and SoftComputing, Vol. 254), Berlin, Heidelberg, Springer, 2010.

**[R5.4]** Eaves, B.C., "Computing Kakutani Fixed Points," *Journal of Applied Mathematics*, Vol. 21, 1971,pp. 236-244.

**[R5.5]** Heilpern, S. "Fuzzy Mappings and Fixed Point Theorem," *Journal of Mathematical Analysis and Applications*, Vol. 83, 1981, pp.566-569.

**[R5.6]** Kacprzyk, J. et. al., (eds.), *Optimization Models Using Fuzzy Sets and Possibility Theory*, Boston, Mass., D. Reidel, 1987.

**[R5.7]** Kakutani, S., "A Generalization of Brouwer's Fixed Point Theorem," *Duke Mathematical Journal*, Vol. 8, 1941, pp. 416-427.

**[R5.8]** Kaleva, O. "A Note on Fixed Points for Fuzzy Mappings", *Fuzzy Sets and Systems*, Vol. 15, 1985, pp. 99-100.

**[R5.9]** Lodwick, Weldon A and Janusz Kacprzyk (eds.), *Fuzzy Optimization: Recent Advances and Applications*, (Studies in Fuzziness and Soft Computing, Vol. 254), Berlin Heidelberg, Springer, 2010,

**[R5.10]** Negoita, C.V. et. al., "Fuzzy Linear Programming and Tolerances in Planning," *Econ. Group Cybernetic Studies*, Vol. 1, 1976, pp. 3-15.

**[R5.11]** Negoita, C.V., and A.C. Stefanescu, "On Fuzzy Optimization," in Gupta, M.M. et. al., (eds.) *Approximate Reasoning In Decision Analysis*, North Holland, New York, 1982. pp. 247-250.

**[R5.12]** Negoita, C.V., "The Current Interest in Fuzzy Optimization," *Fuzzy Sets and Systems*, Vol. 6, #3, 1981, pp. 261-270.

**[R5.13]** Negoita, C.V., et. al., "On Fuzzy Environment in Optimization Problems," in J. Rose et. al., (eds.), *Modern Trends in Cybernetics and Systems*, Springer, Berlin, 1977, pp.13-24.

**[R5.14]** Ralescu, D., "0ptimization in a Fuzzy Environment," in M. M. Gupta et al. (eds.), *Advances in Fuzzy Set Theory and Applications*, New York, North-Holland, 1979, pp.77-91.

**[R5.15]** Warren, R.H., "Optimality in Fuzzy Topological Polysystems," *Jour. Math. Anal.*, Vol. 54, 1976, pp. 309-315.

**[R5.16]** Zimmerman, H.-J., "Description and Optimization of Fuzzy Systems," *Intern. Jour. Gen. Syst.* Vol. 2, #4, 1975, pp. 209-215.

**[R5.17]** Zimmerman, H.J., "Applications of Fuzzy Set Theory to Mathematical Programming," *Information Science*, Vol. 36, #1, 1985, pp. 29-58.

## R6. Fuzzy Probability, Fuzzy Random Variable and Random Fuzzy Variable

**[R6.1]** Bandemer, H., "From Fuzzy Data to Functional Relations," *Mathematical Modelling*, Vol. 6, 1987, pp. 419-426.

**[R6.2]** Bandemer, H. et. al., *Fuzzy Data Analysis*, Boston, Mass, Kluwer, 1992.

**[R6.3]** Kruse, R. et. al., *Statistics with Vague Data*, Dordrecht, D. Reidel Pub. Co., 1987.

**[R6.4]** Chang, R.L.P., et. al., "Applications of Fuzzy Sets in Curve Fitting," *Fuzzy Sets and Systems*, Vol. 2, #1, pp. 67-74.

**[R6.5]** Chen, S.Q., "Analysis for Multiple Fuzzy Regression," *Fuzzy Sets and Systems*, Vol. 25, #1, pp. 56-65.

**[R6.6]** Celmins, A., "Multidimensional Least-Squares Fitting of Fuzzy Model," *Mathematical Modelling*, Vol. 9, #9, pp. 669-690.

**[R6.7]** El Rayes, A.B. et. al., "Generalized Possibility Measures," *Information Sciences*, Vol. 79, 1994, pp. 201-222.

**[R6.8]** Dumitrescu, D., "Entropy of a Fuzzy Process," *Fuzzy Sets and Systems*, Vol. 55, #2, 1993, pp. 169-177.

**[R6.9]** Delgado, M. et. al., "On the Concept of Possibility-Probability Consistency," *Fuzzy Sets and Systems*, Vol. 21, #3, 1987, pp. 311-318.

**[R6.10]** Devi, B.B. et. al., "Estimation of Fuzzy Memberships from Histograms," *Information Sciences*, Vol. 35, #1, 1985, pp. 43-59.

**[R6.11]** Diamond, P., "Fuzzy Least Squares", *Information Sciences*, Vol. 46, #3, 1988, pp. 141-157.

**[R6.12]** Dubois, D. et. al., "Fuzzy Sets, Probability and Measurement," *European Jour. of Operational Research*, Vol. 40, #2, 1989, pp. 135-154.

**[R6.13]** Fruhwirth-Schnatter, S., "On Statistical Inference for Fuzzy Data with Applications to Descriptive Statistics," *Fuzzy Sets and Systems*, Vol. 50, #2, 1992, pp. 143-165.

**[R6.14]** Fruhwirth-Schnatter, S., "On Fuzzy Bayesian Inference," *Fuzzy Sets and Systems*, Vol. 60, #1, 1993, pp. 41-58.

**[R6.15]** Gaines, B.R., "Fuzzy and Probability Uncertainty logics," *Information and Control*, Vol. 38, #2, 1978, pp. 154-169.

**[R6.16]** Geer, J.F. et. al., "Discord in Possibility Theory," *International Jour. of General Systems*, Vol. 19, 1991, pp. 119-132.

**[R6.17]** Geer, J.F. et. al., "A Mathematical Analysis of Information-Processing Transformation Between Probabilistic and Possibilistic Formulation of Uncertainty," *International Jour. of General Systems*, Vol. 20, #2, 1992, pp. 14-176.

**[R6.18]** Goodman, I.R. et. al., *Uncertainty Models for Knowledge Based Systems*, New York, North-Holland, 1985.

**[R6.19]** Grabish, M. et. al., *Fundamentals of Uncertainty Calculi with Application to Fuzzy Systems*, Boston, Mass., Kluwer, 1994.

**[R6.20]** Guan, J.W. et. al., *Evidence Theory and Its Applications*, Vol. 1, New York, North-Holland, 1991.

**[R6.21]** Guan, J.W. et. al., *Evidence Theory and Its Applications*, Vol. 2, New York, North-Holland, 1992.

**[R6.22]** Hisdal, E., Are Grades of Membership Probabilities?," *Fuzzy Sets and Systems*, Vol. 25, #3, 1988, pp. 349-356.

**[R6.23]** Höhle Ulrich , "**A** Mathematical Theory of Uncertainty," in R.R. Yager (ed.) *Fuzzy Set and Possibility Theory: Recent Developments*, New York, Pergamon, 1982, pp. 344 – 355.

**[R6.24]** Kacprzyk, Janusz and Mario Fedrizzi (eds.) *Combining Fuzzy Imprecision with Probabilistic Uncertainty in Decision Making*, New York, Plenum Press, 1992.

**[R6.25]** Kacprzyk, J. et. al., *Combining Fuzzy Imprecision with Probabilistic Uncertainty in Decision Making*, New York, Springer-Verlag, 1988.

**[R6.26]** Klir, G.J., "Where Do we Stand on Measures of Uncertainty, Ambignity, Fuzziness and the like?" *Fuzzy Sets and Systems*, Vol. 24, #2, 1987, pp. 141-160.

**[R6.27]** Klir, G.J. et. al., *Fuzzy Sets, Uncertainty and Information*, Englewood Cliff, Prentice Hll, 1988.

**[R6.28]** Klir, G. J. et. al., "Probability-Possibility Transformations: A Comparison," *Intern. Jour. of General Systems*, Vol. 21, #3, 1992, pp. 291-310.

**[R6.29]** Kosko, B., "Fuzziness vs Probability," *Intern. Jour. of General Systems*, Vol. 17, #(1-3) 1990, pp. 211-240.

**[R6.30]** Manton, K.G. et. al., *Statistical Applications Using Fuzzy Sets*, New York, John Wiley, 1994.

**[R6.31]** Meier, W., et. al., "Fuzzy Data Analysis: Methods and Indistrial Applications," *Fuzzy Sets and Systems*, Vol. 61, #1, 1994, pp. 19-28.

**[R6.32]** Nakamura, A., et. al., "A logic for Fuzzy Data Analysis," *Fuzzy Sets and Systems*, vol. 39, #2, 1991, pp. 127-132.

**[R6.33]** Negoita, C.V. et. al., *Simulation, Knowledge-Based Compting and Fuzzy Statistics*, New York, Van Nostrand Reinhold, 1987.

**[R6.34]** Nguyen, H.T., "Random Sets and Belief Functions," *Jour. of Math. Analysis and Applications*, Vol. 65, #3, 1978, pp. 531-542.

**[R6.35]** Prade, H. et. al., "Representation and Combination of Uncertainty with belief Functions and Possibility Measures," *Comput. Intell.* ,Vol. 4, 1988, pp. 244-264.

**[R6.36]** Puri, M.L. et. al., "Fuzzy Random Variables," *Jour. of Mathematical Analysis and Applications*, Vol. 114, #2, 1986, pp. 409-422.

**[R6.37]** Rao, N.B. and A. Rashed, "Some Comments on Fuzzy Random Variables," *Fuzzy Sets and Systems*, Vol. 6, # 3, 1981, pp.285-292.

**[R6.38]** Sakawa, M. et. al., "Multiobjective Fuzzy linear Regression Analysis for Fuzzy Input-Output Data," *Fuzzy Sets and Systems*, Vol. 47, #2, 1992, pp. 173-182.

**[R6.39]** Schneider, M. et. al., "Properties of the Fuzzy Expected Values and the Fuzzy Expected Interval," *Fuzzy Sets and Systems*, Vol. 26, #3, 1988, pp. 373-385.

**[R6.40]** Slowinski, Roman and Jacques Teghem (eds) *Stochastic versus Fuzzy Approaches to Multiobjective Mathematical Programming under Uncertainty*, Dordrecht, Kluwer, 1990.

**[R6.41]** Stein, N.E. and K Talaki, "Convex Fuzzy Random Variables," *Fuzzy Sets and Systems*, Vol. 6, #3, 1981, pp. 271-284.

**[R6.42]** Sudkamp, T., "On Probability-Possibility Transformations," *Fuzzy Sets and Systems*, Vol. 51, #1, 1992, pp. 73-82.

**[R6.43]** Tanaka, H. et. al., "Possibilistic Linear Regression Analysis for Fuzzy Data," *European Jour. of Operational Research*, Vol. 40, #3, 1989, pp. 389-396.

**[R6.44]** Walley, P., *Statistical Reasoning with Imprecise Probabilities*, London Chapman and Hall, 1991.

**[R6.45]** Wang, G.Y. et. al., "The Theory of Fuzzy Stochastic Processes," *Fuzzy Sets and Systems*, Vol. 51, #2 1992, pp. 161-178.

**[R6.46]** Wang, X. et. al., "Fuzzy Linear Regression Analysis of Fuzzy Valued Variable," *Fuzzy Sets and Systems*, Vol. 36, #1, 19.

**[R6.47]** Zadeh, L. A., "Probability Measure of Fuzzy Event," *Jour. of Math Analysis and Applications*, Vol. 23, 1968, pp. 421 – 427.

## R7.    Ideology and the Knowledge Construction Process

**[R7.1]** Abercrombie, Nicholas et al., *The Dominant Ideology Thesis*, London, Allen and Unwin, 1980.

**[R7.2]** Abercrombie, Nicholas, *Class, Structure, and Knowledge: Problems in the Sociology of Knowledge*, New York, New York University Press, 1980.

**[R7.3]** Aron, Raymond, *The Opium of the Intellectuals*, Lanham, MD, University Press of America, 1985.

**[R7.4]** Aronowitz, Stanley, *Science as Power: Discourse and Ideology in Modern Society*, Minneapolis, University of Minnesota Press, 1988.

**[R7.5]** Barinaga, M. and E. Marshall, *Confusion on the Cutting Edge*, Science, Vol. 257, July 1992, pp. 616-625.

**[R7.6]** Barnett, Ronald, *Beyond All Reason: Living with Ideology in the University*, Philadelphia, PA., Society for Research into Higher Education and Open University Press, 2003.

**[R7.7]** Barth, Hans, *Truth and Ideology*, Berkeley, University of California Press, 1976.

**[R7.8]** Basin, Alberto, and Thierry Verdie, "The Economics of Cultural Transmission and the Dynamics of Preferences," *Journal of Economic Theory*, Vol. 97, 2001, pp. 298-319.

**[R7.9]** Beardsley, Philip L. *Redefining Rigor: Ideology and Statistics in Political Inquiry*, Bevery Hills, Sage Publications, 1980.

**[R7.10]** Bikhchandani, Sushil et al., "A Theory of Fads, Fashion, Custom, and Cultural Change," *Journal of political Economy*, Vol. 100 1992, pp 992-1026.

**[R7.11]** Boyd Robert and Peter J Richerson, *Culture and Evolutionary Process*, Chicago, University of Chicago Press, 1985.

**[R7.12]** Buczkowski, Piotr and Andrzej Klawiter, *Theories of Ideology and Ideology of Theories*, Amsterdam, Rodopi, 1986.

**[R7.13]** Chomsky, Norm, *Manufacturing Consent*, New York, Pantheo Pess, 1988.

**[R7.14]** Chomsky, N., *Problem of Knowledge and Freedom*, Glasgow, Collins, 1972.

**[R7.15]** Cole, Jonathan, R., "Patterns of Intellectual influence in Scientific Research," *Sociology of Education*, Vol. 43, 1968, pp377-403.

**[R7.16]** Cole Jonathan, R. and Stephen Cole, *Social Stratification in Science*, Chicago, University of Chicago Press, 1973.

**[R7.17]** Debackere, Koenraad and Michael A. Rappa, "Institutioal Varations in Problem Choice and Persistence among Scientists in an Emerging Fields," *Research Policy*, Vol. 23, 1994, pp425- 441.

**[R7.18]** Fraser, Colin and George Gaskell (eds.), *The Social Psychological Study of Widespread Beliefs*, Oxford, Clarendon Press, 1990.

**[R7.19]** Gieryn, Thomas, F. "Problem Retention and Problem Change in Science," *Sociological Inquiry*, Vol. 48, 1978, pp96-115.

**[R7.20]** Harrington, Joseph E. Jr, "The Rigidity of social Systems," *Journal of Political Economy*, Vol. 107, pp. 40-64.

**[R7.21]** Hinich, Melvin and Michael Munger, *Ideology and the Theory of Political Choice*, Ann Arbor University of Michigan Press, 1994.

**[R7.22]** Hull, D. L., *Science as a Process: An Evolutionary Account of the Social and Conceptual Development of Science*, Chicago, University of Chicago Press, 1988.

**[R7.23]** Marx, Karl and Friedrich Engels, The German Ideology, New York, International Pub, 1970

**[R7.24]** Mészáros, István , *Philosophy, Ideology and Social Science*: Essay in Negation and Affirmation, Brighton, Sussex, Wheatsheaf, 1986.

**[R7.25]** Mészáros, István *The Power of Ideology*, New York, New York University Press, 1989.

**[R7.26]** Newcomb, Theodore M. et. al., *Persistence and Change*, New York, John Wiley, 1967.

**[R7.27]** Pickering, Andrew, *Science as Practice and Culture*, Chicago, University of Chicago Press, 1992.

**[R7.28]** Therborn, Göran , *The Ideology of Power and the Power of Ideology*, London, NLB Publications, 1980.

**[R7.29]** Thompson, Kenneth, *Beliefs and Ideology*, New York, Tavistock Publication, 1986.

**[R7.30]** Ziman, John, "The Problem of 'Problem Choice'," *Minerva*, Vol. 25, 1987, pp92-105.

**[R7.31]** Ziman, John, *Public Knowledge: An Essay Concerning the Social Dimension of Science*, Cambridge, Cambridge University Press, 1968.

**[R7.32]** Zuckerman, Hrriet, "Theory Choice and Problem Choice in Science," *Sociological Inquiry*, Vol.48, 1978, pp. 65-95.

## R 8.    Information, Thought and Knowledge

**[R8.1]** Aczel, J. and Z. Daroczy, *On Measures of Information and their Characterizations*, New York, Academic Press, 1975.

**[R8.2]** Anderson, J. R., *The Architecture of Cognition*, Cambridge, Mass., Harvard University Press, 1983.

**[R8.3]** Angelov, Stefan and Dimitr Georgiev, "The Problem of Human Being in Contemporary Scientic Knowledge," *Soviet Studies in Philosophy*, Summer, 1974, pp. 49-66.

**[R8.4]** Ash, Robert, *Information Theory*, New York, John Wiley and Sons, 1965.

**[R8.5]** Barlas, Y. and S. Carpenter, "Philosophical Roots of Model Validation: Two Paradigms," *System Dynamic Review*, Vol. 6, 1990, pp148-166.

**[R8.6]** Bergin, J., "Common Knowledge with Monotone Statistics," *Econometrica*, Vol.69, 2001, pp. 1315-1332.

**[R8.7]** Bestougeff, Hélène and Gerard Ligozat, *Logical Tools for Temporal Knowledge Representation*, New York, Ellis Horwood, 1992.

**[R8.8]** Brillouin, L., *Science and information Theory*, New York, Academic Press, 1962.

**[R8.9]** Bruner, J. S., et. al., *A Study of Thinking*, New York, Wiley, 1956.

**[R8.10]** Brunner, K. and A. H. Meltzer (eds.), *Three Aspects of Policy and Policy Making: Knowledge, Data and Institutions*, Carnegie-Rochester Conference Series, Vol. 10, Amsterdam, North-Holland, 1979.

**[R8.11]** Burks, A. W., *Chance, Cause, Reason: An Inquiry into the Nature of Scientific Evidence*, Chicago, University of Chicago Press, 1977.

**[R8.12]** Calvert, Randall, *Models of Imperfect Information in Politics*, New York, Hardwood Academic Publishers, 1986.

**[R8.13]** Cornforth, Maurice, *The Theory of Knowledge*, New York, International Pub. 1972

**[R8.14]** Cornforth, Maurice, *The Open Philosophy and the Open Society,* New York,International Pub. 1970

**[R8.15]** Coombs, C. H., *A Theory of Data*, New York, Wiley, 1964.

**[R8.16]** Dretske, Fred. I., *Knowledge and the Flow of Information*, Cambridge, Mass., MIT Press 1981.

**[R8.17]** Dreyfus, Hubert L., "A Framework for Misrepresenting Knowledge," in Martin Ringle (ed.) *Philosophical Perspectives in Artificial Intelligence,* Atlantic Highlands, N.J., Humanities press, 1979.

**[R8.18]** Fagin R. et al., *Reasoning About Knowledge,* Cambridge, Mass, MIT Press, 1995.

**[R8.19]** Geanakoplos, J., "Common Knowledge," *Journal of Economic Perspectives,*" Vol. 6, 1992, pp53-82.

**[R8.20]** George, F. H., *Models of Thinking*, London, Allen and Unwin, 1970.

**[R8.21]** George, F. H., "Epistemology and the problem of perception," *Mind*, Vol.66, 1957, pp.491-506.

373

**[R8.22]** Harwood, E. C., *Reconstruction of Economics*, Great Barrington, Mass, American Institute for Economic Research, 1955.

**[R8.23]** Hintikka, J., *Knowledge and Belief*, Ithaca, N. Y., Cornell University Press, 1962.

**[R8.24]** Hirshleifer, Jack., "The Private and Social Value of Information and Reward to inventive activity," *American Economic Review*, Vol. 61, 1971, pp.561-574.

**[R8.25]** Kapitsa, P. L., "The Influence of Scientific Ideas on Society," *Soviet Studies in Philosophy*, Fall, 1979, pp.52-71.

**[R8.26]** Kedrov, B. M., "The Road to Truth," *Soviet Studies in Philosophy*, Vol. 4, 1965, pp 3 – 53.

**[R8.27]** Klatzky, R. L., *Human Memory: Structure and Processes*, San Francisco, Ca., W. H. Freeman Pub., 1975.

**[R8.28]** Kreps, David and Robert Wilson, "Reputation and Imperfect Information," *Journal of Economic Theory*, Vol. 27. 1982, pp253-279.

**[R8.29]** Kubát, Libor and J. Zeman (eds.), *Entropy and Information*, Amsterdam, Elsevier, 1975.

**[R8.30]** Kurcz, G. and W. Shugar et al (eds.), *Knowledge and Language*, Amsterdam, North-Holland, 1986.

**[R8.31]** Lakemeyer, Gerhard,and Bernhard Nobel (eds.), *Foundations of Knowledge Representation and Reasoning*, Berlin, Springer-Verlag, 1994.

**[R8.32]** Lektorskii, V. A., "Principles involved in the Reproduction of Objective in Knowledge,", *Soviet Studies in Philosophy*, Vol. 4, #4, 1967, pp. 11-21.

**[R8.33]** Levi, I., *The Enterprise of Knowledge*, Cambridge, Mass. MIT Press 1980.

**[R8.34]** Levi, Isaac, "Ignorance, Probability and Rational Choice", *Synthese*, Vol. 53, 1982, pp. 387-417.

**[R8.35]** Levi, Isaac, "Four Types of Ignorance," *Social Science*, Vol. 44, pp745-756.

**[R8.36]** Marschak, Jacob, *Economic Information, Decision and Prediction: Selected Essays*, Vol. II, Part II, Boston, Mass. Dordrecnt-Holland, 1974.

**[R8.37]** Menges, G. (ed.), *Information, Inference and Decision*, D. Reidel Pub., Dordrecht, Holland, 1974.

**[R8.38]** Michael Masuch and László Pólos (eds.), *Knowledge Representation and Reasoning Under Uncertainty*, New York, Springer-Verlag, 1994.

**[R8.39]** Moses, Y. (ed.), *Proceedings of the Fourth Conference of Theoretical Aspects of Reasoning about Knowledge*, San Mateo, Morgan Kaufmann, 1992.

**[R8.40]** Nielsen, L.T. et al., "Common Knowledge of Aggregation Expectations," *Econometrica*, Vol. 58, 1990, pp. 1235-1239.

**[R8.41]** Newell, A., *Unified Theories of Cognition*, Cambridge, Mass. Harvard University Press, 1990.

**[R8.42]** Newell, A., *Human Problem Solving*, Englewood Cliff, N.J. Prentice-Hall, 1972.

**[R8.43]** Ogden, G. K. and I. A., *The Meaning of Meaning*, New York, Harcourt-Brace Jovanovich, 1923.

**[R8.44]** Planck, Max, Scientific Autobiography and Other Papers, Westport, Conn., Greenwood, 1968.

**[R8.45]** Pollock, J., *Knowledge and Justification*, Princeton, Princeton University Press, 1974.

**[R8.46]** Polanyi, M., *Personal Knowledge*, London, Routledge and Kegan Paul, 1958.

**[R8.47]** Popper, K. R., *Objective Knowledge*, London, Macmillan, 1949.

**[R8.48]** Popper, K. R., *Open Society and it Enemies, Vols. 1 and 2* Princeton, Princeton Univ. Press, 2013

**[R8.49]** Popper, K. R., *The Poverty of Historicism* New York, Taylor and Francis, 2002

**[R8.50]** Price, H. H., *Thinking and Experience*, London, Hutchinson, 1953.

**[R8.51]** Putman, H., *Reason, Truth and History*, Cambridge, Cambridge University Press, 1981.

**[R8.52]** Putman. H., *Realism and Reason*, Cambridge, Cambridge University Press, 1983.

**[R8.53]** Putman, H., *The Many Faces of Realism*, La Salle, Open Court Publishing Co., 1987.

**[R8.54]** Russell, B., *Human Knowledge, its Scope and Limits*, London, Allen and Unwin, 1948.

**[R8.55]** Russell, B., *Our Knowledge of the External World*, New York, Norton, 1929.

**[R8.56]** Samet, D., "Ignoring Ignorance and Agreeing to Disagree," *Journal of Economic Theory*, Vol. 52, 1990, pp. 190-207.

**[R8.57]** Schroder, Harold, M. and Peter Suedfeld (eds.), *Personality Theory and Information Processing*, New York, Ronald Pub. 1971.

**[R8.58]** Searle J., *Minds, Brains and Science*, Cambridge, Mass., Harvard University Press, 1985.

**[R8.59]** Shin, H., "Logical Structure of Common Knowledge," *Journal of Economic Theory*, Vol. 60, 1993, pp. 1-13.

**[R8.60]** Simon, H. A., *Models of Thought*, New Haven, Conn., Yale University Press, 1979.

**[R8.61]** Smithson, M., *Ignorance and Uncertainty, Emerging Paradigms*, New York, Springer-Verlag, 1989.

**[R8.62]** Sowa, John F., *Knowledge Representation: Logical, Philosophical, and Computational Foundations*, Pacific Grove, Brooks Pub., 2000.

**[R8.63]** Stigler, G. J., The Economics of Information," *Journal of Political Economy*, Vol.69, 1961, pp.213-225.

**[R8.64]** Tiukhtin, V. S., "How Reality Can be Reflected in Cognition: Reflection as a Property of All Matter," *Soviet Studies in Philosophy*, Vol.3 #1, 1964, pp.3-12.

**[R8.65]** Tsypkin, Ya Z., *Foundations of the Theory of Learning Systems*, New York, Academic Press, 1973.

**[R8.66]** Ursul, A. D., "The Problem of the Objectivity of Information," in Libor Kubát, and J. Zeman (eds.), *Entropy and Information*, Amsterdam, Elsevier, 1975.pp. 187 – 230.

**[R8.67]** Vardi, M. (ed.), *Proceedings of Second Conference on Theoretical Aspects of Reasoning about Knowledge*, Asiloman, Ca., Los Altos, Ca, Morgan Kaufman, 1988.

**[R8.68]** Vazquez, Mararita, et al., "Knowledge and Reality: Some Conceptual Issues in System Dynamics Modeling," *Systems Dynamics Review*, Vol. 12, 1996, pp. 21-37.

**[R8.69]** Zadeh, L. A., "A Theory of Commonsense Knowledge," in Skala, Heinz J. etal., (eds.), *Aspects of Vagueness*, Dordrecht, D. Reidel Co. 1984. pp 257 – 295.

**[R8.70]** Zadeh, L. A., "The Concept of Linguistic Variable and its Application toApproximate reasoning," *Information Science*, Vol. 8, 1975, pp. 199 – 249 (Also in Vol. 9, pp. 40 – 80).

## R9.    Language and the Knowledge-Production Process

**[R9.1]** Aho, A. V. "Indexed Grammar - An Extension of Context-Free Grammars" *Journal of the Association for Computing Machinery*, Vol. 15, 1968, pp. 647-671.

**[R9.2]** Black, Max (ed.), *The Importance of Language*, Englewood Cliffs, N.J, Prentice-Hall, 1962.

**[R9.3]** Carnap, Rudolff, Meaning and Necessity: A Study in Semantics and Modal Logic, Chicago, University of Chicago Press, 1956.

**[R9.4]** Chomsky, Norm, "Linguistics and Philosophy" in S. Hook (ed.) *Language and Philosophy*, New York, New York University Press, 1968, pp. 51-94.

**[R9.5]** Chomsky, Norm, *Language and Mind*, New York, Harcourt Brace Jovanovich, 1972.

**[R9.8]** Cooper, William S., *Foundations of Logico-Linguistics: A Unified Theory of Information, Language and Logic*, Dordrecht, D. Reidel, 1978.

**[R9.9]** Cresswell, M.J.., *Logics and Languages*, London, Methuen Pub. 1973.

**[R9.10]** Dilman, Ilham, *Studies in Language and Reason*, Totowa, N.J., Barnes and Nobles, Books, 1981.

**[R9.10]** Fodor, Jerry A., *The Language and Thought*, New York, Thom as Y. Crowell Co, 1975

**[R9.12]** Givon, Talmy, *On Understanding Grammar*, New York, Academic Press, 1979

**[R9.13]** Gorsky, D.R., *Definition*, Moscow, Progress Publishers, 1974.

**[R9.14]** Hintikka, Jaakko, The Game of Language, Dordrecht, D. Reidel Pub. 1983.

**[R9.15]** Johnson-Lair, Philip N. *Mental Models: Toward Cognitive Science of Language, Inference and Consciousnes*s, Cambridge, Mass, Harvard University Pres, 1983.

**[R9.16]** Kandel, A., "Codes Over Languages," *IEEE Transactions on Systems Man and Cybernetics*, Vol. 4, 1975, pp. 135-138.

**[R9.17]** Keenan, Edward L. and Leonard M. Faltz, *Boolean Semantics for Natural Languages*, Dordrecht, D. Reidel Pub., 1985

**[R9.18]** Lakoff, G. Linguistics and Natural Logic, *Synthese*, Vol. 22, 1970, pp. 151-271.

**[R9.19]** Lee, E.T., et. al., "Notes On Fuzzy Languages," *Information Science*, Vol. 1, 1969, pp. 421-434.

**[R9.20]** Mackey, A. and D. Merrill (eds.) *Issues in the Philosophy of Language*, New Haven, CT. Yale University Press, 1976.

**[R9.21]** Nagel, T., "Linguistics and Epistemology" in S. Hook(ed.) *Language and Philosophy*, New York, New York University Press, 1969, pp. 180-184.

**[R9.22]** Pike, Kenneth, *Language in Relation to a Unified Theory of Structure of Human Behavior*, The Hague, Mouton Pub., 1969.

**[R9.23]** Quine, W.V. O. *Word and object*, Cambridge, Mass, MIT Press, 1960.

**[R9.24]** Russell, Bernard, *An Inquiry into Meaning and Truth*, Penguin Books, 1970.

**[R9.25]** Tarski, Alfred, *Logic, Semantics and Matamathematics*, Oxford, Clarendon Press, 1956.

**[R9.26]** Whorf, B.L. (ed.), *Language, Thought and Reality*, New York, Humanities Press, 1956.

## R.10. Possible Worlds and the Knowledge Production Process

**[R10.1]** Adams, Robert M., "Theories of Actuality," *Noûs*, Vol. 8, 1974, pp211-231.

**[R10.2]** Allen, Sture (ed.) *Possible Worlds in Humanities, Arts and Sciences*, Proceedings of Nobel Symposium, Vol. 65, New York, Walter de Gruyter Pub. , 1989.

**[R10.3]** Armstrong, D. M., *A Combinatorial Theory of Possibility*. Cambridge University Press, 1989.

**[R10.4]** Armstrong, D.M *A World of States of Affairs*, Cambridge, Cambridge University Press 1997.

**[R10.5]** Bell, J.S., "Six Possible Worlds of Quantum Mechanics" in Allen, Sture (Ed.) *Possible Worlds in Humanities, Arts and Sciences*, Proceedings of Nobel Symposium, Vol. 65, New York, Walter de Gruyter Pub., 1989.pp. 359-373….

**[R10.6]** Bigelow, John. "Possible Worlds Foundations for Probability", *Journal of Philosophical Logic*, 5 (1976), pp. 299-320.

**[R10.7]** Bradley, Reymond and Norman Swartz, *Possible World: An Introduction to Logic and its Philosophy*, Oxford, Bail Blackwell, 1997.

**[R10.8]** Castañeda, H.-N. "Thinking and the Structure of the World", *Philosophia*, 4 (1974), pp. 3-40.

**[R10.9]** Chihara, Charles S. *The Worlds of Possibility: Modal Realism and the Semantics of Modal Logic*, Clarendon, 1998.

**[R10.10]** Chisholm, Roderick. "Identity through Possible Worlds: Some Questions", *Noûs*, 1 (1967), pp. 1-8; reprinted in Loux, *The Possible and the Actual*.

**[R10.11]** Divers, John, Possible *Worlds*, London: Routledge, 2002.

**[R10.12]** Forrest, Peter. "Occam's Razor and Possible Worlds", *Monist*, 65 (1982), pp. 456-64.

**[R10.13]** Forrest, Peter. and Armstrong, D. M. "An Argument Against David Lewis' Theory of Possible Worlds", *Australasian Journal of Philosophy*, 62 (1984), pp. 164-168.

**[R10.14]** Grim, Patrick, "There is No Set of All Truths", *Analysis*, Vol. 46, 1986, pp. 186-191.

**[R10.15]** Heller, Mark. "Five Layers of Interpretation for Possible Worlds", *Philosophical Studies*, 90 (1998), pp. 205-214.

**[R10.16]** Herrick, Paul, *The Many Worlds of Logic,*. Oxford: Oxford University Press 1999.

**[R10.17]** Krips, H. "Irreducible Probabilities and Indeterminism", *Journal of Philosophical Logic*, Vol. 18, 1989, pp. 155-172.

**[R10.18]** Kuhn, Thomas S., "Possible Worlds in History of Science" in Allen, Sture (ed.) *Possible Worlds in Humanities, Arts and Sciences,* Proceedings of Nobel Symposium, Vol. 65, New York, Walter de Gruyter Pub. , 1989. pp. 9-41.

**[R10.19]** Kuratowski, K. and Mostowski, A. *Set Theory: With an Introduction to Descriptive Set Theory,* New York: North-Holland, 1976.

**[R10.20]** Lewis, David, *On the Plurality of Worlds*, Oxford, Basil Blackwell, 1986.

**[R10.21]** Loux, Michael J. (ed.) *The Possible and the Actual: Readings in the Metaphysics of Modality,* Ithaca & London: Cornell University Press, 1979.

**[R10.22]** Parsons, Terence, *Nonexistent Objects,* New Haven, Yale University Press, 1980.

**[R10.23]** Perry, John, "From Worlds to Situations", Journal of Philosophical Logic, Vol. 15, 1986, pp. 83-107.

**[R10.24]** Rescher, Nicholas and Brandom, Robert. *The Logic of Inconsistency: A Study in Non-Standard Possible-World Semantics And Ontology*, Rowman and Littlefield, 1979.

**[R10.25]** Skyrms, Brian. "Possible Worlds, Physics and Metaphysics", *Philosophical Studies*, Vol. 30, 1976, pp. 323-32.

**[R10.26]** Stalmaker, Robert C. "Possible World", *Noûs*, Vol. 10, 1976, pp. 65-75

**[R10.27]** Quine, W.V.O. *Word and Object*, M.I.T. Press, 1960.

**[R10.28]** Quine, W.V.O "Ontological Relativity", *Journal of Philosophy*, 65 (1968), pp. 185-212.

## R11. Rationality, Information, Games, Conflicts and Exact Reasoning

[R11.2] Border, Kim, *Fixed Point Theorems with Applications to Economics and Game Theory*, Cambridge, Cambridge University Press 1985.

[R11.3] Brandenburger, Adam, "Knowledge and Equilibrium Games," *Journal of Economic Perspectives*, Vol.6, 1992, pp. 83-102.

[R13.4] Campbell, Richmond and Lanning Sowden, *Paradoxes of Rationality and Cooperation: Prisoner's Dilemma and Newcomb's Problem*, Vancouver, University of British Columbia Press, 1985.

[R11.6] Gates Scott and Brian Humes, *Games, Information, and Politics: Applying Game Theoretic Models to Political Science*, Ann Arbor, University of Michigan Press, 1996.

[R11.7] Gjesdal, Froystein, "Information and Incentives: The Agency Information Problem," *Review of Economic Studies*, Vol.49, 1982, pp373-390.

[R11.8] Harsanyi, John, "Games with Incomplete Information Played by 'Bayesian' Players I: The Basic Model," *Management Science*, Vol.14, 1967, pp.159-182.

[R11.9] Harsanyi, John, "Games with Incomplete Information Played by 'Bayesian' Players II: Bayesian Equilibrium Points," *Management Science*, Vol.14, 1968, pp.320-334.

[R11.10] Harsanyi, John, "Games with Incomplete Information Played by 'Bayesian' Players III: The Basic Probability Distribution of the Game," *Management Science*, Vol.14, 1968, pp.486-502.

[R11.11] Harsanyi, John, *Rational Behavior and Bargaining Equilibrium in Games and Social Situations*, New York Cambridge University Press, 1977.

[R11.14] Krasovskii, N.N. and A.I. Subbotin, *Game-theoretical Control Problems*, New York, Springer-Verlag, 1988.

[R11.15] Kuhn, Harold (ed.) *Classics in Game Theory*, Princeton, Princeton University Press, 1997.

[R11.16] Lagunov, V. N., *Introduction to Differential Games and Control Theory*, Berlin, Heldermann Verlag, 1985.

[R11.17] Luce, D. R. and H. Raiffa, *Games and Decisions*, New York, John Wiley and Sons, 1957.

[R11.18] Maynard Smith, John, *Evolution and the Theory of Games*, Cambridge, Cambridge University Press, 1982.

**[R11.20]**Myerson, Roger, *Game Theory: Analysis of Conflict*, Cambridge, Mass. Harvard University Press, 1991.

**[R11.21]**Rapoport, Anatol and Albert Chammah, *Prisoner's Dilemma: A Study in Conflict and Cooperation*, Ann Arbor, University of Michigan Press, 1965.

**[R11.22]**Roth, Alvin E., "The Economist as Engineer: Game Theory, Experimentation, and Computation as Tools for Design Economics," *Econometrica*, Vol.70, 2002, pp1341-1378.

**[R11.23]**Shubik, Martin, *Game Theory in the Social Sciences: Concepts and Solutions*, Cambridge, Mass., MIT Press, 1982.

**[R11.24]**Smart, D.R., *Fixes point Theorems*, Cambridge, Cambridge University Press, 1980.

**[R11.26]**Von Neumann, John and Oskar Morgenstern, *The Theory of Games in Economic Behavior*, New York, John Wiley and Sons, 1944.

## R12. Rationality and Philosophy of Exact and Inexact Sciences in the Knowledge Production

**[R12.1]**Achinstein, P., "The Problem of Theoretical Terms," in Brody, Baruch A. (Ed.) *Reading in the Philosophy of Science*, Englewood Cliffs, NJ., Prentice Hall, 1970.

**[R12.2]**Amo Afer, A. G., *The Absence of Sensation and the Faculty of Sense in the Human Mind and Their Presence in our Organic and Living Body, Dissertation and Other essays 1727-1749*, Halle Wittenberg, Jena, Martin Luther Universioty Translation, 1968.

**[R12.3]**Beeson, M. J., *Foundations of Constructive Mathematics*, Berlin/New York, Springer,1985.

**[R12.4]**Benacerraf, P., "God, the Devil and Gödel," *Monist*, Vol. 51, 1967, pp.9-32.

**[R12.5]**Benecerraf, P and H. Putnam (eds.), *Philosophy of Mathematics: Selected Readings*, Cambridge, Cambridge University Press, 1983.

**[R12.6]**Black, Max, *The Nature of Mathematics*, Totowa, Littlefield, Adams and Co. 1965.

**[R12.7]**Blanche, R., *Contemporary Science and Rationalism*, Edinburgh, Oliver and Boyd, 1968.

**[R12.8]**Blanshard, Brand, *The Nature of Thought*, London Allen and Unwin, 1939.

**[R12.9]** Blauberg, I. V., V.N. Sadovsky and E.G. Yudin, Systems Theory: Philosophical and Methodological Problems, Moscow, Progress Publishers, 1977.

**[R12.10]** Braithwaite, R. B., *Scientific Explanation*, Cambridge, Cambridge University Press. 1955.

**[R12.11]** Brody, Baruch A. (ed.), *Reading in the Philosophy of Science*, Englewood Cliffs, N.J., Prentice Hall1970.

**[R12.12]** Brody, Baruch A., "Confirmation and Explanation," in Brody, Baruch A. (ed.) *Reading in the Philosophy of Science*, Englewood Cliffs, N.J., Prentice-Hall, 1970, pp. 410-426.

**[R12.13]** Brouwer, L.E.J., "Intuitionism and Formalism", *Bull of American Math. Soc.*, Vol. 20, 1913, pp81-96.; Also in Benecerraf, P. and H. Putnam (eds.), *Philosophy of Mathematics: Selected Readings*, Cambridge, Cambridge University Press, 1983. pp. 77-89.

**[R12.14]** Brouwer, L.E.J., "Consciousness, Philosophy, and Mathematics," in Benecerraf, P. and H. Putnam (eds.), *Philosophy of Mathematics: Selected Readings*, Cambridge, Cambridge University Press, 1983. Pp. 90-96.

**[R12.15]** Brouwer, L. E. J., Collected *Works, Vol. 1: Philosophy and Foundations of Mathematics* [A Heyting (ed.)],New York, Elsevier, 1975.

**[R12.16]** Campbell, Norman R., *What is Science?*, New York, Dover, 1952.

**[R12.17]** Carnap, R., "Foundations of Logic and Mathematics," in *International Encyclopedia of Unified Science*, Chicago, Univ. of Chicago,1939, pp.143-211.

**[R12.18]** Carnap, Rudolf, "On Inductive Logic," *Philosophy of Science*, Vol. 12, 1945, pp. 72-97.

**[R12.19]** Carnap, Rudolf, "The Methodological Character of Theoretical Concepts," in Herbert Feigl and M. Scriven (eds.) *Minnesota Studies in the Philosophy of Science, Vol. I*, 1956, pp. 38-76.

**[R12.20]** Charles, David and Kathleen Lennon (eds.), *Reduction, Explanation, and Realism*, Oxford, Oxford Unive3rsity Press, 1992.

**[R12.21]** Cohen, Robert S. and Marx W. Wartofsky (eds.), *Methodological and Historical Essays in the Natural and Social Sciences*, Dordrecht, D. Reidel Publishing Co. 1974.

**[R12.22]** Dalen van, D. (ed.), *Brouwer's Cambridge Lectures on Intuitionism*, Cambridge, Cambridge University Press, 1981.

**[R12.23]** Davidson, Donald, *Truth and Meaning: Inquiries into Truth and Interpretation*, Oxford, Oxford University Press, 1984.

**[R12.24a]**Davis, M., *Computability and Unsolvability*, New York, McGraw-Hill, 1958.

**[R12.24b]**Denonn. Lester E. (ed.), *The Wit and Wisdom of Bertrand Russell*, Boston, MA., The Beacon Press, 1951.

**[R12.25]**Dummett, M., "The Philosophical Basis of Intuitionistic Logic," in Benecerraf, P. and H. Putnam (eds.), *Philosophy of Mathematics: Selected Readings*, Cambridge, Cambridge University Press, 1983. pp97-129

**[R12.26]**Feigl, Herbert and M. Scriven (eds.), Minnesota *Studies in the Philosophy ofScience*, Vol. I, 1956.

**[R1.27]**Feigl, Herbert and M. Scriven (eds.), *Minnesota Studies in the Philosophy of Science*, Vol. II, 1958.

**[R12.28]**Garfinkel, Alan, *Forms of Explanation:Structures of Inquiry in Social Science*, New Haven, Conn., Yale University Press, 1981.

**[R12.29]**George, F. H., *Philosophical Foundations of Cybernetics*, Tunbridge Well, Great Britain, 1979.

**[R12.30]** Gillam, B., "Geometrical Illusions," *Scientific American*, January, 1980, pp.102-111.

**[R12.31]** Gödel , Kurt., "What is Cantor's Continuum Problem?" in Benecerraf, P. and H. Putnam (eds.), *Philosophy of Mathematics: Selected Readings* , Cambridge, Cambridge University Press, 1983. pp.470-486.

**[R12.32]**Gorsky, D.R., *Definition,* Moscow, Progress Publishers, 1974.

**[R12.33]**Gray, William and Nicholas D. Rizzo(eds.), *Unity Through Diversity.* New York, Gordon and Breach, 1973.

**[R12.34]**Hart, W. D. (ed.), *The Philosophy of Mathematics*, Oxford, Oxford University Press, 1996.

**[R12.35]**Hartkamper, a and H.- Schmidt, Structure and Approximation in Physical Theories, New York, Plenum Press, 1981

**[R12.36]**Hausman, David, M., *The Exact and Separate Science of Economics,* Cambridge, Cambridge University Press, 1992.

**[R12.37]**Helmer, Olaf and Nicholar Rescher, *On the Epistemology of the Inexact Sciences*, P-1513, Santa Monica, CA, Rand Corporation, October 13, 1958.

**[R12.38]**Hempel, C. G., "Studies in the Logic of Confirmation," *Mind,* Vol.54, Part I, 1945, pp 1-26.

**[R12.39]**Hempel, Carl G., "The Theoretician's Dilemma," in Herbert Feigl and M.Scriven (eds.) *Minnesota Studies in the Philosophy of Science*, Vol.II, 1958, pp. 37-98.

**[R12.40]** Hempel, C. G. and P. Oppenheim, "Studies in the Logic of Explanation," *Philosophy of Science*, Vol. 15, 1948,pp. 135-175. [also in Brody, Baruch A. (ed.) *Reading in the Philosophy of Science*, Englewood Cliffs, NJ., Prentice-Hall,1970, pp. 8-27.

**[R12.41]** Heyting, A., *Intuitionism: An Introduction,* Amsterdam: North-Holland, 1971.

**[R12.42]** Hintikka, Jackko (ed.), *The Philosophy of Mathematics*, London, Oxford University Press, 1969.

**[R12.43]** Hockney D. et al. (eds.), *Contemporary Research in Philosophical Logic and Linguistic Semantics*, Dordrecht-Holland, Reidel Pub., Co. 1975.

**[R12.44]** Hoyninggen-Huene, Paul and F. M. Wuketits, (eds.), *Reductionism and Systems Theory in the Life Science: Some Problems and Perspectives*, Dordrenght, Kluwer Academic Pub. 1989.

**[R12.45]** Ilyenkov, E.V., *Dialectical Logic: Essays on Its History and Theory*, Moscow, Progress Publishers, 1977.

**[R12.46]** Kedrov, B. M., "Toward the Methodological Analysis of Scientific Discovery," *Soviet Studies in Philosophy*, Vol. 11962, pp45 – 65.

**[R12.47]** Kemeny, John G, and P Oppenheim, "On Reduction," in Brody, Baruch A. (ed.) *Reading in the Philosophy of Science*, Englewood Cliffs, NJ., Prentice-Hall,1970, 307-318.

**[R12.48]** Klappholz, K., "Value Judgments of Economics," *British Jour. of Philosophy*, Vol. 15, 1964, pp. 97-114.

**[R 12.49]** Kleene, S.C., "On the Interpretation of Intuitionistic Number Theory," Journal of Symbolic Logic, Vol 10, 1945, pp. 109-124.

**[R12.50]** Kmita, Jerzy, "The Methodology of Science as a Theoretical Discipline," *Soviet Studies in Philosophy*, Spring, 1974, pp. 38 –49

**[R12.51]** Krupp, Sherman R.,(ed.), *The Structure of Economic Science*, Englewood Cliff, N. J., Prentice-Hall, 1966.

**[R12.52]** Kuhn, T., *The Structure of Scientific Revolution*, Chicago, University of Chicago Press, 1970.

**[R12.53]** Kuhn, Thomas, "The Function of Dogma in Scientific Research," in Brody, Baruch A. (ed.) *Reading in the Philosophy of Science*, Englewood Cliffs, NJ., Prentice-Hall,1970 pp.356-374.

**[R12.54]** Kuhn, Thomas, *The Essential Tension: Selected Studies in Scientific Tradition and Change*, Chicago, University of Chicago Press, 1979.

**[R12.55]** Lakatos, I. (ed.), *The Problem of Inductive Logic*, Amsterdam, North Holland, 1968.

**[R12.56]** Lakatos, I., *Proofs and Refutations: The Logic of Mathematical Discovery*, Cambridge, Cambridge University Press, 1976.

**[R12.57]** Lakatos, I., *Mathematics, Science and Epistemology: Philosophical Papers Vol. 2*, edited by J. Worrall and G. Currie, Cambridge, Cambridge Univ. Press, 1978.

**[R12.58]** Lakatos, I., *The Methodology of Scientific Research Programmes*, Vol 1, New York, Cambridge University Press, 1978.

**[R12.59]** Lakatos, Imre and A. Musgrave (eds.)., *Criticism and the Growth of Knowledge*, New York, Cambridge University Press, 1979. Holland, 1979, pp. 153 – 164.

**[R12.60]** Lawson, Tony, *Economics and Reality*, New York, Routledge, 1977.

**[R12.61]** Lenzen, Victor, "Procedures of Empirical Science," in Neurath, Otto et al. (eds.), *International Encyclopedia of Unified Science, Vol. 1 – 10,* Chicago, University of Chicago Press, 1955, pp. 280-338.

**[R12.62]** Levi, Isaac, "Must the Scientist make Value Judgments?," in Brody, Baruch A. (Ed.) *Reading in the Philosophy of Science*, Englewood Cliffs, NJ., Prentice-Hall,1970 pp.559-570.

**[R12.63]** Tse-tung, Mao, On Practice and Contradiction, in Selected works of Mao Tse-tung, Piking, 1937. Also, London, Revolutions, 2008.

**[R12.64]** Lewis, David, *Convention: A Philosophical Study,* Cambridge, Mass., Harvard University Press, 1969.

**[R12.65]** Mayer, Thomas, *Truth versus Precision in Economics*, London, Edward Elgar 1993.

**[R12.66]** Menger, Carl, *Investigations into the Method of the Social Sciences with Special Reference to Economics*, New York, New York University Press, 1985

**[R12.67]** Mirowski, Philip (ed.), *The Reconstruction of Economic Theory*, Boston, Mass.Kluwer Nijhoff, 1986.

**[R12.68]** Mueller, Ian, *Philosophy of Mathematics and Deductive Structure in Euclid's Elements*, Cambridge, Mass., MIT Press, 1981.

**[R12.69]** Nagel, Ernest, "Review: Karl Niebyl, Modern Mathematics and Some Problems of Quantity, Quality, and Motion in Economic Analysis," *The Journal of Symbolic Logic* , 1940, p.74.

**[R12.70]** Nagel, E. et al. (ed.), *Logic, Methodology, and the Philosophy of Science*, Stanford, Stanford University Press 1962.

**[R12.71]** Narens, Louis, "A Theory of Belief for Scientific Refutations," *Synthese,* Vol.145, 2005, pp. 397-423.

**[R12.72]** Narskii, I. S., "On the Problem of Contradiction in Dialectical Logic," *Soviet Studies in Philosophy*, Vol. vi, #4 pp.3-10, 1965

**[R12.73]** Neurath, Otto et al. (eds.), *International Encyclopedia of Unified Science, Vol. 1 – 10*, Chicago, University of Chicago Press, 1955.

**[R12.74]** Neurath Otto, "Unified Science as Encyclopedic," in Neurath, Otto et al. (eds.), *International Encyclopedia of Unified Science, Vol. 1 – 10*, Chicago, University of Chicago Press, 1955, pp.1-27.

**[R12.75]** Planck, Max, *Scientific Autobiography and Other Papers*, Westport, Conn. Greenwood 1971.

**[R12.76]** Planck, Max, "The Meaning and Limits of Exact Science," in Max Planck, *Scientific Autobiography and Other Papers*, Westport, Conn. Greenwood, 1971, pp. 80-120.

**[R12.77]** Polanyi, Michael, "Genius in Science," in Robert S. Cohen, and Marx W. Wartofsky (eds.), *Methodological and Historical Essays in the Natural and Social Sciences*, Dordrecht, D. Reidel Publishing Co. 1974, pp.57-71.

**[R12.78]** Popper, Karl, *The Nature of Scientific Discovery*, New York, Harper and Row, 1968.

**[R12.79]** Putnam, Hilary., "Models and Reality," in Benecerraf, P. and H. Putnam (eds.),*Philosophy of Mathematics: Selected Readings* , Cambridge, Cambridge University Press, 1983. pp. 421-444.

**[R12.80]** Reise, S., *The Universe of Meaning*, New York, The Philosophical Library, 1953.

**[R12.81]** Robinson, R., *Definition*, Oxford, clarendon Press, 1950

**[R12.82]** Rudner, Richard, "The Scientist qua Scientist Makes Value Judgments," *Philosophy of Science*, Vol. 20, 1953, pp 1-6.

**[R12.83]** Russell, B., *Our Knowledge of the External World*, New York, Norton, 1929.

**[R12.84]** Russell, B., *Human Knowledge, Its Scope and Limits*, London, Allen and Unwin, 1948.

**[R12.85]** Russell, B., *Logic and Knowledge: Essays 1901-1950*,New York, Capricorn Books, 1971.

**[R12.86]** Russell, B., *An Inquiry into Meaning and Truth*, New York, Norton, 1940.

**[R12.87]** Russell, Bertrand, *Introduction to Mathematical Philosophy*, London, George Allen and Unwin, 1919.

**[R12.88]** Russell, Bertrand, *The Problems of Philosophy*, Oxford, Oxford University Press, 1978.

**[R12.89]**Rutkevih, M. N., "Evolution, Progress, and the Law of Dialectics," *Soviet Studies in Philosophy*, Vol. IV, #3, PP. 34-43, 1965.

**[R12.90]**Ruzavin, G. I., "On the Problem of the Interrelations of Modern Formal Logic and Mathematical Logic," *Soviet Studies in Philosophy*, Vol. 3, #1, 1964, pp.34- 44.

**[R12.91]**Scriven, Michael, "Explanations, Predictions, and Laws," in Brody, Baruch A. (ed.) *Reading in the Philosophy of Science, Englewood Cliffs*, NJ., Prentice- Hall, 1970, pp.88-104.

**[R12.92]**Sellars, Wilfrid, The Language of Theories," in Brody, Baruch A. (ed.) *Reading in the Philosophy of Science*, Englewood Cliffs, NJ., Prentice-Hall,1970 pp.343-353.

**[R12.93]**Sterman, John, "The Growth of Knowledge: Testing a Theory of Scientific Revolutions with a Formal Model," *Technological Forecasting and Social Change*, Vol. 28, 1995, pp. 93-122.

**[R12.94]**Tsereteli, S. B. "On the Concept of Dialectical Logic,", *Soviet Studies in Philosophy*, Vol. V, #2, pp. 15-21, 1966.

**[R12.95]**Tullock, Gordon, *The Organization of Inquiry*, Indianapolis, Indiana, Liberty Fund Inc. 1966.

**[R12.96]**Van Fraassen, B., *Introduction to Philosophy of Space and Time*, New York, Random House, 1970.

**[R12.97]**Veldman, W., "A Survey of Intuitionistic Descriptive Set Theory," in P.P. Petkov (ed.), Mathematical Logic: Proceedings of the Heyting Conference, New York, Plenum Press, 1990, pp. 155-174.

**[R12.98]**Vetrov, A. A., "Mathematical Logic and Modern Formal Logic," *Soviet Studies in Philosophy*, Vol. 3, #1, 1964 pp. 24 – 33.

**[R12.99]** von Mises, Ludwig, *Epistemological Problems in Economics*, New York, New York University Press, 1981.

**[R12.100]**Wang, Hao, *Reflections on Kurt*Gödel , Cambridge, Mass. MIT Press, 1987

**[R12.101]**Watkins, J. W. N., "The Paradoxes of Confirmation"," in Brody, Baruch A. (ed.) *Reading in the Philosophy of Science*, Englewood Cliffs, NJ., Prentice-hall, 1970 pp. 433-438.

**[R12.102]**Whitehead, Alfred North, *Process and Reality*, New York, The Free Press, 1978.

**[R12.103]**Wittgenstein, Ludwig, *Ttactatus Logico-philosophicus,* Atlantic Highlands, N.J., The Humanities Press Inc.1974.

**[R12.104]**Woodger, J. H., *The Axiomatic Method in Biology*, Cambridge, Cambridge University Press, 1937.

**[R12.105]**Zeman, Jiři', "Information, Knowledge and Time," in Kubát, Libor and J.Zeman (eds.), *Entropy and Information*, Amsterdam, 1975.

## R13 Theory of Planning, the Prescriptive Science and Cost-Benefit Analysis in Trasformations

**[R13.1]**Alexander Ernest R., *Approaches to Planning*, Philadelphia, Pa. Gordon and Breach, 1992.

**[R13.2]**Bailey, J., *Social Theory for Planning*, London, Routledge and Kegan Paul,1975.

**[R13.3]**Burchell R.W. and G. Sternlieb (eds.), *Planning Theory in the 1980's: A Search for Future Directions*, New Brunswick, N. J., Rutgers University Center for Urban and Policy Research, 1978.

**[R13.4]**Camhis, Marios, *Planning Theory and Philosophy*, London, Tavistock Publicationa, 1979.

**[R13.5]**Chadwick, G., *A Systems View of Planning*, Oxford, Pergamon, 1971.

**[R13.6]**Cooke, P., *Theories of Planning and Special Development*, London, Hutchinson, 1983.

**[R13.7]**Dompere, Kofi K., and Taresa Lawrence, "Planning," in Syed B Hussain, *Encyclopedia of Capitalism,* Vol. II, New York, Facts On File, Inc., 2004, pp.649-653.

**[R13.8]**Dompere, Kofi K., *Social Goal-Objective Formation, Democracy and National Interest: A Theory of Political Economy under Fuzzy Rationality*, (Studies in Systems, Decision and Control, Vol. 4), New York, Springer, 2014

**[R13.9]**Dompere, Kofi K., *Fuzziness, Democracy Control and Collective Decision-Choice System: A Theory on Political Economy of Rent-Seeking and Profit-Harvesting*, (Studies in Systems, Decision and Control, Vol. 5), New York, Springer, 2014.

**[R13.10a]**Dompere, Kofi K., *The Theory of Aggregate Investment in Closed Economic Systems*, Westport, CT, Greenwood Press, 1999.

**[R13.10b]**Dompere, Kofi K., *The Theory of Aggregate Investment and Output Dynamics in Open Economic Systems*, Westport, CT, Greenwood Press, 1999.

**[R13.11a]**Faludi, A., *Planning Theory*, Oxford, Pergamon, 1973.

**[R13.11b]** Faludi, A.(ed.), *A Reader in Planning Theory*, Oxford, Pergamon, 1973.

**[R13.12]** Harwood, E.C. (ed.), Reconstruction of Economics, American Institute For Economic Research, Great Barrington, Mass, 1955., Also in John Dewey and Arthur Bently, 'Knowing and the known', Boston, Beacon Press, 1949, p. 269.

**[R13.13]** Kickert, W.J.M., *Organization of Decision-Making A Systems-Theoretic Approach*, New York, North-Holland, 1980.

**[R13.14]** Knight, Frank H. *Risk, Uncertainty and Profit*, Chicago, University of Chicago Press, 1971.

**[R13.15]** Knight, Frank H. *On History and Method of Economics*, Chicago, University of Chicago Press, 1971.

## R14. Social Sciences, Mathematics and the Problems of Exact and Inexact Methods of Thought

**[R14.1]** Ackoff, R.L., *Scientific Methods: Optimizing Applied Research Decisions*, New York, John Wiley, 1962.

**[R14.2]** Angyal, A. "The Structure of Wholes," *Philosophy of Sciences*, Vol.6, #1, 1939, pp 23-37.

**[R14.3]** Bahm, A.J., "Organicism: The Philosophy of Interdependence" *International Philosophical Quarterly*, Vol. VII # 2, 1967.

**[R14.4]** Bealer, George, *Quality and Concept*, Oxford, Clarendon Press, 1982.

**[R14.5]** Black, Max, *Critical Thinking*, Englewood Cliffs, N.J., Prentice-Hall, 1952.

**[R14.6]** Brewer, Marilynn B., and Barry E Collins (eds.) *Scientific Inquiry and Social Sciences*, San Francisco, Ca, Jossey-Bass Pub., 1981.

**[R14.7]** Campbell, D.T., "On the Conflicts Between Biological and Social Evolution and Between Psychology and Moral Tradition", *American Psychologist*, Vol. 30, 1975, pp1103-1126.

**[R14.8]** Churchman, C. W. and P. Ratoosh (eds.) *Measurement: Definitions and Theories*, New York, John Wiley, 1959.

**[R14.9]** Foley, Duncan, "Problems versus Conflicts Economic Theory and Ideology" American Economic Association Papers and Proceedings, Vol. 65, May 1975, pp. 231-237.

**[R14.10]** Garfinkel, Alan, *Forms of Explanation: Structures of Inquiry in Social Science*, New Haven, Conn., Yale University Press, 1981.

**[R14.11]** Georgescu-Roegen, Nicholas, *Analytical Economics*, Cambridge, Harvard University Press, 1967.

**[R14.12]** Gilolispie, C., *The Edge of Objectivity*, Princeton, Princeton University press, 1960.

**[R14.13]** Hayek, F.A., *The Counter-Revolution of Science*, New York, Free Press of Glencoe Inc, 1952.

**[R14.14]** Laudan, L., *Progress and Its Problems: Towards a Theory of Scientific Growth,* Berkeley, CA, University of California Press, 1961.

**[R14.15]** Marx, Karl, *The Poverty of Philosophy,* New York, International Pub. 1971.

**[R14.16]** Phillips, Denis C., *Holistic Thought in Social Sciences*, Stanford, CA, Stanford University Press, 1976.

**[R14.17]** Popper, K., *Objective Knowledge*, Oxford, Oxford University Press, 1972.

**[R14.18]** Rashevsky, N. "Organismic Sets: Outline of a General Theory of Biological and Social Organism," *General Systems*, Vol XII, 1967, pp. 21-28.

**[R14.19]** Roberts, Blaine, and Bob Holdren, *Theory of Social Process*, Ames, Iowa University Press, 1972.

**[R14.20]** Rudner, Richard S., *Philosophy of Social Sciences*, Englewood Cliff, N.J., Prentice Hall, 1966.

**[R14.21]** Simon, H. A., "The Structure of Ill-Structured Problems," *Artificial Intelligence,* Vol. 4, 1973, pp. 181-201.

**[R14.22]** Toulmin, S., *Foresight and understanding: An Enquiry into the Aims of Science,* New York, Harper and Row, 1961.

**[R14.23]** Winch, Peter, *The Idea of a Social Science*, New York, Humanities Press, 1958.

## 15    Tranformation,    Polarity,    Dialectics    and    Categorial Conversion

**[R15.1]** Anovsky, 0mely M.E., *Linin and Modern Natural Science, Moscow*, Progress Pub. 1978

**[R15.2]** Arrow, Kenneth J., "Limited Knowledge and Economic Analysis", American Economic Review, Vol. 64, 1974, pp. 1-10.

**[R15.3]** Berkeley, George, *Treatise Concerning the Principles of Human Knowledge*, Works, Vol. I (edited by A. Fraser), Oxford, Oxford University Press, 1871-1814.

**[R15.4]** Berkeley, George, "Material Things are Experiences of Men or God" in [R1.5], 1967, pp. 658-668.

**[R15.5]** Brody, Baruch A. (ed.), *Readings in the Philosophy of Science*, Englewood Cliffs, NJ., Prentice-Hall Inc., 1970.

**[R15.6** Brouwer, L.E.J., "Consciousness, Philosophy, and Mathematics," in Benecerraf, P. and H. Putnam (eds.), *Philosophy of Mathematics: Selected Readings*, Cambridge, Cambridge University Press, 1983. Pp90-96

**[R15.7]** Brown, B. and J Woods (eds.), *Logical Consequence; Rival Approaches and New Studies in exact Philosophy: Logic, Mathematics and Science*, Vol. II Oxford, Hermes, 2000.

**[R15.8]** Cornforth, Maurice, *Dialectical Materialism and Science*, New York, International Pub.1960

**[R15.9]** Cornforth, Maurice, *Materialism and Dialectical Method*, New York, International Pub.1953

**[R15.10]** Cornforth, Maurice, *Science and Idealism: an Examination of "Pure Empiricism"*, New York International Pub.1947

**[R15.11]** Cornforth, Maurice, *The Open Philosophy and the Open Society: A Reply to Dr. Karl Popper's Refutations of Marxism* New York, International Pub.1968.

**[R15.12]** Cornforth, Maurice, *The Theory of Knowledge*, New York, International Pub.1960

**[R15.13]** Dompere, Kofi K., "On Epistemology and Decision-Choice Rationality" in R. Trapple (ed.), *Cybernetics and System Research*, New York, North Holland, 1982, pp. 219-228.

**[R15.14]** Dompere, Kofi K. and M. Ejaz, *Epistemics of Development Economics: Toward a Methodological Critique and Unity*, Westport, CT, Greenwood Press, 1995.

**[R15.15]** Dompere, Kofi K., *The Theory of Categorial Conversion: Analytical Foundations of Nkrumaism*, Working Monograph on Mathematics, Philosophy, Economic and Decision Theories, Wasgington, D.C., Howard University, 2013.

**[R15.16]** Dompere, Kofi K., Nkrumaism: Socio-Political Philosophy and Ideology, Working Monograph on Mathematics, Philosophy, Economic and Decision Theories, Wasgington, D.C., Howard University, 2013.

**[R15.17]** Dompere, Kofi K., *Polyrhythmicity: Foundations of African Philosophy*, London, Adonis and Abbey Pub, 2006.

**[R15.18]** Dompere, Kofi K., "On Epistemology and Decision-Choice Rationality, in R. Trappl (ed.), *Cybernetics and System Research,* New York, North-Holland, 1982, pp219-228.

**[R15.19]** Engels, Frederick, *Dialectics of Nature*, New York, International Pub., 1971

**[R15.20]**Engels, Frederick, *Origin of the Family, Private Property and State*, New York, International Pub., 1971.

**[R15.21]** Ewing, A.C., "A Reaffirmation of Dualism" in [R1.5], pp. 454-461.

**[R15.22]**Fedoseyer, P.N. *et al., Philosophy in USSR: Problems of Dialectical Materialism*, Moscow, Progress Pub., 1977

**[R15.23]**Kedrov, B. M., "On the Dialectics of Scientific Discovery," *Soviet Studies in Philosophy* , Vol. 6 1967, pp16 – 27.

**[R15.24]**Lenin, V. I. *Materialism and Empirio-Criticism: Critical Comments on Reactionary Philosophy*, New York, International Pub., 1970.

**[R15.25]**Lenin, V. I. *Collected Works Vol. 38: Philosophical Notebooks*, New York, International Pub., 1978

**[R15.26]**Lenin, V. I., *On the National Liberation Movement*, Peking, Foreign Language Press, 1960

**[R15.27]**Hegel, George, *Collected Works*, Berlin, Duncher und Humblot, 1832 – 1845 [also *Science of Logic*, translated by W. H. Johnston and L. G. Struther , London, 1951].

**[R15.28]**Hempel, Carl G. and P. Oppenheim, "Studies in the Logic of Explanation," in [R15.5], pp. 8 – 27.

**[R15.29]**Ilyenkov, E.V., *Dialectical Logic: Essays on its History and Theory*, Moscow, Progress Pub. 1977.

**[R15.30]**Keirstead, B.S., "The Conditions of Survival," American Economic Review, Vol. 40, #2, pp.435- 445.

**[R15.31]**Kühne, Karl, *Economics and Marxism, Vol.I: The Renaissance of the Marxian System*, New York, St Martin's Press, 1979.

**[R15.32]**Kühne, Karl, *Economics and Marxism, Vol.II: The Dynamics of the Marxian System*, New York, St Martin's Press, 1979.

**[R15.33]** March, J. C., "Bounded Rationality, Ambiguity and Engineering of Choice," *The Bell Journal of Economics*, Vol. 9 (2), 1978

**[R15.34]**Marx, Karl, *Contribution to the Critique of Political Economy*, Chicago, Charles H. Kerr and Co. 1904.

**[R15.35]** Marx, Karl, *Economic and Philosophic Manuscripts of 1884,* Moscow, Progress Pub., 1967.

**[R15.36]**Marx, Karl, *The Poverty of Philosophy*, New York, International Publishers, 1963.

**[R15.37]**Marx, Karl, Economic and Philosophic Manuscripts of 1844, Moscow, Progress Pub, 1967

**[R15.38]**Niebyl, Karl, H., "Modern Mathematics and Some Problems of Quantity, Quality and Motion in Economic Analysis," *Philosophy of Science*, Vol 7, # 1, January, 1940, pp. 103 – 120.

**[R15.39]**Price, H. H., *Thinking and Experience*, London, Hutchinson, 1953.

**[R15.40]**Putman, H., *Reason, Truth and History*, Cambridge, Cambridge University Press, 1981.

**[R15.41]**Putman. H., *Realism and Reason*, Cambridge, Cambridge University Press, 1983.

**[R15.42]** Robinson, Joan, *Economic Philosophy*. New York, Anchor Books, 1962.

**[R15.43]**Robinson, Joan, *Freedom and Necessity: An Introduction to the Study of Society*, New York, Vintage Books, 1971.

**[R15.44]**Robinson, Joan, *Economic Heresies: Some Old-Fashioned Questions in Economic Theory*, New York, Basic Books, 1973.

**[R15.45]**Schumpeter, Joseph A., *The Theory of Economic Development*, Cambridge, Mass. Harvard University Press, 1934.

**[R15.46]**Schumpeter, Joseph A., *Capitalism, Socialism and Democracy*, New York, Harper & Row, 1950.

**[R15.47]**Schumpeter, Joseph A., "March to Socialism,*"American Economic Review*, Vol. 40 May 1950, pp. 446 456.

**[R15.48]**Schumpeter, Joseph A., "Theoretical Problems of Economic Growth" *Journal of Economic History* Vol. 8, Supplement 1947, pp. 1-9.

**[R15.49]**Schumpeter, Joseph A., "The Analysis of Economic Change," *Review of Economic Statistics*, Vol. 17, 1935, pp 2-10.

**R16    Vagueness, Approximation and Reasoning in the Knowledge Development and Categorial Conversion**

**[R16.1]**Adams, E. w., and H. F. Levine, "On the Uncertainties Transmitted from Premises to Conclusions in deductive Inferences," *Synthese* Vol. 30, 1975, pp. 429 – 460.

**[R16.2]**Arbib, M. A., *The Metaphorical Brain*, New York, McGraw-Hill, 1971.

**[R16.3]** Bečvář, Jiři , " Notes on Vagueness and Mathematics," in Skala, Heinz J. et al., (eds.), *Aspects of Vagueness*, Dordrecht, D. Reidel Co. 1984, pp.1-11.

**[R16.4]**Black, M, "Vagueness: An Exercise in Logical Analysis," *Philosophy of Science*, Vol. 17, 1970, pp141-164.

**[R16.5]** Black, M. "Reasoning with Loose Concepts," *Dialogue*, Vol. 2, 1973, pp. 1-12.

**[R16.6]** Black, Max, *Language and Philosophy*, Ithaca, N.Y.: Cornell University Press. 1949

**[R16.7]** Black, Max, *The Analysis of Rules*, in Black, Max [] *Models and Metaphors: Studies in Language and Philosophy*, 1962 pp. 95-139.

**[R16.8]** Black, Max Models *and Metaphors: Studies in Language and Philosophy*,
Ithaca, NewYork: Cornell University Press. 1962.

**[R16.9]** Black, Max Margins *of Precision*, Ithaca: Cornell University Press,1970.

**[R16.10]** Boolos, G. S. and R. C. Jeffrey, *Computability and Logic,* New York, Combridge University Press, 1989.

**[R16.11]** Cohen, P. R., *Heuristic Reasoning about uncertainty: An Artificial Intelligent Approach*, Boston, Pitman, 1985.

**[R16.12]** Darmstadter, H., "Better Theories," *Philosophy of Science*, Vol. 42, 1972, pp. 20 – 27.

**[R16.13]** Davis, M., *Computability and Unsolvability*, New York, McGraw-Hill, 1958.

**[R16.14]** Dummett, M., "Wang's Paradox," *Synthese,* Vol. 30, 1975, pp301 – 324.

**[R16.15]** Dummett, M., *Truth and Other Enigmas*, Cambridge, Mass. Harvard University Press, 1978.

**[R16.16]** Endicott, Timothy, *Vagueness in the Law*, Oxford, Oxford University Press, 2000.

**[R16.17]** Evans, Gareth, "Can there be Vague Objects?," *Analysis,* Vol. 38, 1978, p. 208.

**[R16.18]** Fine, Kit, "Vagueness, Truth and Logic," *Synthese,* Vol.54, 1975, pp. 235-259.

**[R16.19]** Gale, S., "Inexactness, Fuzzy Sets and the Foundation of Behavioral Geography," *Geographical Analysis*, Vol.4, #4, 1972, pp.337-349.

**[R16.20]** Ginsberg, M. L. (ed.), *Readings in Non-monotonic Reason*, Los Altos, Ca., Morgan Kaufman, 1987.

**[R16.21]** Goguen, J. A., "The Logic of Inexact Concepts," *Synthese*, Vol. 19, 1968/69, pp. 325 – 373.

**[R16.22]** Grafe, W., "Differences in Individuation and Vagueness," in A. Hartkamper and H. –J. Schmidt, *Structure and Approximation in Physical Theories*, New York, Plenum Press, 1981. pp.113–122.

394

**[R16.23]** Goguen, J. A, "The Logic of Inexact Concepts" *Synthese*, Vol.19, 1968-196**9**.

**[R16.24]** Graff, Delia and Timothy (eds.), *Vagueness*, Aldershot, Ashgate Publishing, 2002.

**[R16.25]** A. Hartkämper and H.J. Schmidt (eds.), *Structure and Approximation in Physical Theories*, New York, Plenum Press, 1981

**[R16.26]** Hersh, H.M. et. al., "A Fuzzy Set Approach to Modifiers and Vagueness in Natural Language," *J. Experimental*, Vol. 105, 1976, pp. 254-276.

**[R16.27]** Hilpinen, R., "Approximate Truth and Truthlikeness," in M. Pprelecki et al. (eds.) *Formal Methods in the Methodology of Empirical Sciences*, Wroclaw, Reidel, Dordrecht and Ossolineum, 1976 pp. 19 – 42.

**[R16.28]** Hockney D. et al. (eds.), *Contemporary Research in Philosophical Logic and Linguistic Semantics*, Dordrecht-Holland, Reidel Pub. Co. 1975.

**[R16.29]** Höhle Ulrich et al (eds.), *Non-Clasical Logics and their Applications to FuzzySubsets: A Handbook of the Mathematical Foundations of Fuzzy Set Theory*, Boston, Mass. Kluwer, 1995.

**[R16.30]** Katz, M., "Inexact Geometry," *Notre-Dame Journal of Formal Logic*, Vol.21, 1980, pp. 521-535.

**[R16.31]** Katz, M., "Measures of Proximity and Dominance," *Proceedings of the Second World Conference on Mathematics at the Service of Man*, Universidad Politecnica de Las Palmas, 1982, pp. 370 – 377.

**[R16.32]** Katz, M., "The Logic of Approximation in Quantum Theory," *Journal of Philosophical Logic*, Vol. 11, 1982, pp. 215 – 228.

**[R16.33]** Keefe, Rosanna, *Theories of Vagueness*, Cambridge, Cambridge University Press, 2000.

**[R16.34]** Keefe, Rosanna and Peter Smith (eds.) *Vagueness: A Reader*, Cambridge, MIT Press, 1996.

**[R16.35]** Kling, R., "Fuzzy Planner: Reasoning with Inexact Concepts in a Procedural Problem-solving Language," *Jour. Cybernetics*, Vol. 3, 1973, pp. 1-16.

**[R16.36]** Kruse, R.E. et. al., *Uncertainty and Vagueness in Knowledge Based Systems: Numerical Methods*, New York, Springer-Verlag 1991.

**[R16.37]** Ludwig, G., "Imprecision in Physics," in A. Hartkämper and H.J. Schmidt (eds.), *Structure and Approximation in Physical Theories*, New York, Plenum Press, 1981, pp. 7 – 19.

395

**[R16.38]** Kullback, S. and R. A. Leibler, "Information and Sufficiency," *Annals of Math. Statistics*, Vol. 22, 1951, pp. 79 – 86.

**[R16.39]** Lakoff, George, "Hedges: A Study in Meaning Criteria and Logic of Fuzzy Concepts," in, Hockney D. et al. (eds.), *Contemporary Research in Philosophical Logic and Linguistic Semantics*, Dordrecht-Holland, Reidel Pub. Co. 1975, pp. 221-271.

**[R16.40]** Lakoff, G., "Hedges: A Study in Meaning Criteria and the Logic of Fuzzy Concepts," *Jour. Philos. Logic*, Vol. 2, 1973, pp. 458-508.

**[R16.41]** Levi, I., *The Enterprise of Knowledge*, Cambridge, Mass. MIT Press 1980.

**[R16.42]** Łucasiewicz, J., *Selected Works: Studies in the Logical Foundations of Mathematics,* Amsterdam, North-Holland, 1970.

**[R16.43]** Machina, K.F., "Truth, Belief and Vagueness," *Jour. Philos. Logic*, Vol.5, 1976, pp. 47-77.

**[R16.44]** Menges, G., et. al., "On the Problem of Vagueness in the Social Sciences," in Menges, G. (ed.), *Information, Inference and Decision*, D.Reidel Pub., Dordrecht, Holland, 1974, pp. 51-61.

**[R16.45]** Merricks, Trenton, "Varieties of Vagueness," *Philosophy and Phenomenological Research*, Vol.53, 2001, pp. 145-157.

**[R16.46]** Mycielski, J., "On the Axiom of Determinateness," *Fund. Mathematics*, Vol. 53, 1964, pp. 205 – 224.

**[R16.47]** Mycielski, J., "On the Axiom of Determinateness II," *Fund. Mathematics,* Vol.59, 1966, pp. 203 – 212.

**[R16.48]** Naess, A., "Towards a Theory of Interpretation and Preciseness," in L. Linsky (ed.) *Semantics and the Philosophy of Language*, Urbana, Ill. Univ. of Illinois Press, 1951.

**[R16.49]** Narens, Louis, "The Theory of Belief," *Journal of Mathematical Psychology*, Vol. 49, 2003, pp 1-31.

**[R16.50]** Narens, Louis, "A Theory of Belief for Scientific Refutations," *Synthese,* Vol.145, 2005, pp 397-423.

**[R16.51]** Netto, A. B., "Fuzzy Classes," *Notices, Amar, Math. Society*, Vol. 68T- H28, 1968, pp. 945.

**[R16.52]** Neurath, Otto et al. (eds.), *International Encyclopedia of Unified Science, Vol.1 – 10,* Chicago, University of Chicago Press, 1955.

**[R16.53]** Niebyl, Karl, H., "Modern Mathematics and Some Problems of Quantity, Quality and Motion in Economic Analysis," *Science*, Vol 7, # 1, January, 1940, pp. 103 – 120.

**[R16.54]** Orlowska, E., "Representation of Vague Information," *Information Systems*, Vol. 13, #2, 1988, pp. 167-174.

**[R16.55]** Parrat, L. G., *Probability and Experimental Errors in Science*, New York, John Wiley and Sons, 1961.

**[R16.56]** Raffman. D., "Vagueness and Context-sensitivity," *Philosophical Studies*, Vol.81, 1996, pp. 175-192.

**[R16.57]** Reiss, S., *The Universe of Meaning*, New York, The Philosophical Library, 1953.

**[R16.58]** Russell, B., "Vagueness," *Australian Journal of Philosophy*, Vol.1, 1923 pp. 84-92.

**[R16.59]** Russell, B., *An Inquiry into Meaning and Truth*, New York, Norton, 1940.

**[R16.60]** Shapiro, Stewart, *Vagueness in Context*, Oxford, Oxford University Press, 2006.

**[R16.61]** Skala, H. J. "Modelling Vagueness," in M. M. Gupta and E. Sanchez, *Fuzzy Information and Decision Processes*, Amsterdam North-Holland, 1982, pp 101 – 109.

**[R16.62]** Skala, Heinz J. et al., (eds.), *Aspects of Vagueness*, Dordrecht, D. Reidel Co. 1984.

**[R16.63]** Sorensen, Roy, *Vagueness and Contradiction*, Oxford, Oxford University Press, 2001.

**[R16.64]** Tamburrini, G. and S. Termini, "Some Foundational Problems in Formalization of Vagueness," in M. M. Gupta et al (eds.), *Fuzzy Information and Decision Processes*, Amsterdam, North Holland, 1982, pp. 161-166.

**[R16.65]** Termini, S. "Aspects of Vagueness and Some Epistemological Problems Related to their Formalization," in Skala, Heinz J. et al., (eds.), *Aspects of Vagueness,* Dordrecht, D. Reidel Co. 1984, pp.205 – 230.

**[R16.66]** Tikhonov, Andrey N. and Vasily Y. Arsenin., *Solutions of Ill-Posed Problems*, New York, John Wiley and Sons, 1977.

**[R16.67]** Tversky, A. and D. Kahneman, "Judgments under Uncertainty: Heuristics and Biases," *Science,* Vil 185 September 1974, pp. 1124-1131.

**[R16.68]** Ursul, A. D., "The Problem of the Objectivity of Information," in Kuba't, Libor and J. Zeman (eds.), *Entropy and Information*, Amsterdam, Elsevier, 1975.pp. 187-230.

**[R16.69]** Vardi, M. (ed.), *Proceedings of Second Conference on Theoretical Aspects of Reasoning about Knowledge*, Asiloman, Ca, Los Altos, Ca, Morgan Kaufman, 1988.

**[R16.70]** Verma, R.R., "Vagueness and the Principle of the Excluded Middle," *Mind*, Vol. 79, 1970, pp. 66-77.

**[R16.71]** Vetrov, A. A., "Mathematical Logic and Modern Formal Logic," *Soviet Studies in Philosophy*, Vol. 3, #1, 1964 pp. 24 – 33.

**[R16.72]** von Mises, Richard, *Probability, Statistics and Truth*, New York, Dover Pub. 1981.

**[R16.73]** Williamson, Timothy, *Vagueness*, London, Routledge, 1994.

**[R16.74]** Wiredu, J.E., "Truth as a Logical Constant With an Application to the Principle of the Excluded Millde," *Philos. Quart.*, Vol. 25, 1975, pp. 305-317.

**[R16.75]** Wright, C., "On Coherence of Vague Predicates," *Synthese*, Vol. 30, 1975. pp. 325 – 365.

**[R16.76]** Wright, Crispin, "The Epistemic Conception of Vagueness," *Southern Journal of Philosophy*, Vol. 33, Supplement, 1995, pp. 133-159.

**[R16.77]** Zadeh, L. A., A Theory of Commonsense Knowledge," in Skala, Heinz J. et al., (eds.), *Aspects of Vagueness*, Dordrecht, D. Reidel Co. 1984., pp 257 – 295.

**[R16.78]** Zadeh, L. A., "The Concept of Linguistic Variable and its Application to Approximate reasoning," *Information Science*, Vol. 8, 1975, pp. 199 – 249 (Also in Vol. 9, pp. 40 – 80.

## R17 Vagueness and Fuzzy Game Theory in Categotial Conversion and Philosophical Consciencism

**[R17.1]** Aubin, J.P. "Cooperative Fuzzy Games", Mathematics of Operations Research, Vol.6, 1981, pp1 – 13.

**[R17.2]** Aubin, J.P. Mathematical Methods of Game and Economics Theory, New York, North Holland.

**[R17.3]** Butnaria, D., "Fuzzy Games: A description pf the concepts," Fuzzy sets and systems Vol.1, 1978, pp. 181 – 192.

**[R17.4]** Butnaria, D., "Stability and shapely value for a n – persons Fuzzy Games," Fuzzy sets and systems, Vol. 4, #1, 1980, pp 63 – 72.

**[R17.5]** Nurmi, H.., "A Fuzzy Solution to a Majority Voting Game, "Fuzzy sets and systems, Vol.5, 1981 pp187-198.

**[R17.6]** Regade., R. K., " Fuzzy Games in the Analysis of Options," jour. Of Cybernetics, Vol. 6, 1976, pp 213 – 221.

**[R17.7]** Spillman, B. et al., "Coalition Analysis with Fuzzy Sets," Kybernetes, Vol. 8, 1979, pp. 203-211.

**[R17.8]**Wernerfelt, B., "Semifuzzy Games" Fuzzy sets and systems, Vol. 19, 1986, pp 21 – 28

# INDEX